Reviews in Modern Astronomy

4

Gerhard Klare (Ed.)

Reviews in Modern Astronomy 4

With 138 Figures

Springer-Verlag

Berlin Heidelberg New York
London Paris Tokyo
Hong Kong Barcelona
Budapest

Edited on behalf of the *Astronomische Gesellschaft* by

Dr. *Gerhard Klare*

Landessternwarte, Königstuhl,
W-6900 Heidelberg 1, Fed. Rep. of Germany

ISBN-13:978-3-642-76752-4 e-ISBN-13:978-3-642-76750-0
DOI: 10.1007/978-3-642-76750-0

Softcover reprint of the hardcover 1st edition 1991

Typesetting: Camera ready by author
56/3140-543210 – Printed on acid-free paper

Preface

The yearbook series *Reviews in Modern Astronomy* of the Astronomische Gesellschaft (AG) was established three years ago in order to bring the scientific events of the meetings of the society to the attention of the worldwide astronomical community. *Reviews in Modern Astronomy* is devoted exclusively to the invited reviews, the Karl Schwarzschild lectures, and the highlight contributions from leading scientists reporting on recent progress and scientific achievements at their research institutes.

Volume 4 comprises all eighteen contributions which were presented during the fall meeting of the AG at Freiburg/Breisgau in September 1990. They cover problems in solar research and the solar system as well as the first results of the ROSAT and Hipparcos space missions, stellar and extragalactic studies, and Supernova 1987A in the Large Magellanic Cloud.

The Karl Schwarzschild Medal was awarded to Professor Eugene Parker[1]. His lecture entitled "Convection, Spontaneous Discontinuities, and Stellar Winds and X-Ray Emission" begins this volume.

Heidelberg, May 1991 *G. Klare*

---

[1] The complete text of the laudation will be published in Mitteilungen der AG Vol. 74, 1991.

Contents

Convection, Spontaneous Discontinuities, and Stellar Winds and X-Ray Emission

E.N. Parker

Depts. of Astronomy & Astrophysics, Enrico Fermi Institute, The University of Chicago, Chicago, IL 60637, USA

1 Introduction

The subject of this lecture is the origin of the solar corona, but I would like to begin with a brief recollection of the basic ideas on which present concepts of solar and stellar activity are based. First of all, there is the star itself, first represented by Emden's gaskügeln. His simple polytrope models are the first scientific representation of the static equilibrium of a self-gravitating star. The polytropes are not precise models of a star, but they capture the basic theoretical principles of the balance of thermal pressure against gravity, with a high central temperature declining toward the surface.

The next milestone was the recognition of convection by Prof. Karl Schwarzschild, in his statement of the criterion for the onset of convection that now bears his name, viz. that convection arises where the temperature gradient exceeds the adiabatic gradient. Convection is the basic cause of the activity of all solitary stars, such as the sun. Where there is convection there are the familiar suprathermal effects of flares, coronae, etc. Thus Schwarzschild laid the first foundation block for modern inquiries into the nature of stellar activity.

Historically the next major step was the realization that the sun, and the stars, are composed mainly of hydrogen and helium. That fact is certainly not obvious from inspection of the solar spectrum, and it was a major revelation when first pointed out. We take it for granted today, as if it had been known from the beginning of time. With the introduction of Rosseland's mean opacity, then, the modern development of the theory of stellar interiors got underway.

The nuclear origin of stellar energy completes the list of fundamental ideas on which understanding of the steady state of a star is currently based.

In the nineteen thirties, Edlen, Grotrian, and Lyot established that the corona of the sun – visible during lunar eclipses of the sun – is really an extended suprathermal atmosphere of 10^6 K. That fundamental step was critized as absurd by some of the most distinguished names of the time. But it was not long before *L. Biermann* [1], *H. Alfven* [2] and *M. Schwarzschild* [3] pointed out that the suprathermal corona is a direct consequence of the convection beneath the visible surface of the sun. The convection is a heat engine, doing mechanical work that is dissipated in the chromosphere and corona to provide the elevated temperatures. The heat engine is, of course, subject to the well known thermodynamic limitations of the Carnot cycle. But the work produced by the heat engine can be used to heat a tenuous atmosphere to any desired temperature.

In practice the heat input to the tenuous outer atmosphere of the sun causes the upward expansion of the gas, to a pressure scale height 2 kT/Mg where M is approximately the mass of a hydrogen atom (fully ionized in the corona) and g is the gravitational acceleration ($\sim 2.7 \times 10^4$ cm/sec^2). The density increases until outward expansion into space, and/or thermal emission of electromagnetic radiation consumes the heat input, and a steady state is reached.

Biermann, Alfven, and Schwarzschild expounded the basic principles in the context of waves (acoustic waves and Alfven waves) created in the subphotospheric convection and propagating upward into the chromosphere and corona, where dissipation of the waves supplies the heat to maintain the elevated temperature. The principal point of this lecture is to note that the convective frequencies are too low to create waves that heat the corona in the detailed manner currently observed. Instead of wave oscillations, it is suggested that the mechanical work done by the convection is of a quasi-static nature so far as the active corona is concerned. This is only a quantitative variation of the physical principle enunciated by Biermann, Alfven, and Schwarzschild. The basic idea remains unchanged.

Now the corona of the sun, and by inference the coronas of most stars, consist of the active X-ray corona on the one hand and the coronal hole on the other hand. These two coronal conditions are qualitatively distinct and have, so far as can be determined, entirely different causes and effects. The active X-ray corona is the source of the X-ray emission of the star, whereas the coronal hole is the source of the high speed streams in the solar wind, i.e. the mass loss from the sun, creating the heliosphere [4]. Presumably the same applies to other stars. Consider first the causes of the X-ray corona.

2 The Active X-Ray Corona

The active X-ray corona consists of gas at $2 - 3 \times 10^6$ K forcibly confined in the 10^2 gauss (10^{-2} tesla) bipolar magnetic fields of active regions. The energy input [5] is of the order of 10^7 ergs/cm^2 sec, which causes the number density N to increase to about 10^{10} atom/cm^3 so that the thermal bremsstrahlung emission balances the heat input. The pressure scale height is then 10^{10} cm or more, to be compared with the horizontal scale of $10^9 - 10^{10}$ cm of the bipolar magnetic fields of the small and large active regions. The sound speed is of the order of 2×10^7 cm/sec and the Alfven speed is approximately 10 times larger at 2×10^8 cm/sec. The gas pressure is only 2×10^{-2} of the pressure of the confining magnetic field.

Now the extensive studies of wave generation and propagation [6], [7], to reference only a few of the many scientists that have worked on this complex problem, indicate that only the Alfven waves, guided along the magnetic field, reach the corona. The acoustic waves, the internal gravity waves, the fast and slow mode MHD waves, etc. are either dissipated before reaching the corona (contributing to chromospheric heating) or are refracted back into the sun. The theoretical problem then is to show that the necessary flux of Alfven waves is produced by the convection and is dissipated upon their arrival in the corona.

We expect Alfven waves from the subphotospheric convection to have periods comparable to the 300 sec characteristic time of the photospheric convection – the granules – with characteristic dimensions ℓ of 10^3 km and velocities $v \cong 1$ km/sec. Shorter periods down to 50 sec are detected, but at reduced power levels. The granules are evidently strongly turbulent, with Reynolds numbers of 10^{10} or more, so that one may reasonably expect something of the form of the Kolmogoroff turbulent spectrum to apply. This predicts that the kinetic energy density $\frac{1}{2}\rho v^2$ at any period τ scales directly as τ.

The fundamental difficulty with wave heating is that the wavelengths of the principal Alfven wave flux in the corona are too large to match to the bipolar magnetic fields of the X-ray corona. A period of 10^2 sec is associated with a half-wavelength of 10^5 km, which is as large as the larger bipolar fields. The smaller bipolar fields at 10^4 km are so small that a wave period of 10^2 sec represents a quasi-static displacement of the entire bipole rather than a wave within the bipole. Yet the heat input to the small bipole is observed to be about the same as to the large bipole, both experiencing 10^7 ergs/cm^2 sec. So evidently it is not Alfven waves that supply the principle heat input. That is to say, the Alfven waves that might reasonably be expected from the photospheric convection have periods that are too long to account for an energy input that is independent of the scale of the magnetic bipolar regions over the range $10^9 - 10^{10}$ km. One could, of course, postulate a wave flux of 10^7 ergs/cm^2 sec with periods of, say 0.1 sec, appealing to surface resonant absorption [8] to dissipate the waves (with length 200 km) equally effectively over all scales. But so large a wave flux at such short periods requires conditions within the convective zone so totally unlike what is presently believed that it merely substitutes one mystery for another.

In fact, it appears that the clue to the heat input of the active X-ray corona lies in the quasi-static nature of the convective deformation of the magnetic bipoles. The continuing (presumably random) displacement of the photospheric footpoints of the magnetic field progressively winds and interweaves the field lines into a complicated topology, sketched in Fig. 1. The basic magnetostatic theorem states that the magnetostatic equilibrium of almost all field topologies (i.e. all but the most carefully tailored magnetic fields) necessarily involves tangential discontinuities as an intrinsic part of the balance between the magnetic pressure and the magnetic tension along the field. It follows that the random winding of the field lines causes the development of many internal tangential discontinuities in the bipole magnetic field of an active region on the sun.

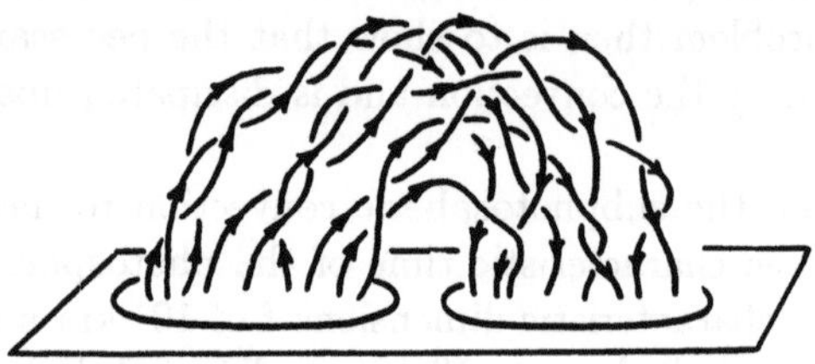

Fig.1. A schematic drawing of the interwoven field of a bipolar magnetic region, arising form the random shuffling of the footpoints of the field by the photospheric convection.

The tangential discontinuity is a surface lying along the field lines across which the field magnitude is continuous (required for balance of the magnetic pressure $B^2/8\pi$) while the direction of the field changes discontinuosly. The tangential discontinuity represents a current sheet. The thickness of the current sheet is limited only by the small resistivity η of the fluid and is identically zero in the limit of vanishing η. Thus, no matter how small is η, there is resistive dissipation. In fact, both laboratory experience with Tokamaks and numerical simulations [9], [10], [11] suggest that the dissipation is largely in the form of bursts of rapid reconnection.

We have suggested [12]–[16] that it is this sporadic explosive dissipation at the tangential discontinuities in the bipolar fields on the sun that creates the active X-ray corona. The heating occurs in bursts, which are estimated to involve individually $10^{23} - 10^{25}$ ergs. Such a burst is too small to be readily observable in the tenuous gas of the corona and we refer to the invidiaul burst as a *nanoflare*, as 10^{-9} of a large flare of 10^{32} ergs/cm^2 sec. On this basis, the X-ray corona is a sea of nanoflares [17]. More generally, the X-ray emission of solitary middle and lower main sequence stars, is simply the net radiative output of a captive sea of nanoflares. Recent calculations [18] show that the net effect of impulsive heating of the corona is a theoretical emission spectrum with the basic properties of the observed emission.

3 The Magnetostatic Theorem

To avoid confusion and misunderstanding consider the idealized situation sketched in Fig. 2(a), in which a uniform magnetic field B_0 extends in the z-direction from the boundary $z = 0$ to the boundary $z = L$, the region $0 < z < L$ being filled with a fluid of zero electrical resistivity. To proceed formally, suppose that at time $t = 0$ the fluid throughout $0 \leq z \leq L$ is caused to move with the velocity $\mathbf{v}$ given by

$$v_x = +kz\partial\psi/\partial y, \quad v_y = -kz\partial\psi/\partial x, \quad v_z = 0, \tag{1}$$

where $\psi = \psi(x, y, kzt)$ is a bounded, smooth, continuous, differentiable, and generally well behaved function of its three arguments. This velocity leaves the footpoints fixed at $z = 0$ while it creates an arbitrary continuous interweaving of the field lines throughout $0 < z < L$, in association with the more or less random continuous mixing and swirling of the footpoints at $z = L$. The magnetic field is mapped into

$$B_x = +B_0kt\partial\psi/\partial y, \quad B_y = -B_0kt\partial\psi/\partial x, \quad B_z = B_0, \tag{2}$$

and, in view of the continuity of ψ and its derivatives, is continuous throughout $0 < z < L$. In the general case the field lines are randomly interwoven, in the manner sketched in Fig. 2(b).

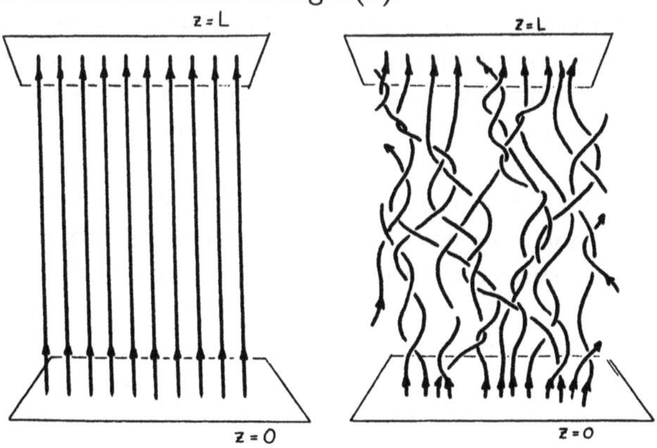

(a) (b)

Fig. 2. (a) A sketch of the initial uniform magnetic field B_0 through $0 < z < L$. (b) A sketch of the continuous field of equation (2).

Suppose that at time $t = t_0$ the fluid motion is switched off, the footpoints of the field at $z = 0, L$ are held fixed, and the fluid is released so that the field relaxes into the lowest available energy state, thereby achieving magnetostatic equilibrium. It may be convenient to introduce some slight viscosity to achieve the final static state. The magnetostatic theorem asserts that the continuous field described by equation (2) develops internal tangential discontinuities (in almost all cases) as it relaxes to equilibrium [19]. The discontinuities are a consequence of the mutual winding and wrapping of the field lines.

This, perhaps surprising, property of magnetostatic fields can be understood on a relatively simple level. For instance, if the fluid pressure p is maintained uniform at the boundaries $z = 0, L$, then since $\mathbf{B} \cdot \nabla p = 0$, it is uniform throughout $0 < z < L$, and the static equilibrium of the field is described by the familiar force–free equation

$$\nabla \times \mathbf{B} = \alpha \mathbf{B}, \tag{3}$$

where

$$\alpha = (\mathbf{B} \cdot \nabla \times \mathbf{B})/B^2 \tag{4}$$

is the torsion coefficient. Since $\nabla \cdot \mathbf{B} = 0$, the divergence of equation (3) yields the familiar result,

$$\mathbf{B} \cdot \nabla \alpha = 0, \tag{5}$$

that the torsion is uniform along each field line. This fact introduces an apparent contradiction, for in the random winding of the field lines, described by the arbitrary function ψ, an elemental flux bundle may wind first one way and then the other around the neighboring bundles. The torsion within the bundle is fixed by the average torsion along the entire length of the bundle. Hence the bundle cannot adapt its local torsion to the varying torsion of its winding about the neighboring flux bundles. The result is the formation of tangential discontinuities where the flux bundle winds around its neighbors. The tangential discontinuity represents the incompatibility between α and the local torsion of the winding of the flux bundle around its neighbors. The discontinuity does not violate equation (5) because the discontinuity contains no flux, it being the surface between two adjacent regions of magnetic field.

Another way to see the necessity for tangential discontinuities is to consider the fragility of a continuous field [20]. Suppose, for instance, that by suitable choice of ψ we construct the continuous equilibrium field described by the vector potential $A = (B_0/k)\ \mathit{sinkx\ sinky}$ where the field is given by

$$B_x = +\partial A/\partial y,\ B_y = -\partial A/\partial y,\ B_z^2 = B_0^2 + 2k^2 A^2 \tag{6}$$

and satisfies the Grad-Shafranov equation,

$$\nabla^2 A + B_z'(A) B_z(A) = 0, \tag{7}$$

for static equilibrium. The field represents a close-packed rectangular array of alternately twisted flux bundles, whose cross section is sketched in Fig. 3(a).

So long as the physical manipulation is mathematically precise, it can produce the foregoing continuous field. But in fact no physical manipulation in the real world has mathematical precision in any aspect. The point is, then, that the perfect continuous field sketched in Fig. 3(a) is slightly deformed in several ways, with the result that it contains tangential discontinuities. For instance, it may be that we err in giving the counterclockwise cells slightly more torsion than the clockwise cells. Then the counterclockwise cells expand slightly at the expense of the clockwise cells. The X-type neutral points in the transverse field (B_x, B_y) in Fig. 3(a) are deformed into pairs of Y-type points, sketched in Fig. 3(b). That is to say, the corners of the cells no longer match perfectly, so that cells with the same torsion come into contact over a finite width, necessarily producing tangential discontinuities, represented by the heavy lines in Fig. 3(b).

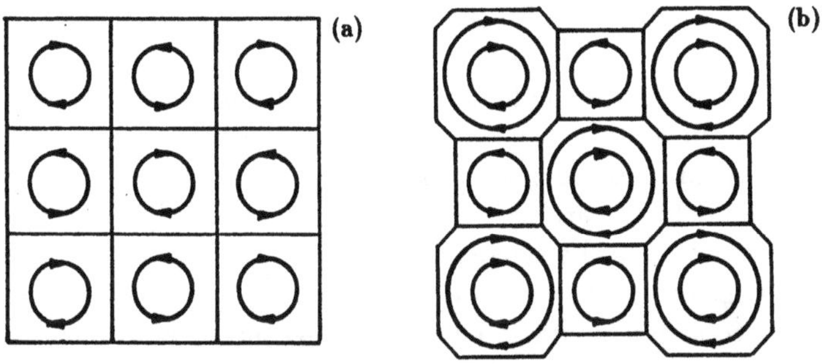

Fig. 3. (a) Cross section z = *constant* of the field from equation (6). (b) Cross section of the field when the counterclockwise twist is slightly stronger than the clockwise twist.

The discontinuities form at the separatrices between regions of different topology, and one finds that almost all deformations of a continuous field cause the formation of tangential discontinuities where the different topological regions slide over each other [20].

Another way to understand the creation of tangential discontinuities is to note that equation (3) has three sets of characteristics. Two are imaginery, as with Laplace's equation, $\nabla^2\phi = 0$, indicating an elliptic equation, with the property that any continuous field throughout the interior of the region $0 < z < L$ is completely determined by specification of the field on the boundary. But B_z can be pushed around on $z = 0$ and $z = L$ so as to achieve any desired variation with x and y, without, however, changing the topology of the wrapping and interweaving of the field throughout the interior. So on physical grounds the solutions to equation (3) cannot be uniquely determined by the boundary conditions. The equation avoids this contradiction through the third set of characteristics, which are real. These characteristics are the field lines, along which there may be tangential discontinuities. The presence of discontinuities voids the unique determination of the internal field by the field at the boundaries. This situation is illustrated by a specific example of a field extending a very long distance with a random pattern of winding [21].

When it comes to formal mathematical solution of equation (3), the conventional methods of analysis are effective only in the linear and quasilinear cases of this generally nonlinear equation, and they automatically provide continuous solutions. There are infinitely many continuous solutions, of course, but they constitute a set of measure zero compared to all possible solutions, most of which involve tangential discontinuities. In special cases, however, the integration can be carried forward under less restrictive conditions leading directly to tangential discontinuities [22].

A particularly useful mathematical device for illustrating and understanding the formation of tangential discontinuities is the *optical analogy* [23][24] noting that any magnetic field $\mathbf{B}(\mathbf{r})$ described by equation (3) can be written as $-\nabla\phi$ in any two dimensional space represented by a flux surface. It follows that the pattern of the field lines in any flux surface is the same as the optical ray paths in an index of refraction proportional to the field magnitude $B = |\mathbf{B}|$. Hence the field is refracted around a local maximum in B, so that a sufficiently concentrated maximum causes a bifurcation in the field, with the field lines passing around, rather than over, the maximum. This gap in the field pattern, centered on the maximum in B, is the singular feature that produces the tangential discontinuity. The gap in the flux surface permits the otherwise separate regions of field on opposite sides of the flux surface to come into contact. The fields are generally not precisely parallel where they meet in the gap, producing a tangential discontinuity.

4 Energy Considerations

Consider the energy input to the active X-ray corona through the nanoflares at the tangential discontinuities [16][17]. Fig. 4 is a sketch of an idealized situation in which the footpoint at $z = 0$ of a single elemental flux bundle is transported in a random walk with a velocity v, of the order of 0.5 km/sec, representing the displacement of the footpoint by the photospheric convection. The footpoint at $z = L$ is held fixed. Suppose that the length ℓ of each random step is larger than the average separation of magnetic fibrils, so that the effective path length vt takes the flux bundle on a looping journey among the neighboring flux bundles (shown without footpoint displacement in Fig. 4) so that the inclination θ of the flux bundle to the vertical is given by

$$\tan\theta = vt/L. \tag{8}$$

The transverse (horizontal) component of the field produced in this way is

$$B_\perp = B_0 \tan\theta. \tag{9}$$

The Maxwell stress pulling back on the advancing footpoint is $B_\perp B_0/4\pi$ dynes/cm^2, so the rate $\mathcal{E}$ at which the footpoint motion does work on the field is

$$\mathcal{E} = vB_\perp B_0/4\pi = (v^2t/L)B_0^2/4\pi \text{ ergs/cm sec.} \tag{10}$$

The X-ray corona requires $\mathcal{E} = 10^7$ ergs/cm^2 sec, as already noted, in a field $B_0 = 10^2$ gauss. If the effective value of v is 0.5 km/sec, it follows that $B_\perp \cong \frac{1}{4}B_0$, i.e. $\tan\theta \cong 0.25$ and $\theta \cong 14°$. Only a modest inclination of the field is required, then, to supply the necessary energy input.

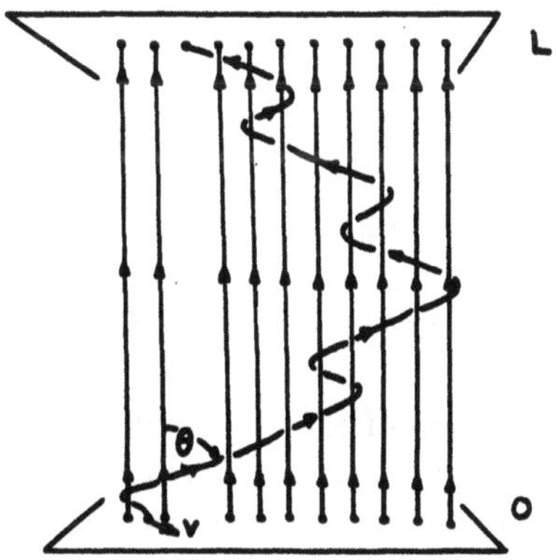

Fig. 4. A sketch of a single flux bundle whose footpoint at $z = 0$ meanders among the neighboring flux bundles with a velocity v.

The transverse component $B_\perp$ is responsible for creating the tangential discontinuities, so suppose that when $B_\perp$ exceeds $\frac{1}{4}B_0$ by some amount, the otherwise slow resistive reconnection of $B_\perp$ across the discontinuities provides a burst of rapid reconnection [9], [11], [17] substantially reducing $B_\perp$ and lapsing back to slow reconnection again while $B_\perp$ builds up to $\frac{1}{4}B_0$, etc. In this way an average steady state is reached in which the average value of $\mathcal{E}$ is 10^7 ergs/cm^2 sec and the average $B_\perp$ is $\frac{1}{4}B_0$. The individual burst of reconnection is the nanoflare mentioned earlier.

Now suppose that rapid reconnection is not as effective as assumed. Suppose $B_\perp$ must increase beyond $\frac{1}{4}B_0$ before triggering a burst of rapid reconnection. The result is, obviously, an energy dissipation $\mathcal{E}$ that is larger than 10^7 ergs/cm^2 sec because $B_\perp$ on the right hand side of equation (10) is larger. So with the tangential discontinuities the energy input $\mathcal{E}$ is *inversely* proportional to the effectiveness of the dissipation. This is the reverse of most other heating mechanisms.

This is perhaps the appropriate point to emphasize that it remains for observations to establish the effective value of v in equation (10), and it remains for observation to tell us more about the character of the individual nanoflare, which we estimate to be in the range $10^{23} - 10^{25}$ ergs/cm^2 sec. Its existence is already known from observational detection of many tiny bursts of energy on the sun, producing small jets, transient x-ray emission, etc. [25]–[28].

One can hope that with the rapid advance of instrumentation on ground based telescopes and in spacecraft, much more can be learned about the basic process outlined here. Only when the basic processes are substantially observed will we be sure that we have the correct picture. Plausibility is a good place to start, of course, but we should not be content to let things rest there.

5 The Coronal Hole

The coronal hole is qualitatively different from the active X-ray corona because only one end of the field is attached to the sun. The field is weak, 5–10 gauss, and is forced open by the expanding gas to form the solar wind, so that the other "end" of the field lines is at infinity. It follows that any transverse field $B_\perp$ produced by the motion of the footpoints simply propagates to infinity along the field and the field remains continuous. Hence tangential discontinuities do not develop in the mean field as a consequence of the random motion of the footpoints of the field in the photospheric convection.

The coronal hole has a gas density of the order of $N = 10^8$ atoms/cm^3 and a magnetic field of the order of 10 gauss, so the Alfven speed is again 2×10^8 cm/sec, as in the active corona. The gas temperature is typically $1.5 - 2 \times 10^6$ K. Radiative losses are negligible because of the small density of the gas, so that the principal heat input is balanced by the cooling of the gas as it expands outward into space. Withbroe and Noyes [5] estimate that the energy input is approximately 5×10^5 ergs/cm^2 sec, and a more detailed analysis by Withbroe [29] shows that most of this input occurs in the first one or two solar radii above the surface of the sun. On the other hand, some fraction, of the order of $1 - 2 \times 10^5$ ergs/cm^2 sec, is introduced to maintain the total gas pressure beyond the sonic point in the wind to provide the further acceleration to the 500-700 km/sec observed in the fast streams of wind from the coronal hole.

The traditional idea has been that the coronal hole is heated by the dissipation of Alfven waves from the photospheric convection. But a recent quantitative review [30] of the known mechanisms for damping Alfven waves (e.g. phase mixing, resonant absorption, anomalous resistivity, thermal conduction, etc.) produces significant dissipation of photospheric Alfven waves ($\tau \sim 10^2$ sec) only over distances of 5-10 $R_\odot$. There is no known mechanism that dumps most of the energy in 1-2 $R_\odot$. The only evident alternative is the idea put forth some years ago by Martin [31]–[35], that the principal heating in a coronal hole is the activity in the *network fields*, i.e. in the small ($1 - 5 \times 10^3$ km) magnetic features in the boundaries of supergranule cells, particularly concentrated in the downdrafts at the junctions of three boundaries.

Small weak magnetic fields continually emerge in the updrafts at the centers of supergranules. These *intranetwork* fields become increasingly concentrated as they are carried by the outflow into the boundaries of the supergranules where they interact with the network fields. Porter and Moore estimate a mean energy dissipation of 5×10^5 ergs/cm^2 sec in the microflaring observed in the network, suggesting a comparable heat input to the ambient coronal gas. The theoretical burst of magnetic reconnection that gives rise to the microflare produces a violent reconfiguration of the local magnetic fields, as well as high speed jets of fluid (at the Alfven speed). The characteristic Alfven transit time across the characteristic scale of, say, 2×10^3 km is of the order of 1 sec, suggesting that the agitation produces waves of short period. McKenzie [35] points out that waves of this period would preferentially heat the minor ions by gyro resonance, consistent with observation. Much of the disturbance must be dissipated locally, providing the principal heating of the coronal hole close to the sun. It is not unreasonable to suppose that some modest fraction of the Alfven waves survive to distances of several solar radii to provide the energy and pressure gradient beyond the sonic point in the wind that accelerates the wind to the final high speeds of 500-700 km/sec. Alfven waves of photospheric origin may also contribute to the energy and pressure in the distant coronal hole, because, as already noted, they are expected to dissipate somewhere beyond 5 $R_\odot$. The total energy input to the distant coronal hole is estimated at $1 - 2 \times 10^5$ ergs/cm^2 sec, it will be recalled.

It is evident that the idea, that the heating of coronal holes originates in the network fields, needs extensive observational study to determine its importance. The observations are difficult because the small scale ($\sim 2 - 5 \times 10^3$ km) of the magnetic features and the relatively small magnetic fluxes ($10^{17} - 10^{18}$ Maxwells) lie at the limits of resolution of present instrumentation. There is no way to guess what will appear when, and if, one day the spatial resolution is significantly improved. But the observational study, for all its difficulty, is of prime importance because the activity of the network fields appears to be the principal heat source for the high speed streams in the solar wind. It follows, then, that the activity of the network fields is the primary architect of the heliosphere, as well as the basic energy source for planetary magnetic activity, just as the convective manipulation of the magnetic fields of active regions is the basic source of stellar X-ray emission.

6 High Speed Solar Wind

The high speed streams of solar wind are produced by the expansion of gas from the coronal holes, reaching velocities sometimes as high as 600-700 km/sec far the sun compared with 300 km/sec in the quiet wind. Noting that the gravitational escape velocity from the surface of the sun is approximately 650 km/sec, it is evident that the final asymptotic kinetic energy density (ergs/gm) of the fast solar wind is about equal to the energy expended to lift the matter out of the gravitational potential well of the sun. That is to say, the final kinetic energy of the fast wind is about half of the total energy producing the wind, whereas in the quiet wind at 300 km/sec, the kinetic energy is only about a fifth of the total energy producing the wind.

Now a solar wind of 600-700 km/sec can be understood as the thermal expansion of an isothermal corona at $1.5 - 2 \times 10^6$ K. Fig. 5 is a plot of wind velocity as a function of radial distance from the sun [4] for simple radial expansion. In fact the outflow broadens more rapidly than radial flow as the coronal holes expand transversely to fill in beyond the closed active regions. The result is a somewhat faster expansion, nearer sonic point and ultimately a faster, more tenuous stream in the solar wind.

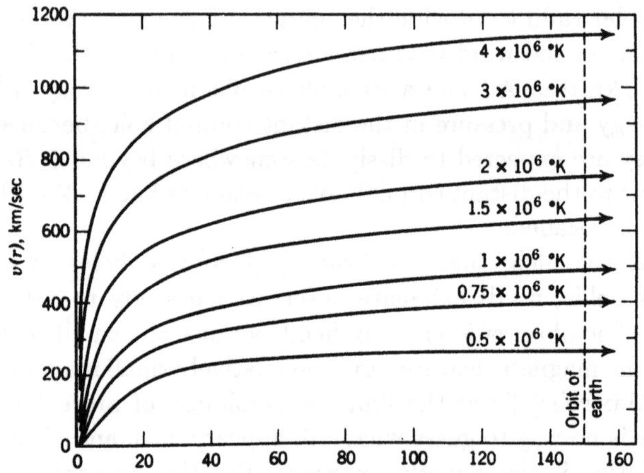

Fig. 5. A plot of the radial expansion velocity of an isothermal atmosphere (at the indicated temperature) around the sun as a function of radial distance (in units of 10^6 km) (from [4]).

As a first approximation, then, the fast stream can be understood on the basis of a flux of Alfven waves in the solar wind that maintains the thermal temperature of the gas in the neighborhood of 1.5×10^6 K to a distance of $10^2\ R_\odot$ [36]. However, it has been pointed out [37]–[39] that a wave flux of this intensity at large radial distance from the sun contributes a magnetic pressure and a "turbulent" pressure of its own, adding significantly to the thermal pressure $p = 2NkT$. Thus the total effective pressure P can be written

$$P = p + \frac{1}{2}\rho\langle(\delta v)^2\rangle + \langle(\delta B)^2\rangle/8\pi$$

where δv is the longitudinal (radial) velocity fluctuation of the waves and δB is the magnetic variation in the wave. It is the gradient dP/dr of this total pressure that accelerates the expanding coronal gas. In addition the transverse wave velocity fluctuation δw may contribute a centrifugal force $\rho\langle(\delta w)^2\rangle/r$ to the acceleration. The equation of motion for radial expansion is, then,

$$N(r)Mv(r)\frac{dv}{dr} = \frac{dP}{dr} - \frac{GM_{\odot}N(r)M}{r^2} + \frac{N(r)M\langle(\delta w)^2\rangle}{r}$$

for the gas with number density $N(r)$. This equation approximates to the familiar static barometric equation from the base of corona out to the sonic point, where v reaches the local speed of sound [36], [40]. In this quasi-static region the wave pressure is negligible, and the support and acceleration of the gas is entirely thermal. Presumably the temperature declines somewhat with increasing distance, whereas the transverse wave velocity δw increases as $N(r)^{-\frac{1}{4}}$ (if wave dissipation is neglected), so that there may be a cross over somewhere beyond the sonic point with $NM\langle(\delta w)^2\rangle/r$ and $(d/dr)(\delta B)^2/8\pi$ becoming larger than the thermal pressure gradient dp/dr. If the waves are dissipated sufficiently slowly that the temperature falls and the wave disturbance δw exceeds the characteristic ion thermal velocity, then the distant acceleration is more a result of the pressure contributed by the wave than by the thermal energy of the gas. In this way the Alfven waves produced in the photosphere and in the network activity provide both extra thermal energy, so as to maintain p, and an extra pressure of their own, to boost the wind velocity to the final velocity of the high speed stream. It should be emphasized that this function, of providing an "afterburner" to produce the high speed streams, is an essential part of any final quantitative model of the coronal hole and its heat source. The ultimate high speeds of 500-700 km/sec cannot be reached without the energy transport of the Alfven waves into the distant coronal hole. To what extend wave pressure compliments thermal pressure remains to be determined. It should be noted that the coronal expansion, against the gravitational field of the sun, past the sonic point to speeds of, say, 300 km/sec is a thermal process, in which waves play only the role of transporting energy and maïntaining the temperature of the gas. Some curious statements are sometimes found in the literature nowadays to the effect that the solar wind is driven by wave momentum rather than by thermal pressure gradients. The fact is that the supersonic wind is created by thermal expansion, and the final boost to create high speed streams is some (model dependent) combination of thermal and wave pressure.

A semi-quantitative treatment of the relative importance of wave pressure and thermal pressure can be constructed from the expression

$$I_w(r) = N(r)M\langle(\delta w)^2\rangle V_A \quad \text{ergs/cm}^2\text{ sec} \tag{11}$$

for the energy flux in an Alfven wave with Alfven speed $V_A = B(r)/[4\pi MN(r)]^{\frac{1}{2}}$ and mean square transverse velocity $\langle(\delta w)^2\rangle$. Suppose that the cross sectional area of a flux bundle, along which the corona is expanding, varies as $f(r)^2$ with $f(r) \sim r/a$ for a purely radial field. Then at a radial distance r the field density is

$$B(r) = B(a)/f(r)^2, \tag{12}$$

and conservation of matter requires

$$N(r) = N(a)v(a)/f(r)^2 v(r), \tag{13}$$

where $v(r)$ is the wind velocity and $r = a$ is some reference level in the corona. If a fraction $\gamma(r)$ of the initial wave intensity has been dissipated at a radial distance r, then

$$I_w(r) = I_w(a)[1-\gamma(r)]/f(r)^2 \tag{14}$$

Solve equation (11) for $\langle(\delta w)^2\rangle$ and use equations (12)–(14) to express the result in terms of f, v, and γ, obtaining

$$f(r) = \frac{\langle(\delta w)^2\rangle}{u(r)^2}\left[\frac{v(a)}{v(r}\right]^{\frac{1}{2}}\frac{T(r)}{T(a)}\frac{V_A(a)p(a)}{I_w(a)[1-\gamma(r)]}, \tag{15}$$

where u is the characteristic thermal velocity such that $p = 2NkT = NMu^2$ for ionized hydrogen, and $p(a) = 2N(a)kT(a)$. Let the reference level $r = a$ represent the base of the coronal hole ($a = 1R_\odot$), where $N(a) \cong 10^8$ atoms/cm^3, $B(a) = 10$ gauss, and $V_A(a) = 2\times10^8$ cm/sec. The Alfven waves that heat the wind beyond the sonic point have an energy flux of the order of $I_w(a) \cong 1\times10^5$ ergs/cm^2 sec, so that (15) becomes

$$f(r) = \frac{80\langle(\delta w)^2\rangle}{[1-\gamma(r)]u(r)^2}\left[\frac{v(a)}{v(r)}\right]^{\frac{1}{2}}\frac{T(r)}{T(a)}. \tag{16}$$

Thus $u^2/\langle(\delta w)^2\rangle = 80$ at $r = a$ where $\gamma = 0$. So the wave pressure is negligible. But this ratio diminishes with increasing r/a, reaching one where

$$f(r) = \frac{80}{[1-\gamma(r)]}\left[\frac{v(a)}{v(r)}\right]^{\frac{1}{2}}\frac{T(r)}{T(a)}. \tag{17}$$

As an example, suppose that $T(r)$ has fallen to half of $T(a)$, and $\gamma(r)$ is also equal to one half. Then with $v(r) = 400$ km/sec and $v(a) = 10$ km/sec, the result is $f(r) = 12$. For a purely radial expansion, then, the wave pressure and the thermal pressure become equal at $r/a = 12$.

Equation (13) can be used to write (15) as

$$\left[\frac{N(a)}{N(r)}\right]^{\frac{1}{2}} = \frac{\langle(\delta w)^2\rangle}{u^2}\left[\frac{T(r)}{T(a)}\right]\frac{V_A(a)p(a)}{I_w(a)[1-\gamma(r)]}, \quad (18)$$

so that the thermal pressure and the wave pressure become equal where $N(a)/N(r) \cong 7 \times 10^3$, or $N(r) = 1.4 \times 10^4$ atoms/cm^3.

Discussion on a more quantitative basis requires a precise model for the geometry $f(r)$ of the coronal hole and the radial temperature variation $T(r)$. The present simple relationship is sufficient to illustrate the general idea.

7 Summary

There are a number of ideas on the nature of coronal heating and the creation of the high speed streams of solar wind to be found in the literature. The foregoing exposition expresses our own evaluation of the available ideas. It is essential in any theory of the corona that all the basic observational facts should be confronted and expounded, insofar as possible, particularly if there is serious difficulty in meeting them. There are serious gaps in the views expressed here that can be filled in only by future observation. As already remarked, we can be certain of nothing until there is observational verification or support, because it is well known from past experience that Nature is much more imaginative than the human mind. The conjectures and speculations of the theoretician are shooting at a target in an n-dimensional (parameter) space, where n is a large number. The probability of hitting the bulls eye through apriori reasoning is slight.

It should be noted that in every case the crucial observations to test the theory of coronal heating involve measurements on small-scales to detect small events. The random walk of the photospheric magnetic fibrils is an essential part of the theory of the active X-ray corona, as is also the individual nanoflare and the collective sea of nanoflares that makes up a coronal X-ray loop. All of these must be studied closely if the theory is to be properly tested.

The detailed behavior and activity of the network fields, and the associated coronal waves and wave dissipation, must be determined by observation if we are to be clear on the formation and operation of the coronal hole and the formation of the high speed streams in the solar wind.

This will all take time and a lot of ingenuity and hard work. The degree of success will depend upon the development of improved instrumentation, with the ground based LEST and the orbiting OSL as two essential developments to lead the investigation. The terrestrial effects of the solar wind and the UV and X-rays alone justify the effort on a purely practical or economic basis. The scientific importance of understanding the mass loss and X-ray emission from ordinary stars goes without saying. In particular, the field of X-ray astronomy can make only limited scientific progress until the reasons for the X-ray emission are clearly understood.

References

1. L. Biermann: Naturwissenschaften **33** 118 (1946)
2. H. Alfven: Mon. Not. Roy. Astron. Soc. **107** 211 (1947)
3. M. Schwarzschild: Astrophys. J. **107** 1 (1948)
4. E.N. Parker: Interplanetary Dynamical Processes, New York Interscience Division of John Wiley, 1963, pp. 113-130
5. G.L. Withbroe and R.W. Noyes: Annual Rev. Astron. Astrophys. **15** 363 (1977)
6. D.E. Osterbrock: Astrophys. J. **134** 347 (1961)
7. R.F. Stein and J. Leibacher: Annual Rev. Astron. Astrophys. **12** 407 (1974)
8. M.A. Lee and B. Roberts, Astrophys. J. **301** 430 (1986)
9. V.M. Vasyliunas: Rev. Geophys. Space Phys. **13** 303 (1975)
10. T. Tajima, F. Brunel, and J. Sakai: Astrophys. J. Letters **258** L45 (1982)
11. J. Sakai, T. Tajima, and F. Brunel: Solar Phys. **91**, 103 (1984)
12. W.M. Glencross: Astrophys. J. Letters **199** L53 (1975)
13. E.N. Parker: Cosmical Magnetic Fields, Oxford, Clarendon Press (1979) p. 391
14. W.M. Glencross: Astron. Astrophys. **83** 65 (1980)
15. E.N. Parker: Astrophys. J. **244** 644 (1981)
16. E.N. Parker: Astrophys. J. **264** 642 (1983)
17. E.N. Parker: Astrophys. J. **330** 474 (1988)
18. P.A. Sturrock, W.W. Dixon, J.A. Klimchuck, and S.K. Antiochos: Astrophys. J. Letters **356** L31 (1990)
19. E.N. Parker: Geophys. Astrophys. Fluid Dyn. **34** 243 (1986); Geophys. Astrophys. Fluid Dyn. **35** 277 (1986)
20. E.N. Parker: Geophys. Astrophys. Fluid Dyn. **22** 195 (1982); Geophys. Astrophys. Fluid Dyn. **23** 85 (1983)
21. E.N. Parker: Geophys. Astrophys. Fluid Dyn. **45** 159 (1989)
22. E.N. Parker: Geophys. Astrophys. Fluid Dyn. **52** 183 (1990)
23. E.N. Parker: Astrophys. J. **244** 631 (1981)
24. E.N. Parker: Geophys. Astrophys. Fluid Dyn. **45** 169 (1989); Geophys. Astrophys. Fluid Dyn. **46** 105 (1989)
25. S.R. Habbal and G.L. Withbroe: Solar Phys. **69** 77 (1981)
26. G.E. Brueckner and J.D. Bartoe: Astrophys. J. **272** 329 (1983)
27. R.P. Lin, R.A. Schwartz, S.R. Kane, R.M. Pelling, and K.C. Hurley: Astrophys. J. **283** 421 (1984)
28. J.G. Porter, R.L. Moore, E.J. Reichmann, O. Engvold and K.L. Harvey: Astrophys. J. **323** 380 (1987)
29. G.L. Withbroe: Astrophys. J. **325** 442 (1988)
30. E.N. Parker: Astrophys. J. (in press) (1990)
31. S.F. Martin: Small-Scale Dynamical Processes in Quiet Stellar Atmospheres, National Solar Observatory/Sacramento Peak, ed. S.L. Keil (1984) p. 30
32. S.F. Martin: Solar Phys. **117** 243 (1988)
33. S.F. Martin: Solar Photosphere: Structure, Convection and Magnetic Fields, IAU Symp., ed. J.O. Stenflo (1990) p. 129
34. J.G. Porter and R.L. Moore: Proc. Ninth Sacramento Peak Summer Symp. (National Solar Observatory/Sacramento Peak) ed. R.C. Altrock (1987) p. 125
35. J.F. McKenzie, Seventh Quadrennial STP Symp. The Hague (1990)

36. E.N. Parker: Space Sci. Rev. **4** 666 (1965)
37. A.J. Hundhausen: Coronal Holes and High Speed Wind Streams, Boulder, Colorado, Colorado Associated University Press, ed. J.B. Zirker (1977) p. 225
38. T.E. Holzer and E. Leer: J. Geophys. Res. **85** 4665 (1980)
39. T.E. Holzer and T. Fla, and E. Leer: Astrophys. J. **275** 808 (1983)
40. E.N. Parker: Astrophys. J. **139** 72 (1964)

The Sun as a Prototype in the Study of Stellar Magnetic Activity

C.J. Schrijver

European Space Agency (ESTEC), P.O. Box 299,
2200 AG Noordwijk, The Netherlands

1. Introduction

The proximity of the Sun to the Earth allows detailed, spatially-resolved observations. The complexity of the solar atmosphere has not ceased to surprise us ever since the telescope, and in particular the spectrograph, has been used to observe the Sun. Details show up down to scales below an arcsecond, while future high-resolution telescopes probably hold more surprises in store. Not only the high degree of structuring is surprising, but also the high temperature of the outer atmosphere (i.e. everything above the photosphere), reaching values in excess of 10^6K in the outermost regions. The atmospheric plasma is evidently non-radiatively heated by the dissipation of energy which is being delivered either actively or passively (or both) to the outer layers by the complex magnetic field which at the same time serves to confine the heated plasma. Observers structured the confusion contained in the observed phenomena by introducing names: chromosphere, coronal mass ejection, ephemeral region, facula, flare, granule, network, plage, pore, prominence, spicule, spot, umbra, ... The list of terms is a long one indeed with new ones still being added. Continued observations revealed the solar cycle with its associated "butterfly diagram," the differential rotation, etc. Observations of the "objects" which are described by these terms create the mental framework within which solar physicists communicate.

Even with all its surprises, the Sun appears to be an ordinary star: non-radiatively heated atmospheres, shaped by magnetic fields, are common to most stars on the cool side of the Hertzsprung–Russell diagram. Astrophysicists began to use the mental framework developed by the solar physicists to interpret the far less detailed stellar observations. The positive effect of this is that many observational results from stellar studies can be modelled in what thus far seems to be a realistic representation of physical reality. But there is the danger that the mental framework turns into a cage which restrains our imagination from looking for new ideas to interpret observational data of structures which seem to have no counterpart on the Sun.

This review outlines some of the areas where the Sun appears to be a useful archetype of a magnetically active star, and some areas where this may not be the case. It is not meant to be complete in any sense. Conclusions drawn from the minds of many and refined in series of publications will often simply be stated with only the minimum of references. This paper should present a brief and basic summary of a selection of topics in the field of stellar magnetic activity. With magnetic activity I mean all phenomena, particularly those in the stellar atmospheres, which are directly or indirectly caused by the presence of a magnetic field in stars similar to the Sun (see Section 3 for the required properties for this similarity), although large–scale flaring processes are excluded.

This review concentrates on the peculiar circumstance that relationships between radiative flux densities from the outer atmospheres of a large variety of cool stars are often distinctly non–linear while exhibiting surprisingly little scatter despite the large range of fundamental properties of the stellar sample. This is all the more astounding in view of the complex and variable structure of the solar outer atmosphere. Why this is so, and how it may be explained by combining solar and stellar observations, constitutes the core of this review. The range of topics covered serves also to convey the complexity of the stellar outer atmosphere which nevertheless results in such simple flux-flux relationships.

2. Solar Magnetic Activity

2.1 Photospheric Magnetic Fields

2.1.1 Manifestation of Magnetic Flux

Most of the magnetic flux passing through the photosphere is concentrated in patches of high field strength embedded in relatively field-free plasma (although a weak intranetwork field appears to exist [1]). The basic building block of the photospheric magnetic field is the flux tube [2, 3], which is formed by a convective collapse which follows the surfacing of strands of magnetic field through the photosphere. The mechanism which drives this collapse is the following: The plasma rising with the magnetic field cools quickly by radiation as soon as the photosphere is reached. Observations indicate that the magnetic field strength in the emerging tubes is already high, and close to the equipartition value of 500 Gauss [4], so that the energy which is lost by radiation cannot be replenished by the impeded convective motions. The radiatively cooled plasma flows down along the field lines, and the subsequent concentration of the emerged magnetic field by the "convective collapse" causes field strengths to reach values in excess of 1500 Gauss. The downflow stops when the increased magnetic pressure balances the gas-pressure difference between the exterior and the relatively cool interior of the tube.

The size of the magnetic elements determines the overall contrast: The result of the cooling of the interior of the flux tube is that the interior of spots and pores appears dark. In small flux concentrations, however, another effect dominates: the lower gas pressure lowers the level at which $\tau_{5000} = 1$ within the tube so that hotter, and brighter, subphotospheric layers contribute significantly to the brightness of the tubes depending on the size and the viewing angle (reviewed in [5]). The observational properties of patches of magnetic flux thus depend on the total magnetic flux which they contain; a hierarchy of increasing magnetic flux results in flux fibers, faculae or network clusters, magnetic knots or micropores, pores, and sunspots (e.g. [6]). Typical observed field strengths range from about 1200 Gauss in the smallest elements up to 3200 Gauss in large sunspots.

The magnetic elements in the photosphere commonly surface in bipolar patterns referred to as "active regions." Active regions come in many shades of complexity, with sizes ranging over several orders of magnitude if small emerging flux regions are included in the definition. Active regions have a tendency to appear in the immediate vicinity of an existing region or at a site where an active region surfaced previously. Such clusters of active regions are called "active nests," and typically live some 6 to 15 months. In large nests the total magnetic flux remains constant to within a factor of two, and the magnetic polarities are closely balanced [7].

2.1.2 Evolution of Magnetic Flux

The magnetic flux tubes are embedded in the solar convection. The granulation operates on a scale of about one thousand kilometers. Its characteristics were succinctly summarized by *Stein* [8]: "Fast, cool, thin, twisting, isolated downdrafts [merging at larger depths] embedded in slow, warm, diverging, smooth upflows [much of which never reaches the surface]." This flow pattern twists and tangles the magnetic flux tubes, which may cause field discontinuities with associated energy dissipation in the higher regions of the atmosphere (e.g. *Parker*, these proceedings).

The supergranulation displaces magnetic flux over relatively large distances. The first description of the supergranular velocity field was given by *Hart* [9, 10], soon followed by the statistical studies by *Leighton* et al. [11] and *Simon* and *Leighton* [12]. Approximately 10,000 supergranules cover the solar surface, with a typical diameter of ≈ 28,000 km [13, 14, 15]. Life time estimates for supergranules range from ≈ 20 hrs. [12, 16, 17]) up to ≈ 40 hrs. [18, 19, 20]. These estimates are often "deformation time scales," i.e. correlation times over which the feature(s) used to trace the supergranule (for instance the area of downdraft, or the chromospheric or magnetic network) is displaced by its own characteristic dimension. The cells may in fact exist up to two days or more (e.g. [21, 22]). Network cells forming the enhanced network appear to live longer than cells in regions with a low flux tube density [4], perhaps as long as 3 days [23] or more [24].

Evolutionary changes of supergranules, including their formation and disappearance, have been suggested to cause a random–walk dispersal of the flux tubes [25]. In this process, the flux – that has been transported to the areas of downflow immediately after surfacing in the photosphere – shifts so that it always remains at the cell boundaries. The decay of magnetic plages, however, appears to defy a description by a simple, uniform diffusion process: for one thing plages do not gradually weaken and become less well defined as the magnetic flux leaks out into the surrounding network. Instead they maintain a more-or-less clear periphery at ≈ 50 Gauss, while the average magnetic flux density within a 50 Gauss perimeter is 100 ± 20 Gauss, regardless of the size or evolutionary stage of the region, with the possible exception of the emergence phase. The mean value of about 100 Gauss implies a magnetic filling factor in the photosphere of $\approx 5 - 10\%$.

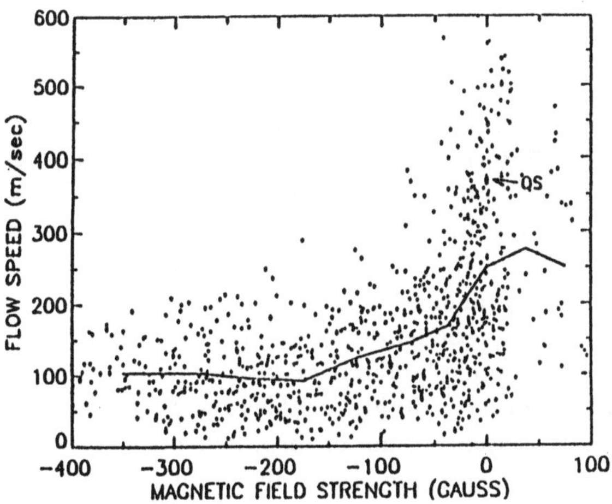

Fig. 1. Horizontal flow speed v_{phot} of the photospheric plasma vs. magnetic flux density $|\varphi|$. The solid curve represents the mean flow speed $\langle v_{\text{phot}} \rangle$ as a function of $|\varphi|$. The point labelled QS shows $\langle v_{\text{phot}} \rangle$ for the quiet Sun with $|\varphi| \approx 0$ (from [26])

This so-called "plage phenomenon" may be caused by an interaction between the local flux–tube density and the size of the network cells (e.g. [4]): within dense plages no cell structure is seen except for some isolated, relatively empty patches with diameters up to several thousand kilometers, while rings of supergranular sizes are only found in the outskirts of active regions where the magnetic flux density is relatively low. Hence the suggestion that convective cells appear wherever the local magnetic flux density is sufficiently low, while the cells are larger where the magnetic flux density is lower. This is supported by observations showing that the characteristic diameter for the active–region "network" is $\simeq$14,000 km [27], and plage cell sizes inferred from maps of horizontal flow patterns are typically 10,000 km to 15,000 km [28], suggesting that not just the network pattern but indeed the flow pattern itself is modified. Observations by the Solar Optical Universal Polarimeter [29] flown on Spacelab 2, in combination with a powerful technique to track the motions of granules [30], suggest that the presence of magnetic flux also affects the flow velocity in the centers of the cells as well as the cell size; average horizontal velocities for granules caught in the supergranular flow lie around 300-400 m/s and 100-150 m/s in quiet and magnetic Sun, respectively (cf Fig. 1).

These changes in the convective properties as a function of the mean magnetic flux density can slow the diffusion of flux away from plages, while a sufficiently pronounced decrease of cell size with increasing magnetic flux density could explain the rapid transition from plage to network. The slower dispersal of flux into the network also offers an explanation for the fact that a dominant fraction of the active-region flux appears to cancel near the polarity-inversion line within the bipolar plage [31], leaving a reduced amount to disperse into the network [32].

Once magnetic flux has left the active region, it is further dispersed by the supergranulation and large-scale flow patterns such as the differential rotation, the meridional circulation, the torsional oscillation, and other flow patterns occurring on intermediate scales (e.g. [33]). The dispersion of magnetic flux over the solar surface on a scale much larger than a typical plage can indeed be modelled accurately by a random–walk diffusion process, with differential rotation and meridional flow taken into account (e.g. [34]).

Motions along supergranular lanes appear to contribute substantially to the random walk of magnetic features (e.g. [22, 28, 35]). If the flux tubes move along the supergranule boundaries with a mean velocity of 100 m/s [36] a supergranule lifetime of 50 hrs. enables the flux tubes to travel some 18,000 km without experiencing major displacements caused by the evolution of supergranules. The process of flux displacement along supergranular lanes could therefore form a major contribution to the random walk of magnetic flux tubes (implications for the decay of plages are discussed by [32]).

Eventually flux tubes will either experience a chance encounter with a flux tube of opposite polarity and be (partially) cancelled, or they may drift to high latitudes. These continuous processes of surfacing and cancelling of magnetic flux result in a wide range of flux–tube filling factors. The process which leads to the disappearance of flux from the photosphere is still largely unknown. However, because of the low electrical resistivity of the plasma, the magnetic field cannot free itself from the gas to escape, but must be transported together with the gas: patches of opposite polarity must be moved together to start a convective downdraft which can pull the flux loop

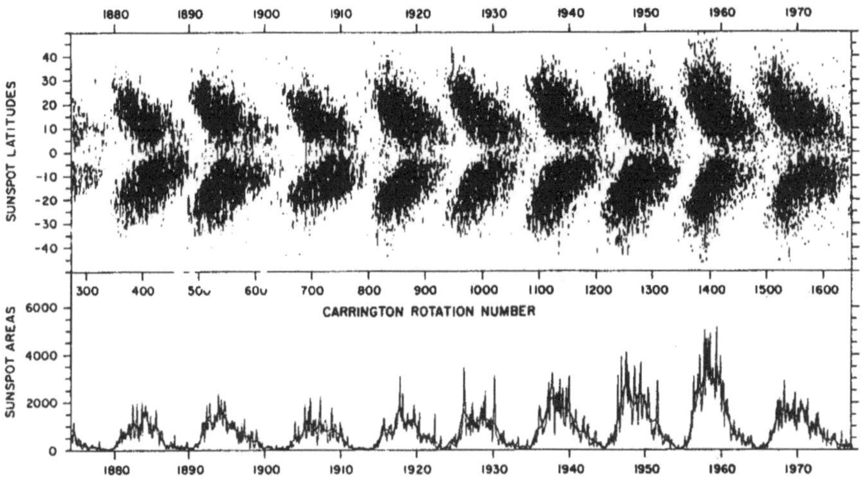

Fig. 2. Butterfly diagram displaying the observed latitudes of sunspots as a function of time (compiled by J.A. Eddy, HAO, from [37])

downward, which can either be a retraction of the complete loop or part of the field may be detached from the subphotospheric field (for further details see review by *Zwaan* [6]. The observational effect is that of a "cancellation" of photospheric flux, which is, of course, a purely descriptive term.

2.1.3 The Solar Cycle

The amount of magnetic flux in the solar photosphere, the occurence of sunspots, the latitude at which active regions appear, ... in fact just about every aspect of the observed patterns in photospheric magnetic activity is variable on a time scale of about 11 years. This phenomenon, referred to as the "solar cycle," varies in duration and amplitude, ostensibly without defining a predictable pattern. The number of active regions visible on the surface varies strongly throughout the cycle, although the vast majority of the active regions follow well-studied patterns: Active regions surface generally in bands on either side of the equator, extending up to some 40 degrees in latitude. The average latitude at which active regions surface shifts from about 30 degrees down to about 10 degrees as the solar cycle progresses, but individual regions surface in a range of ≈20 degrees around the average latitude, thus giving rise to the "butterfly diagram" (Fig. 2). The majority of active regions, with the exception of the smallest regions, are oriented nearly east-west, with a small inclination which puts the polarity preceding in the solar rotation closer to the equator. Most active regions also obey Hale's polarity law: active regions on either side of the equator have the same polarity leading in rotation, while these polarities are opposite on

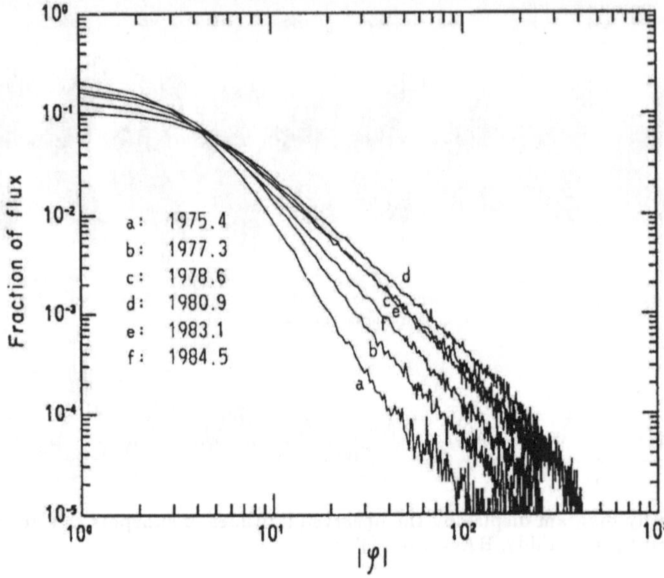

Fig. 3. Distribution function $h_t(|\varphi|)d|\varphi|$ (derived from solar synoptic magnetograph maps with a resolution of 1° in heliographic longitude and 1/90 in sine latitude; the bin width for the histogram is 1 Gauss) for absolute values of magnetic flux densities at different times in the solar cycle (see legend). The time-dependent distribution function is shown after averaging over sets of six consecutive solar rotations (from [40]).

the northern and southern hemispheres, as well as alternate in successive cycles. The evolution of plages appears to be independent of the phase in the solar cycle, and the shape of the distribution function for areas of emerging bipolar regions does not change with the phase in the activity cycle but is merely offset by a multiplicative factor (e.g. [38, 39]), except perhaps for the smallest regions.

The surfacing and subsequent dispersion of magnetic flux changes the distribution of magnetic flux over the solar surface. Of particular interest to this review (see Section 5) is the distribution function for magnetic flux densities[1], presented in Fig. 3 for different phases in the solar cycle. The temporal changes in $h_t(|\varphi|)$ are smooth: an increase of the tail at high values of $|\varphi|$, caused by an increase in the emergence of bipolar regions, is coupled with a decreased coverage by areas with the lowest flux densities, with related changes anywhere between these two extremes. As a result, the distribution function $h_t(|\varphi|)$ *cannot* be split into two distinct distribution

[1] The mean magnetic flux density $|\varphi|$ is the absolute value of the magnetic flux as measured in one resolution element of the magnetograph, expressed per unit area, i.e. in units of weber/cm^2. This is equivalent to the average value $|\varphi|$ of the product of the surface filling factor f and the intrinsic magnetic field strength $B \approx$ 1–2kG.

functions, one for a "typical quiet Sun" and one for the variable active–region and enhanced–network component, with time–dependent area coverage factors (see also [39], and 41]: the entire solar surface is affected by variations in activity, and a simple separation into "quiet" and "active" components cannot be made. Note, however, that the tails of $\mathbf{h}_t(|\varphi|)$ at different times t are parallel, reflecting the similarity of plage properties at different phases in the cycle (see Section 2.1.2).

2.2 Atmospheric Radiative Fluxes

2.2.1 Chromosphere, Transition–Region, and Corona

The non-radiatively heated solar atmosphere emits radiation from different temperature regimes. Throughout this review the solar and stellar outer atmospheres are divided into three regimes, defined as follows: the chromosphere, with temperatures $T \lesssim 2\,10^4$K, the transition region, with $5\,10^4 \lesssim T \lesssim 3\,10^5$K, and the corona, with $T \gtrsim 10^6$K. Typical chromospheric diagnostics are the Ca II H+K and Mg II h+k resonance lines. Transition–region diagnostics are mainly found in the UV, such as the C II and C IV resonance lines. One of the most important diagnostics for coronae is the soft X–ray flux. These three regions, each highly structured and certainly not with a simple stratified temperature structure, are particularly bright over magnetic plages, but in the case of the Sun the emission of the Sun-as-a-star is dominated by emissions from the "quiet" network in all but the coronal soft X–rays [39].

The transition region between chromosphere and corona should not be viewed as merely the classical conduction-heated interface between regions of low and high temperature: it extends over several thousand kilometers, much more than would be expected in the classical case of thermal conduction, and has a complicated structure. The region may be heated, in addition to heating by classical thermal conduction, by ambipolar diffusion of ionized and neutral hydrogen [42], by turbulence [43], or by other mechanisms.

A direct comparison of atmospheric radiative losses from different temperature regimes is difficult because of the complex topology, with perspective distortions confusing the problem. In order to circumvent this one may study, for instance, the relationships between integrated radiative losses from different temperature regimes of entire active regions and of large sections of quiet regions [44]. The relationships between the total intensities of plages in several UV lines or soft X–rays are tight, but these relationships mainly reflect changes in area of the plages, because the average intensities of active regions (i.e. the total intensity divided by the total area), as observed in a number of UV lines, cover only a small range: less than a factor of two in chromospheric lines, and less than a factor of four in transition–region and coronal lines. This phenomenon reflects the plage phenomenon

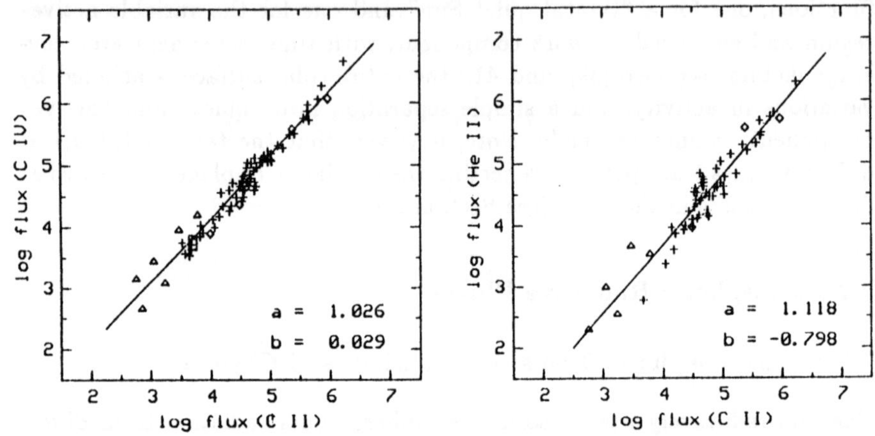

Fig. 4. Comparison of line fluxes (in $\mathrm{erg\,cm^{-2}\,s^{-1}}$) for magnetically active late-type stars and for spatially resolved solar regions. Symbols: × stellar data from low-resolution IUE spectra, ◇ solar data obtained with the $2'' \times 60''$ NRL Slit Spectrograph on Skylab, □ disk-integrated solar data obtained with the Solar Mesosphere Explorer. Power-law fits are shown, together with the correlation coefficient and the slope and intercept of the best-fit line. (From [46])

described in Section 2.1.3. Because the area-averaged properties of plages differ little, they are of little direct value for the study of the interdependence of radiative flux densities (but see [45]).

When looking at chromospheric lines and – to a lesser extent – at transition-region lines the perspective effects are not too important, and a direct point-by-point comparison of the radiative losses may be attempted. Such a comparison is shown in Fig. 4 for emissions integrated over a $2'' \times 60''$ (or $\approx 1500 \times 47000$ km) slit [46]. Some of the relationships are slightly non-linear. The agreement with stellar data is striking (see Sections 3.1 and 5).

2.2.2 Radiative and Magnetic Flux Densities

The observed structures in the solar atmosphere agree well with computed magnetic field lines, as estimated from the observed photospheric boundary conditions, revealing a direct link between the field and the heating. Relationships between radiative and magnetic flux densities from the solar outer atmosphere have been determined; these are generally non-linear, except perhaps for the coronal emission (reviewed in [45], and Figs. 5 and 6). When studying chromospheric emission a minimum or "basal" flux – unrelated to magnetic activity – has to be corrected for (see Section 3.1).

The relationship between $F_{\mathrm{Ca\ II}}$ and $|\varphi|$ is shown in Fig. 5. The figure relates the observed radiative and magnetic fluxes for a resolution of about

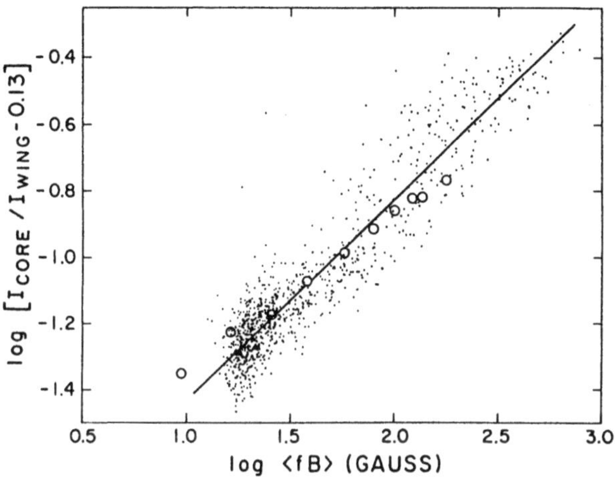

Fig. 5. The Ca II K line–core intensity ratio I_{core}/I_{wing} (I_{core} the intensity in a 1.04Å band centered on the Ca II K line, and I_{wing} the intensity in a 2.08Å wide window 7.39Å to the red) after subtraction of the minimal observed line–core intensity ratio (0.13) as observed in the plage and surrounding quiet Sun, vs. the magnetic flux density $|\varphi| \equiv \langle fB \rangle$ for a resolution of 14.″4 (from [48]). The power–law fit has an exponent of 0.6. The circles represent interval–averaged values from Skumanich et al. [49] for a quiet solar region.

14″. The original data have a resolution of 2.″4, but these exhibited a large scatter. Lowering the resolution decreased the scatter about the mean relationship markedly, without affecting its characteristics. Schrijver et al. [48] argue that 14″ represents a typical length scale within which the mean magnetic flux density does not change markedly, but which permits averaging over a number of flux tubes so as to smooth out both temporal changes and effects of the relative positions of the flux tubes within the resolution elements (see Section 2.2.3 on a possible interaction between flux tubes).

The implications of the relationship between $F_{\mathrm{Ca\,II}}$ and $|\varphi|$ for the mechanism(s) heating the chromosphere are rather indirect, because the Ca II H, K lines are optically thick (but see Section 2.2.3). The relationship between $F_{C\ IV}$ and $|\varphi|$ is also non–linear (Fig. 6), but here the implications for the mechanism heating the effectively thin transition region are more direct: the amount of energy radiated in the C IV doublet varies weaker than linearly with the magnetic flux density. Since the transition region probably loses most of its energy by radiation, the net heating rate of the transition region appears to vary non–linearly with the magnetic flux density.

Establishing the relationship between coronal soft X–ray and magnetic flux densities is greatly hampered by the severe projection effects which occur when studying the extended coronal loops, as well as complicated by the fact that individual bright loops are seen in what must be a continuous magnetic field. A clear constraint on the relationship is formed by the contrast in the soft X–ray flux density between a typical plage and a typical quiet Sun which is a factor of ≈ 30 [44], while the magnetic flux density

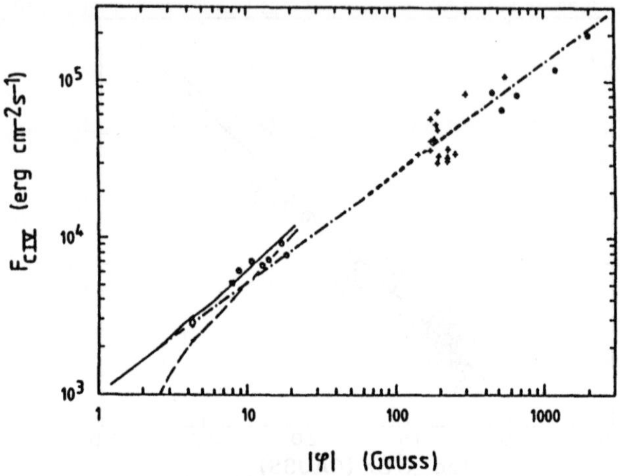

Fig. 6. **Relationship between flux densities $F_{C\ IV}$ in the C IV λ 1548 + 1551 doublet and magnetic flux densities $|\varphi|$ (from [45]). The full–drawn curve represents the relation derived from distribution functions of C IV intensities and magnetic flux densities for solar quiet regions, with an asterisk denoting the point above which the relation is determined by less than 10% of the data, and the curve ending at the point where less than 1% of the data determine the relation. After correction for a possible basal flux level of 10^3 erg cm^{-2} s^{-1} (see Section 3.2) the long–dashed curve results. The power–law fit derived for active–region data is shown by the heavy, short–dashed line segment. The extension of this line segment, which is also the best–fit relationship $F_{C\ IV} = .10^3 \cdot |\varphi|^{0.70 \pm 0.03}$ is represented by the dashed–dotted line. Symbols: ◇ the mean values for a quiet region, ○ fluxes for the Sun–as–a–star (observed by SME), + average values for active regions, • stellar (IUE) data (i.e. for hemisphere–averaged flux densities $\langle F_{CIV} \rangle$ and $\langle |\varphi| \rangle$).**

differs by a factor of ≈ 25 [47]. This suggests that the relationship between F_X and $|\varphi|$ is nearly linear, at least when looking on a scale which is large enough to average over entire active regions.

One phenomenon which should not be neglected in a description of magnetic activity is the coronal hole: if active regions surface in a suitable configuration, a large area may be formed where one polarity dominates in the network .The field lines in the outermost layers of the corona over these "monopolar areas" (which also occur at high latitudes during sunspot minimum, causing polar coronal holes) can be open to the interplanetary space. The coronal brightness is greatly reduced (cf. Fig. 9), although the chromospheric and transition–region flux densities differ only slightly between the area of the coronal hole and ordinary quiet regions. The open fields guide the solar wind. The relative rigidity of this flow out to a considerable distance from the Sun enables an efficient loss of angular momentum. On the time scale of the solar evolution this results in a significant "magnetic braking".

2.2.3 Magnetic Topology in the Lower Chromosphere

The complex topology of radiating features in the upper chromosphere and above is caused by highly non–local heating even though the magnetic field is almost ubiquitous. In the lower chromosphere the topology of the field itself is such that magnetic and non–magnetic parts of the atmosphere are interspersed: The field in the solar photosphere is concentrated mainly into flux tubes with kG field strengths. These are concentrated in a network along the edges of the supergranular cells, so that the magnetic field covers only a small fraction of the solar surface. The gas pressure decreases with increasing height above the surface, and consequently the flux tubes expand until they cover almost the entire solar surface (cf. Fig. 7). The lower chromosphere appears to be divided into regions with a high temperature in the magnetic part of the atmosphere, and with a low temperature in the non-magnetic atmosphere [53, 54]. Depending on the local non-radiative heating, the temperature stabilizes either well below 4000K, caused by an efficient cooling by bands of molecular CO, or well above 5000K, where UV resonance lines control the energy balance. *Ayres* and co–workers (e.g. [53, 54]) suggested that most of the gas at chromospheric altitudes is in fact cool, while the classical chromosphere is concentrated in small-scale, intensely-heated bright points.

The effect of this thermal bifurcation is that magnetic flux tubes expand much more rapidly with height than in a smoothly layered homogeneous atmosphere (cf. Fig. 7); the flux tubes merge and form a magnetic canopy [55] with an almost horizontal base which does not exceed approximately 800-1000 km [50], independent of the magnetic filling factor. The expected presence of waves in the solar atmosphere, generally accompanied with pressure fluctuations, would yield a canopy base which is continuously rising and falling. *Solanki* and *Steiner* [50] call to mind the picture of water lilies: the firm stem is left largely unaffected by motions in the water, while the broad, supple leaf rides water surface waves.

If the topology of the chromosphere is such that a dominant contribution to the Ca II K line-core emission originates around the canopy height, the emission would depend strongly on the flux-tube filling factor, even if the atmospheric structure in the flux tube and outside would not change [56]: as the filling factor increases, the emission increases rapidly at first, soon beginning to level off as the free expansion of the flux tubes is impeded by neighbouring flux tubes, and eventually saturating at the maximum emission that can be expected from the flux–tube atmosphere. This would cause a non-linearity of the relationship between magnetic and chromospheric radiative fluxes, such as shown in Fig. 5, but more studies are required to determine whether this geometric effect dominates, or whether changes in the atmospheric structure also play a crucial part.

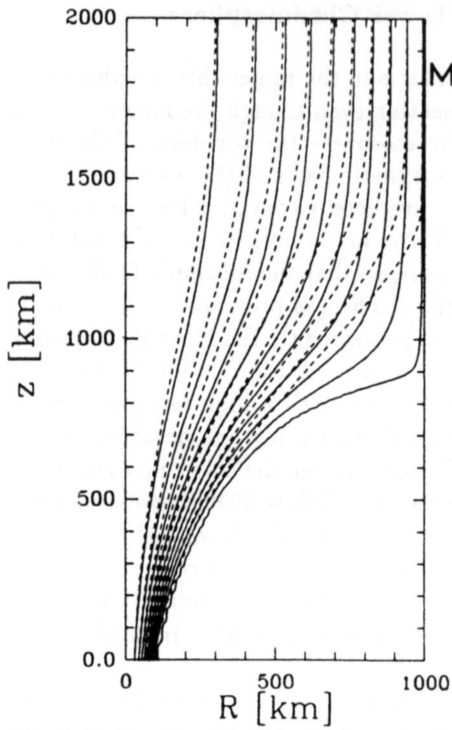

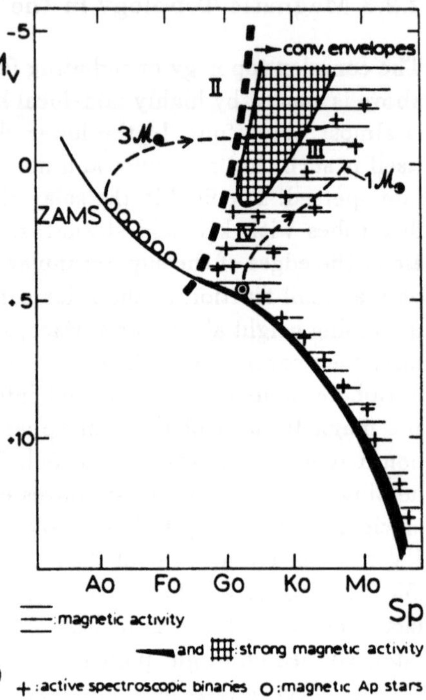

Fig. 7. (left) Selected field lines of a model for magnetic flux tubes [50]. The dashed lines show the tube structure in an atmosphere with equal temperature inside and outside the tube. The temperature profile was taken from the C′ model [51]. The solid curves correspond to the field lines belonging to a flux tube surrounded by a cool radiative equilibrium model, with the temperature within the flux tube again given by the C′ model. The field strength at the base was taken to be 1500 Gauss, the radius at the base 100 km, and the photospheric magnetic filling factor 1%.

Fig. 8. (right) Schematic distribution of magnetically active stars over the Hertzsprung–Russell diagram. Solar–like magnetic activity is found in the region of stars with convective envelopes (bounded to the left by the dashed bold line). Dashed thin lines mark approximate evolutionary tracks for stars of 1 and 3 solar masses (from [52]).

3. Stellar Magnetic Activity

3.1 Photospheric Magnetic Fields

Similar to the sunspots contained in the solar photosphere, starspots appear to occur in stellar photospheres (e.g. [57]). The occurence of large, dark starspots results in a measurable variation of the visual magnitude of the star as spots move onto and off from the visible hemisphere as the star rotates and as the spots evolve. Careful analysis of spectral line profiles makes possible what has become known as "Doppler imaging" [58]: as the spots move across the disk their contribution to line profiles will move from one side of the rotationally broadened line to the other as the line-of-sight velocity changes as a function of the position on the disk. Line-profile studies therefore constrain the size and location (mainly longitude, but also latitude) of the spots or spot groups (the resolution is, of course, rather low). In the most active stars, substantial coverage by starspots has been inferred, up to several tens of percents.

The past decade has seen considerable advances in our ability to measure magnetic fields on cool stars (see, for instance, review by *Saar* [59]), which is hampered by the pepper–and–salt nature of the field, largely cancelling the polarization of the emitted radiation which is present on small length scales. New techniques for observations and analysis have been employed, producing a considerable body of data on the mean magnetic field strength in stellar atmospheres and the hemisphere–averaged fractional surface filling factor for the field. The observations support the theory outlined in Section 2.1.1, in that the field strength increases with decreasing $T_{\rm eff}$ such that the square of the field strength roughly equals the gas pressure, suggesting that it is primarily the external gas pressure which balances the magnetic pressure, while repeated measurements suggest that the field strength appears to be constant while the filling factor may vary. Another conclusion is that the filling factor increases with decreasing rotation period provided the periods are not too short (cf. Section 3.3). Combination of measurements of stellar magnetic and radiative flux densities will allow studies similar to those described for the spatially resolved Sun in Section 2.2.2. The presently available data sets are still limited, but they indicate a good agreement with the spatially resolved solar data (cf. Fig. 6, and [45]).

3.2 Atmospheric Radiative Fluxes

The spectroscopic signatures which revealed the non-radiative heating in the solar atmosphere are also seen on stars: chromospheric and transition-region emission lines and coronal soft X–ray emissions. In many respects these emissions are reminiscent of solar activity. Not only are the relative emission strengths from different temperature regimes comparable (see below), but other striking similarities are found: rotational modulation (e.g. [60]), evidence for starspots (see Section 3.1), cycle–like modulations (e.g. [61]), ... Current data suggest that the required property which makes a star "similar to the Sun" within the context of magnetic activity is the existence of a convective envelope immediately beneath the photosphere (Fig. 8).

Diagrams in which the atmospheric radiative flux densities are plotted as a function of colour $B-V$ show two striking properties (see, for instance, the reviews [52, 63]) a) a well–defined lower boundary to the observed flux which decreases sharply with increasing $B-V$, and b) a large range in

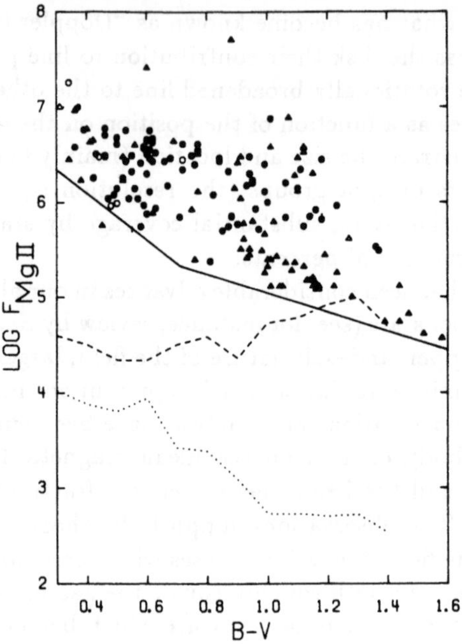

Fig. 9. Flux density $F_{\mathrm{Mg\ II}}$ in the Mg II resonance lines at the stellar surface (in $\mathrm{erg\,cm^{-2}\,s^{-1}}$) vs. colour. Symbols: ○ dwarfs, △ (sub)giants. The estimated detection limit is represented by the dashed and dashed–dotted lines for dwarfs and giants, respectively. The solid line denotes the minimum observed flux level $F_{\mathrm{Mg\ II,min}}$ (closely resembling the basal flux which appears to be a constant contribution to stellar radiative fluxes, varying with stellar colour while independent of rotation rate). Figure from [62]

observed fluxes, even for stars of comparable $B-V$ and luminosity class (see example in Fig. 9). Detailed studies suggest that the observed fluxes contain two distinct components: 1) an "active" component, associated with the magnetic activity of the stars, and 2) a "basal" component which depends markedly only on the stellar colour, perhaps varying slightly with the surface gravity. This basal component is most pronounced in the chromospheric emission, while it is very low or non-existent in coronal emission. The basal component may reflect acoustic heating in the outer atmospheres of stars, caused by the dissipation of sound waves generated in the stellar convection zone and travelling upward into the outer atmosphere [64].

If the observed fluxes are corrected, where appropriate, for the basal emission, colour-independent relationships are found between emissions from different temperature regimes in stellar outer atmospheres (see example in Fig. 10). Within the range of late–F–, G–, and K–type stars which obey the main flux-flux relationships it is worth noting that the most active stars are brighter than the average solar plage and the least active stars (in general the effectively single giants, which have suffered an increase in the moment of inertia due to the increase in radius and which have lost angular

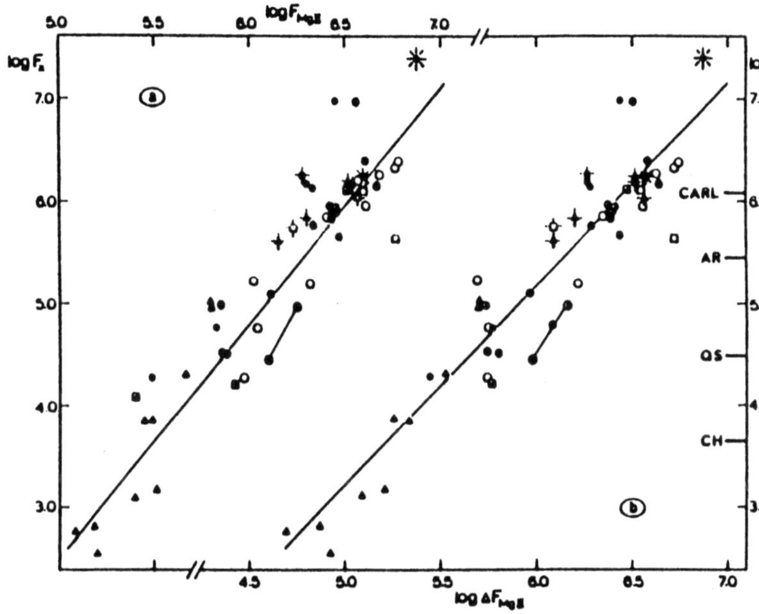

Fig. 10. Soft X-ray flux density, F_X, vs. the Mg II h+k flux density, $F_{Mg\ II}$, (left) and excess flux density $\Delta F_{Mg\ II} = F_{Mg\ II} - F_{Mg\ II,min}$ (right), with $F_{Mg\ II,min}$ as in Fig. 9. Symbols: ⊙ average Sun, ◯ luminosity class LC V and IV-V, □ LC IV, △ LC III and II-III. Open symbols for $B - V < 0.6$, half-filled $0.6 \leq B - V < 0.9$, filled $B - V \geq 0.9$. Typical X-ray flux densities of solar coronal holes (CH), quiet regions (QS), active regions (AR) and compact active-region loops (CARL, the brightest known non-flaring features on the Sun) are displayed on the right-hand side. Close binaries are identified by bars around the symbols, * refers to overactive stars (see Section 4.2). Note that the X-ray and Mg II fluxes were measured up to a decade apart, so that the intrinsic stellar variability contributes to the scatter in the diagram. The uncertain coronal temperature structure also contributes to the scatter, because of the related uncertainty in the countrate-to-flux conversion factor. Figure from [52].

momentum by a stellar wind , [65]) are fainter than a typical coronal hole (cf. Fig. 10). Yet these stars at the extremes of stellar magnetic activity obey the same flux-flux relationships defined by stars with levels of activity closer to that of the Sun.

The relationships between radiative fluxes after correction for the basal fluxes (leaving the "excess" fluxes) from different regimes of the atmosphere are simple power laws, with the value of the exponent differing more from unity with increasing difference in the formation temperature for the two radiative fluxes being compared. There is little or no colour dependence, no clearly observed dependence on luminosity class, age, binarity ... [62], (but see Section 4.1). In fact, the data suggest that once energy is pumped into the magnetic field, it is distributed over the different temperature regimes from chromosphere to corona in such a way that it is impossible to tell apart different types of stars. The structure of the corona, however, may form an exception: X-ray spectroscopy of quiescent stellar coronae revealed that stellar coronae do differ from star to star, depending on luminosity class and rotation rate [66]. Temperatures range from below 1 MK to around 25 MK (even in seemingly quiescent coronae). More active stars appear to have a stronger $\approx$ 25MK component in their coronae, while giants generally appear to have hotter coronae than dwarfs, except for the most active dwarfs. The 25 MK component on stars may not have a quiescent solar counterpart, although observations made with the HXIS instrument on the *Solar Maximum Mission* reveal very weak, small, but long-lived features with temperatures in excess of 10 MK [67]. The nature of these hot features is still unclear. With this one exception, temperatures as high as 25 MK are observed on the Sun only in flares. The stellar 25 MK component may, of course, be related to flares, in which case the generally featureless X-ray light curves which was observed for some of the sources would imply a large number of small flares.

3.3 Activity and Rotation

The excellent similarity in the chromospheric, transition–region, and coronal emissions suggests that one "activity parameter" determines the heating of the outer atmospheres of F–, G–, and K–type stars [68, 69, 70, 71, 72]. One of the most important stellar properties which fix the value of the activity parameter is the stellar rotation rate: with increasing rotation rate the activity level also increases, until $P \lesssim 1$ day (see Section 4.2). It remains as yet unclear how this rotation rate is best expressed (rotation period, Rossby number – i.e. period scaled by the overturn time of convection at some point in the convection zone (mostly at the base) –, or some other scaling), but some basic conclusions are clear: for sufficiently narrow color bins the relation between rotation period and atmospheric radiative flux density is tight, mildly dependent on stellar colour, not significantly dependent on surface gravity, and valid even for cool components in binaries provided the components are not too close (cf. Section 4.2 on "overactivity"). There are some exceptions to this rule, see Section 4.1.

4. Extremes of Stellar Magnetic Activity

4.1 Single Stars

Although most cool stars obey the flux–flux and rotation–activity relationships, some types of stars deviate from these relationships. Here I briefly discuss the F-, and M-type main sequence stars, and the T Tauri stars.

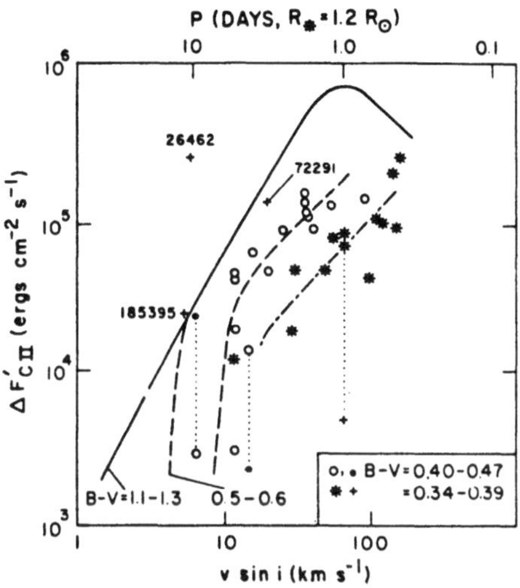

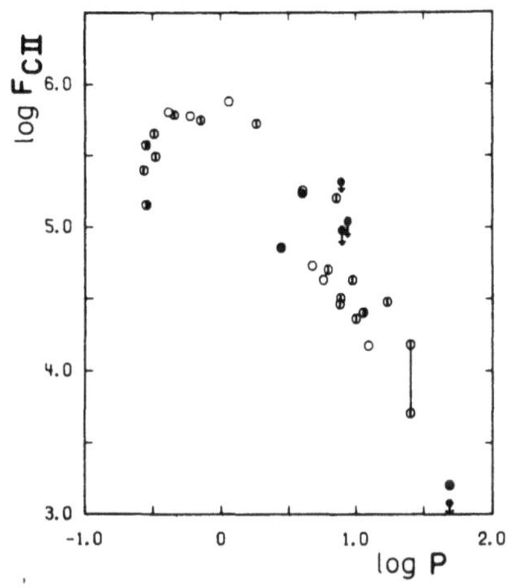

Fig. 11. (left) The weighted average of C II and C IV excess flux $\Delta F'_{\mathrm{C\,II}} \equiv 0.5(F_{\mathrm{C\,II}} + 0.63\,F_{\mathrm{C\,IV}}) - F'_{\mathrm{C\,II,min}}$ vs. the projected equatorial rotation velocity $v\,sin(i)$. The mean relationship for stars with $B - V > 0.5$ is marked by the solid curve [73]. Some deviating points (o,+) are also shown if the observed C II flux is used, i.e. without averaging with the C IV flux; in that case HD 185395 would lie on the basal flux level. The reason for the deviation of HD 26462 and HD 72291 from the relationship defined by similar stars is unknown: a large inclination angle of the rotation axis (resulting in a small $sin(i)$) or a possible companion (contaminating the signal) could be the cause (from [74]).

Fig. 12. (right) The C II flux density versus rotation period P for dwarf stars (from [73]). Stars with $P < 0.7$ days in the sample are W UMa type binaries. Symbols: open $B-V < 0.6$, with bar $0.6 \leq B-V < 0.8$, half-filled $0.8 \leq B-V < 1.0$, filled $B-V \geq 0.8$.

For early F–type main sequence stars rotation and activity are still related, but the radiative losses of the active atmosphere of a star at a given rotation period decrease strongly for stars hotter than $B - V = 0.45$ (≈F5), possibly associated with the disappearance of the convective envelope (Figs. 8 and 11). Note that at this point also the basal flux, which is relatively strong in F–type stars (cf. Fig. 9), appears to decrease: if a convective envelope is thick enough to produce acoustic heating, it can apparently also support magnetic activity, although with an efficiency that depends on the envelope's thickness [74].

At the cool end of the HR diagram we find the M–type dwarfs. These cool stars deviate systematically from flux-flux relationships valid for warmer stars [75]. The flux densities in the Ca II H+K line cores in the M–type spectra, for instance, are up to 60 times smaller than observed from stars of earlier spectral type with the same X-ray flux density. The departure from the flux-flux relationships seems to be a function of stellar colour and possibly of metallicity. The departure decreases with increasing temperature of formation of the line. The reason for this deviation remains unclear [75].

The T Tauri stars form another group of stars which deviate from the main flux-flux relationships. These stars are too bright in the diagnostics used to measure stellar magnetic activity given their rotation periods, except in soft X-rays. The large widths of the UV lines suggests that the surplus of the emission originates in some inner part of an accretion disk [76].

4.2 Binary Systems

For rotation periods smaller than one day the atmospheric radiative flux densities decrease with increasing rotation rate [77], although one should note that the data in Fig. 12 for rotation periods below 0.7 days are from contact binaries of the W UMa type, so the inversion in the activity-rotation relation may be an effect typical of these peculiar objects. The W UMa contact binaries also depart from some of the flux-flux relationships [71].

While most late–F–, G–, and K–type stars obey a well–defined activity-rotation relationship (Fig. 13), some stars are much more active that the rotation rate predicts [78, 79]. This "overactivity" generally occurs in close binaries. The level of overactivity depends on the properties of the secondary star (see [80] for details), while the bulk of the emission comes from the coolest and most evolved component or from both if they are comparable. These systems do not deviate systematically from flux–flux relationships defined by single stars. It is conceivable that tidal interaction by the companion of the cool component causes the overactivity. *Zwaan* [52] speculates

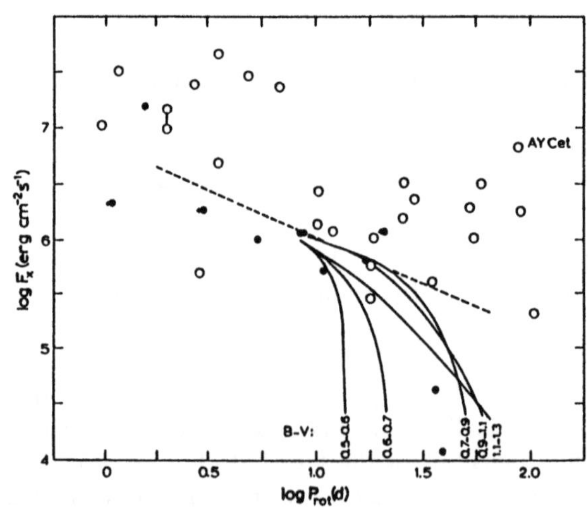

Fig. 13. Rotation–activity diagram: plotted are the surface flux density in soft X–rays, F_X, versus rotation period P_{rot}. The small dots represent selected single stars, while relatively close binaries are shown as open circles. The curves show the colour–dependent rotation–activity relationships derived using a sample of about 300 single stars and wide binaries [79]. Figure from [80].

that whereas in a single star the convective mantle spins down by the loss of angular momentum ("magnetic braking"), the cool component of a relatively close binary is forced to co-rotate because of the tidal interaction, so that the convective mantle is forced to spin up in order to adjust to the loss of orbital and spin angular momentum of the system, while the stellar surface keeps losing angular momentum by the stellar wind. Hence the shear in the convection zone may differ from that in a single star, which could result in a changed efficiency with which the magnetic field is generated.

5. The Solar–Stellar Connection

One of the areas in which the combined study of solar and stellar data has proved of enormous value is in exploring the origin of the non–linear flux-flux relationships found for stellar data. Section 3.2 discusses the high degree of correlation between the radiative losses in various emission lines from stellar outer atmospheres, and notes that several significantly non–linear relationships were found between chromospheric or transition–region flux densities and magnetic flux densities. Astoundingly, the relations between fluxes derived from spatially–resolved solar data are (almost) the same as the relations between surface–averaged fluxes from stars of different levels of activity (see also [45, 46 , 48, 81], and Fig. 4). If stellar atmospheres would consist merely of ensembles of identical active regions with only the number of these regions changing as a function of the level of activity, linear relations should result between the flux densities. Hence either the structure of the active regions changes with the global level of activity, or the mixture of distinct components in the atmosphere changes with activity, or both.

The emission from a magnetically active star is determined by a summation of the radiative losses over the visible hemisphere. These radiative flux densities F_i in turn depend on the associated magnetic flux densities $|\varphi|$. Hence, the total stellar emission is a convolution of the relation between F_i and the associated photospheric magnetic flux density $|\varphi|$ (as discussed in Section 2.2.2), with a distribution function $\mathbf{h}_t(|\varphi|)$ for the magnetic flux density (this was studied by [40], whose work is summarized below). The time–dependent distribution function $\mathbf{h}_t(|\varphi|)$ (Fig. 3) appears to be only a function of the total magnetic flux present in the solar photosphere – i.e. the solar level of activity –, and does not exhibit significant hysteresis effects associated with the phase in the cycle: after at most six months – the resolution of $\mathbf{h}_t(|\varphi|)$ – the solar photosphere has effectively "forgotten" that it ever had a different mean level of activity.

If there is a relationship

$$F_i = a_i \, |\varphi|^{b_i} \tag{1}$$

between a radiative flux density, F_i, and the mean magnetic flux density, $|\varphi|$, observed with a resolution that is poor enough to smooth out small-scale temporal and spatial fluctuations (see Section 2.2.2), the fluxes from the Sun–as–a–star throughout the cycle can be approximated by integrating over the surface, using $\mathbf{h}_t(|\varphi|)$:

$$\langle F_i \rangle_t \approx \left\{2.5^{b_i(b_i-1)}\right\} a_i \left\langle |\varphi| \right\rangle_t^{b_i}. \tag{2}$$

The relation between two disk–averaged stellar radiative flux densities $\langle F_i \rangle_t$ and $\langle F_j \rangle_t$ can be determined by applying Eq. (2) to appropriate pairs of constants (a_i, b_i) and (a_j, b_j). For two radiative flux densities, F_i and F_j, for which spatially resolved data define relationships $F_i = a_i \; |\varphi|^{b_i}$ and $F_j = a_j \; |\varphi|^{b_j}$, the flux–flux relationship for spatially resolved data transforms into a relationship between mean stellar flux densities, $\langle F_i \rangle$ and $\langle F_j \rangle$, like this:

$$F_i = A_{ij} \, F_j^{B_{ij}} \Rightarrow \langle F_i \rangle = \left\{2.5^{b_i b_j \cdot (B_{ij}-1)}\right\} A_{ij} \, \langle F_j \rangle^{B_{ij}}. \tag{3}$$

with $B_{ij} = b_i/b_j$ and $A_{ij} = a_i/a_j^{B_{ij}}$. If, for example, $b_i = 0.6$, as is the case for the relation between ΔF_{CaII} and $|\varphi|$ [69], $2.5^{b_i(b_i-1)}$ equals 0.8, which explains why the flux–flux relationships for mean stellar fluxes and for spatially–resolved solar fluxes are nearly the same. Apparently the time–dependent distribution function of magnetic flux densities, as determined from solar synoptic maps, has the remarkable property that power–law relations between two radiative flux densities for spatially resolved data are transformed into power–law relations for surface–averaged data with the same power–law index, and only a small change in the constant of proportionality. Because the "photospheric memory" for magnetic activity is short compared to the time scale of the activity cycle (hysteresis effects appear to be absent, see above), the Sun in different stages of activity resembles stars of different mean levels of activity. Hence the transformation of relations between fluxes for spatially resolved data to surface–averaged fluxes for the Sun is probably also valid for stars. The Sun–as–a–star follows the relations between radiative flux densities defined by other cool stars. Therefore the distribution function of magnetic flux densities over the surface of stars like the Sun, but at different levels of activity, probably resembles that for the Sun in different phases of the cycle. The process which determines the functional form of the distribution function $\mathbf{h}_t(|\varphi|)$ is therefore of fundamental importance. This process is clearly related with the location where and the frequency with which flux surfaces in the photosphere (cf. Section 2.1.1), with the way the flux diffuses over the solar surface (cf. Section 2.1.2), and with the way in which magnetic flux disappears from the photosphere (cf. Section 2.1.2). Understanding the non–linearity of relations between stellar radiative flux densities now requires understanding of the nature of the non–linearity of the relation between the radiative and magnetic flux densities for spatially resolved data (cf. Section 2.2.2).

6. Conclusions

The combination of solar and stellar data, partly described in this review, has resulted in a model for stellar magnetic activity as presented in Fig. 14. Solar observations have disclosed the properties of the outer layers of the convection zone, differential rotation (with helioseismology now even revealing the rotation as a function of depth), the large- and small-scale topology of the photospheric magnetic field (including such things as the active latitudes and the cool, CO-cooled atmosphere between flux tubes), and some of the details of non-radiative atmospheric heating. Stellar observations have revealed the dependence of the activity level on rotation rate, the possibilities to change the spin-angular momentum by mass loss by a stellar wind in a magnetosphere or by tidal interaction in close binaries, the existence of a basal flux, and the peculiar atmospheres of M-type dwarf stars, T Tauri stars, and close binaries with at least one cool component, and the effects of a shallow convection zone in F-type stars. Not only do the solar observations provide a stepping stone towards the structure of stellar outer atmospheres: combined observations suggest that the non-linearity of stellar and solar flux-flux relationships are brought about by both the dispersal of magnetic

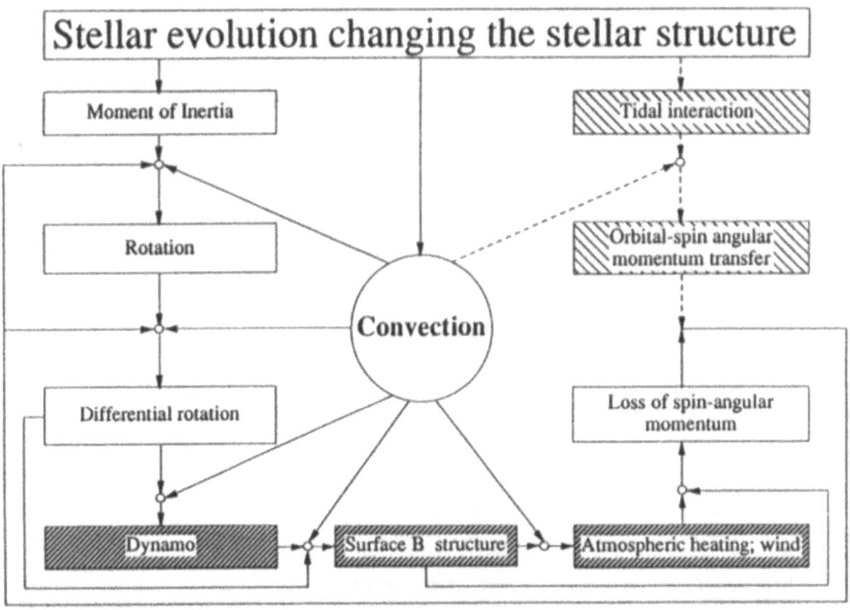

Fig. 14. Diagram illustrating the interactions responsible for driving the dynamo and the surface phenomena associated with magnetic activity in evolving cool stars, including the effects caused by tidal interaction in relatively close binaries. The thickly shaded boxes represent effects associated with the magnetic field, the lightly shaded boxes, connected by dashed arrows, those acting in close binaries. Some of the major interactions are shown between the boxes (from [80]).

flux over the stellar surface, and an intrinsically non-linear relationship between magnetic flux density and local radiative losses, affected by non-local flux tube interactions. This also appears to eliminate stellar gravity, effective temperature, etc. in the mechanism heating the outer atmospheres, at least where the relative strength of the different atmospheric temperature regimes is concerned. Theoretical studies suggest that convection and (differential) rotation interact to drive a "dynamo" which generates the magnetic field. Only by the combination of solar observations, stellar observations, and theory could the diagram of Fig. 14 be constructed, revealing the essential part played by the convection in many different processes. Figure 14 contains only the basic building blocks of stellar magnetic activity, and needs to be refined and possibly expanded in the future when new observational and theoretical areas and techniques are explored.

References

1 W.C. Livingston, J.W. Harvey: Bull. Am. Astron. Soc. **7**, 346 (1975)
2 H.C. Spruit: Solar Phys. **50**, 269 (1976)
3 H.C. Spruit: Solar Phys. **55**, 3 (1977)
4 C. Zwaan: Solar Phys. **60**, 213 (1978)
5 C. Zwaan, L.E. Cram: in "FGK Stars and T Tauri Stars," ed. by L.E. Cram and L.V. Kuhi (NASA, 1989) p. 215
6 C. Zwaan: Ann. Rev. Astron. Astrophys. **25**, 83 (1987)
7 V. Gaizauskas, K.L. Harvey, J.W. Harvey, C. Zwaan: Astrophys. J., **265**,1056 (1983)
8 R. Stein: in "Mechanisms of Chromospheric and Coronal Heating", ed. by P. Ulmschneider, E. Priest, and R. Rosner, (Springer–Verlag, 1990) in press
9 A.B. Hart: Mon. Not. Royal Astron. Soc. **114**, 17 (1954)
10 A.B. Hart: Mon. Not. Royal Astron. Soc. **116**, 38 (1956)
11 R.B. Leighton, R.W. Noyes, G.W. Simon: Astrophys. J. **135**, 474 (1962)
12 G.W. Simon, R.B. Leighton: Astrophys. J. **140**, 1120 (1964)
13 J. Singh, M.K.V. Bappu: Solar Phys. **71**, 161 (1981)
14 R. Brune, H. Wöhl: Solar Phys. **75**, 75 (1982)
15 G. Küveler: Solar Phys. **88**, 13 (1983)
16 E.H. Rogers: Solar Phys. **13**, 57 (1970)
17 T.J. Janssens: Solar Phys. **11**, 222 (1970)
18 R.C. Smithson: Solar Phys. **29**, 365 (1973)
19 S.P. Worden, G.W. Simon: Solar Phys. **46**, 73 (1976)
20 T.L. Duvall, Jr.: Solar Phys. **66**, 213 (1980)
21 J.M. Mosher: Ph.D. Thesis, California Inst. Tech. (1977)
22 H. Wang: Ph.D. Thesis, California Inst. Tech. (1988)
23 N.R. Sheeley, Jr.: Solar Phys. **9**, 347 (1969)
24 W.C. Livingston, F.Q. Orrall: Solar Phys. **39**, 301 (1974)
25 R.B. Leighton: Astrophys. J. **140**, 1547 (1964)
26 A.M. Title, R.A. Shine, T.D. Tarbell, K.P. Topka, G.B. Scharmer: in "Solar Photosphere: Structure, Convection, and Magnetic Fields," ed. by J.O. Stenflo (Kluwer Academic Publishers, Dordrecht, 1989) p. 49
27 B. Foing, R.M. Bonnet, M. Bruner: Astron. Astrophys. **162**, 292 (1986)
28 G.W. Simon, A.M. Title, K.P. Topka, T.D. Tarbell, R.A. Shine, S.H. Ferguson, H. Zirin, and the SOUP team: Astrophys. J. **327**, 964 (1988)
29 A.M. Title, T.D. Tarbell, G.W. Simon, and the SOUP team: Adv. Space Res. **6**, No. 8, 253 (1986)

30 L.J. November, G.W. Simon, T.D. Tarbell, A.M. Title, S.H. Ferguson: in "Second Workshop on Theoretical Problems in High Resolution Solar Physics," NASA Conf. Publ. 2483, p. 121 (1986)
31 K.L. Harvey: private communication (1988)
32 C.J. Schrijver: Solar Phys. **122**, 193 (1989)
33 R.S. Bogart: Solar Phys. **110**, 23 (1987)
34 N.R. Sheeley, Jr., C.R. DeVore, J.P. Boris: Solar Phys. **98**, 219 (1985)
35 S.F. Martin, S.H.B. Livi, J. Wang: Australian J. Phys. **38**, 929 (1985)
36 T.D. Tarbell: private communication (1987)
37 H. Zirin: "Astrophysics of the Sun" (Cambridge University Press, Cambridge, 1989)
38 F. Tang, R. Howard, J.M. Adkins: Solar Phys. **91**, 75 (1984)
39 C.J. Schrijver: Astron. Astrophys. **189**, 163 (1987)
40 C.J. Schrijver, K.L. Harvey: Astrophys. J. **343**, 481 (1989)
41 A. Skumanich, J.L. Lean, O.R. White, W.C. Livingston: Astrophys. J. **282**, 776 (1982)
42 E.H. Avrett, J. Fontenla: in "Mechanisms of Chromospheric and Coronal Heating," ed. by P. Ulmschneider, E. Priest, and R. Rosner (Springer–Verlag, 1990) in press
43 P.S. Cally: in "Mechanisms of Chromospheric and Coronal Heating," ed. by P. Ulmschneider, E. Priest, and R. Rosner (Springer–Verlag, 1990) in press
44 C.J. Schrijver, C. Zwaan, C.W. Maxson, R.W. Noyes: Astron. Astrophys. **149**, 123 (1985)
45 C.J. Schrijver: in "Mechanisms of Chromospheric and Coronal Heating," ed. by P. Ulmschneider, E. Priest, and R. Rosner (Springer–Verlag, 1990), in press
46 A. Cappelli, M. Cerruti–Sola, C.C. Cheng, R. Pallavicini: Astron. Astrophys. **213**, 226 (1989)
47 C.J. Schrijver: Astron. Astrophys. **180**, 241 (1987)
48 C.J. Schrijver, J. Coté, S.H. Saar, C. Zwaan: Astrophys. J., **337**, 964 (1989)
49 A. Skumanich, C. Smythe, E.N. Frazier: Astrophys. J. **200**, 747 (1975)
50 S.K. Solanki, O. Steiner: Astron. Astrophys. **234**, 519 (1990)
51 P. Maltby, E.H. Avrett, M. Carlsson, O. Kjeldseth–Moe, R.L. Kurucz, R. Loeser: Astrophys. J. **306**, 284 (1986)
52 C. Zwaan: in "Mechanisms of Chromospheric and Coronal Heating," ed. by P. Ulmschneider, E. Priest, and R. Rosner (Springer–Verlag, 1990), in press
53 T.R. Ayres, L. Testerman: Astrophys. J. **245**, 1124 (1981)
54 T.R. Ayres, L. Testerman, J.W. Brault: Astrophys. J. **304**, 542 (1986)
55 R.G. Giovanelli, H.P. Jones: Solar Phys. **79**, 267 (1982)
56 S.K. Solanki, O. Steiner, H. Uitenbroek: Astron. Astrophys., in press (1990)
57 S.S. Vogt: in "Activity in Red-Dwarf Stars", ed. by P.B. Byrne and M. Rodonò (D. Reidel, Dordrecht, 1983), p. 137
58 S.S. Vogt, G.D. Penrod: in "Activity in Red-Dwarf Stars", ed. by P.B. Byrne and M. Rodonò (D. Reidel, Dordrecht, 1983) p. 379
59 S.H. Saar: in "Solar Photosphere: Structure, Convection, and Magnetic Fields," ed. by J.O. Stenflo (Kluwer Academic Publishers, Dordrecht, 1989), p. 427
60 S.L. Baliunas, A.H. Vaughan: Ann. Rev. Astron. Astrophys., **23**, 379 (1985)
61 S.L. Baliunas: in "Cool Stars, Stellar Systems, and the Sun," ed. by M. Zeilik and D.M. Gibson, (springer–Verlag, 1986) p. 3
62 R.G.M. Rutten, C.J. Schrijver, A.F.P. Lemmens, C. Zwaan: subm. to Astron. Astrophys. (1990)
63 C.J. Schrijver: in "Cool Stars, Stellar Systems, and the Sun," ed. by J.L. Linsky and R.E. Stencel (Springer–Verlag, 1987), p. 135
64 P. Ulmschneider: in "Mechanisms of Chromospheric and Coronal Heating," ed. by P. Ulmschneider, E. Priest, and R. Rosner (Springer–Verlag, 1990), in press
65 R.G.M. Rutten, E. Pylyser: Astron. Astrophys. **191** 227 (1988)
66 C.J. Schrijver: in "New Windows to the Universe," ed. by M. Vazquez (Cambridge University Press, 1990), in press
67 A. Schadee: Solar Phys. **89**, 287 (1983)
68 B.J. Oranje, C. Zwaan, F. Middelkoop: Astron. Astrophys. **177**, 131 (1982)
69 C.J. Schrijver: Astron. Astrophys. **127**, 289 (1983)
70 C. Zwaan: in "Solar and Stellar Magnetic Fields: Origins and Coronal Effects," ed. by J.O. Stenflo (Reidel, Dordrecht, 1983), p. 85

71 B.J. Oranje: Astron. Astrophys. **154**, 185 (1986)
72 G. Basri: Astrophys. J. **316**, 377 (1987)
73 R.G.M. Rutten, C.J. Schrijver: Astron. Astrophys. **177**, 155 (1987)
74 F.M. Walter, C.J. Schrijver: in "Cool Stars, Stellar Systems, and the Sun," ed. by J.L. Linsky and R.E. Stencel (Springer-Verlag, 1987), p. 262
75 R.G.M. Rutten, C.J. Schrijver, C. Zwaan, D.K. Duncan, R. Mewe: Astron. Astrophys. **219**, 239 (1989)
76 A.F.P. Lemmens, C. Zwaan, R.G.M. Rutten: Astron. Astrophys., in press (1990)
77 O. Vilhu, S.M. Rucinsky: Astron. Astrophys. **127**, 5 (1983)
78 G. Basri, R. Laurent, F.M. Walter: Astrophys. J. **298**, 761 (1985)
79 R.G.M. Rutten: Astron. Astrophys. **177**, 131 (1987)
80 C.J. Schrijver, C. Zwaan: submitted to Astron. Astrophys. (1990)
81 S.H. Saar, C.J. Schrijver: in "Cool Stars, Stellar Systems, and the Sun," ed. by J.L. Linsky and R.E. Stencel (Springer-Verlag, 1987), p. 38

Hydrodynamics of the Solar Photosphere: Model Calculations and Spectroscopic Observations

M. Steffen and B. Freytag

Institut für Theoretische Physik und Sternwarte der Universität, Olshausenstr. 40, W-2300 Kiel 1, Fed. Rep. of Germany

1. Introduction

A close-up view of the solar photosphere under good conditions of atmospheric seeing shows a characteristic cellular pattern covering the entire surface (except for sunspots), the so called ***solar granulation***. Bright isolated elements, the granules, are surrounded by a network of interconnected dark lanes. In contrast to W. Herschel, who first reported the existence of some small-scale inhomogeneities on the solar surface in 1801, we now have a basic physical understanding of the origin of the solar granulation. The essential foundation was laid by Unsöld [1], who pointed out that because of the sharp increase with depth of opacity and ionization of hydrogen, the Schwarzschild criterion for convective instability should be satisfied in the deep photospheric layers. According to current mixing length models of the solar atmosphere (Kurucz [2]), the dividing level between stable (radiative) and unstable (convective) layers is located somewhere between $\tau_{5000} = 0.8$ and $\tau_{5000} = 1$. Consequently, granulation is identified with convection cells in the uppermost layers of the hydrogen convection zone, forming the observed continually changing cellular pattern at the interface between optically thick and optically thin conditions. ***Granules*** represent hot rising gas elements while the ***intergranular lanes*** consist of cooler sinking gas. This picture has been verified spectroscopically, confirming the convective origin of the solar granulation.

Although granulation is basically understood as a convective phenomenon, our quantitative knowledge about its dynamical and thermal structure is still incomplete. From current observational material it seems impossible to derive a detailed quantitative picture of granular convection by a purely empirical analysis. To make progress empirically, extensive spectroscopic observations with very high spatial resolution ($\approx 0.1''$) would be required. Lacking such observations, we have to rely on theoretical models. While local mixing length theory may be adequate to model the mean stratification of deep convective layers, it certainly cannot provide a reasonable description of the details of photospheric convection. Apart from the problem that the mixing length is a free parameter which has to be calibrated somehow, such models give only an average stratification, while we are interested in the effect of horizontal inhomogeneities. Moreover, they assume an abrupt transition between stable and unstable conditions, while actually convective motions must be expected to extend into the stable atmospheric layers to some degree. Although several non-local mixing length theories have been developed (e.g. Shaviv and Salpeter [3], Roxburgh [4]), there is no commonly accepted formalism to model this so called *convective overshoot* in a simple way.

The only possibility to take into account the complex physics of stellar photospheric convection seems to be a direct numerical simulation of the photospheric gas flows. This approach requires the use of powerful computers and generates large amounts of data.

In the following we shall describe such detailed hydrodynamical calculations for the solar photosphere. Based on the resulting physically self-consistent, parameter-free models we calculate synthetic spectra in order to be able to make a direct comparison with various spectroscopic observations. We conclude that our time-dependent radiation-hydrodynamics models seem to be a reasonable representation of the real solar photosphere, which, on the other hand, is obviously quite different from the picture suggested by classical mixing length models.

2. Modelling the Solar Photosphere

2.1 Physics of Photospheric Convection

Before setting up numerical simulations of convective flows in stellar photospheres, it is important to realize the complexity of the physics involved. In the following we emphasize some important aspects.

The basic processes of photospheric convection are governed by time-dependent, non-linear hydrodynamics in 3 dimensions. Gravity causes the gas to be stratified, implying large variations of gas pressure and density with depth. Apart from the stratification, compressibility of the medium will affect the characteristics of the convective flow as well as lead to additional features like pressure waves.

The stellar gas is a plasma, consisting of different elements (essentially hydrogen and helium) which begin to ionize at different temperatures. Ionization has important consequences for the energy balance in the plasma and increases the degree of convective instability.

While in the deep, optically thick layers radiation can be described as a diffusive process, this is no longer possible in the optically thin photospheric regions. Depending on the magnitude of the frequency-dependent opacity, the radiation field eventually is controlled by non-local transfer beyond different levels in the photosphere. Radiative cooling rates can be substantial, causing a strong coupling between hydrodynamics and radiative transfer.

The very low viscosity of the stellar gas gives rise to another complexity. In view of the correspondingly high Reynolds number ($Re \approx 10^9$) we must expect the flow to be fully turbulent. Among other things this implies the existence of an enormous range of spatial scales, between some 1000 km for the largest granular dimensions and a few cm, where small-scale turbulent energy is eventually dissipated.

Finally, in convective stellar atmospheres magnetic fields may play an important role. At least where locally concentrated, they may even dominate the gas dynamics.

From this brief overview it is clear that it is impossible to model the complete physics of the problem in a single numerical simulation. Rather, it is necessary to introduce simplifications and approximations. The hope is, however, that the simplified problem can still account for the essential features of photospheric convection.

2.2 Numerical Simulations

For our numerical simulations of photospheric convection we have adopted the following major simplifications of the problem. First, we ignore magnetic fields altogether. Second, we restrict the flow to 2 dimensions, assuming axial symmetry in cylindrical coordinates (r: parallel to the stellar surface, z: upwards, antiparallel to the direction of gravity). Finally, every numerical simulation of turbulent flows has to treat the effect of turbulence on subgrid scales in some schematic way. For this purpose we use the concept of *turbulent viscosity* according to Smagorinsky [5] (see also Deardorff [6], [7]). Application of this scheme results in a so called "large eddy simulation approach", where one resolves only the largest (granular) scales, without trying to model the various smaller scales in detail. The idea is that the large-scale behaviour of a turbulent system becomes insensitive to the details of the physics at the smaller scales if Re is sufficiently large (see Chan and Sofia [8] for a discussion of this concept in the context of numerical simulations of turbulent stellar convection).

Within the framework of these simplifications we try to include as much realistic background physics as possible. Details about the input physics used for our numerical simulations have been described elsewhere (Steffen et al. [9] and Steffen [10]). In the following we briefly reiterate the main points.

To account for the basic physics of compressible convection in a stratified, turbulent atmosphere, we solve the Navier Stokes equations, prescribing the conservation of mass and momentum, together with some energy equation appropriate for the conditions in stellar atmospheres. Turbulence, represented by subgrid scale eddy viscosity, has several effects. *Turbulent exchange of momentum* gives rise to an additional force in the momentum equation and leads to *turbulent dissipation* of large-scale kinetic energy into small-scale (turbulent) kinetic energy and eventually into heat, a process which is taken into account in our energy equation. A second, more important effect on the energy equation is due to the *turbulent exchange of heat*. The corresponding heat flux is modelled as being proportional to the local entropy gradient.

The most important term in the energy balance of the photosphere is the *exchange of radiation*. Hence, a realistic description of radiative transfer effects is essential. Avoiding approximations like the diffusion or Eddington approximation, we use a system of several thousand rays covering the entire model volume under different angles of inclination. Tracing each ray from the top to the bottom of the model, assuming the lateral boundary of the cylindrical computational domain to be reflective, we solve the equation of radiative transfer using a modified Feautrier technique. Assuming LTE, there is no coupling between the rays and the Feautrier scheme can be vectorized efficiently. Although we have recently modified the code to take into account the frequency-dependence of the radiation field in a simplified way [10], the results reported here refer to models calculated in the grey approximation, using a realistic Rosseland mean opacity $\kappa_{Ross}(P, T)$.

The gas is taken to be a mixture of hydrogen and helium with variable degree of ionization, and a representative neutral metal, with relative abundances according to solar composition. Excitation and ionization of H, He, He^+ and the formation of H_2 molecules are calculated in LTE (Boltzmann and Saha formulae). With this realistic equation of state there are no longer simple relations for e.g. density and entropy as a function of pressure and temperature, and the mean molecular weight and specific heats cannot be regarded as constant (C_p and C_v vary by a factor of 20 or more in the solar photosphere!).

Our models are designed to simulate *photospheric convection* and cannot comprise the entire solar convection zone. Including only the uppermost superadiabatic layers in the computational domain makes the problem of specifying a reasonable lower boundary condition difficult since simple assumptions like a rigid wall are not realistic in this situation. To restrict the flow as little as possible, we have adopted an open lower boundary, allowing a free flow of gas out of and into the model. The gas pressure at the bottom is continuously adjusted in such a way as to conserve the total mass within the model volume. The temperature of the inflowing gas is regulated such that the time averaged radiative surface flux conforms to the specified effective temperature. The flow velocity at the bottom can then be computed from the hydrodynamical relations, giving the flow the freedom to chose the regions of up and downflows wherever it likes.

The upper boundary is treated as a closed top by requiring the vertical velocity to vanish, while the horizontal velocity has a zero z-derivative at the top. At the axis of symmetry as well as at the side walls the horizontal velocity and the r-derivatives of all other variables are taken to be zero.

3. Results of the Numerical Simulations

3.1 Steady State Models

Depending on the adopted size of the computational domain, the numerical simulations lead to *steady state* and *instationary* solutions, respectively. The flow develops towards a steady state whenever the model diameter is smaller than a critical upper limit, roughly 1500 km. An example of a steady state solution in a cell measuring 1050 km in diameter is displayed in Fig. 1.

The properties of the cylindrical steady state flows have been described in detail by Steffen et al. [9]. Here we just summarize the main characteristics.

Fast narrow converging *downflows* occur at the axis of symmetry with downflow velocities near the speed of sound ($\approx$ 8 km/s), and at the lateral boundary with somewhat lower velocities.

The *upflows* are slower, not faster than 2 to 3 km/s. In the ascending regions, which are distinctly broader than the downflows, the flow diverges on its way up.

Convection generates large horizontal *temperature differences* between up and downflows in order to carry the solar flux. A maximum temperature difference of 4800 K (at constant geometrical height) is found about 70 km below the visible surface where the average temperature is approximately 9200 K. Another conspicious feature is the occurrence of very steep *temperature gradients* at the boundary between convectively unstable and stable regions. The main reason is efficient radiative cooling once the ascending hot gas becomes optically thin.

The convective "overshoot layers" extend about 300 km above $\tau = 1$, but the velocity and temperature fluctuations induced in these layers are not very large. Note that the temperature fluctuations change sign in the overshoot layers, i.e. it is relatively hotter above the downflows (intergranular lanes) and cooler above the upflows (granules), a consequence of adiabatic cooling of expanding gas rising in convectively stable layers.

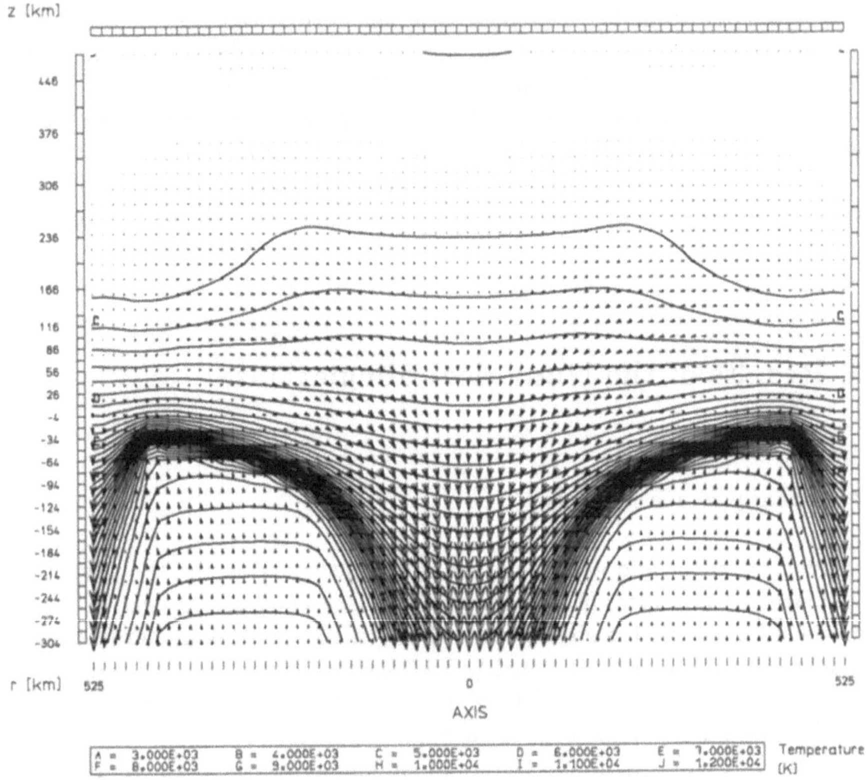

Fig. 1. Steady state flow in a granular convection cell measuring 1050 km in diameter. This simulation uses 40 vertical grid points (spacing $\Delta z = 15$ km in the lower and $\Delta z = 35$ km in the upper part) and 36 equidistant horizontal grid points ($\Delta r = 15$ km). $z = 0$ corresponds to $\tau_{ross} \approx 1$. Temperature contours are shown in steps of 200 K, while arrows indicate the velocity field. Maximum temperature is 11960 K, maximum flow velocity is 8.6 km/s.

In the highest layers of the model, the stratification is essentially static, plane-parallel and in radiative equilibrium. Here temperature is almost constant with height and the boundary temperature is that of a grey radiative equilibrium atmosphere, $T_{bound} \approx 4700$ K.

3.2 Instationary Models

If the model diameter is increased beyond a critical limit ($D \gtrsim 1500$ km) the resulting solutions become highly instationary. While many of the basic characteristics of granular convection found in the steady state models are retained, the important new feature is that now the topology of the convective flow pattern changes with time. These changes may be interpreted as a continuous splitting and merging of "granules" on time scales of the order of 10 to 30 minutes, comparable to observed granular lifetimes. Although the

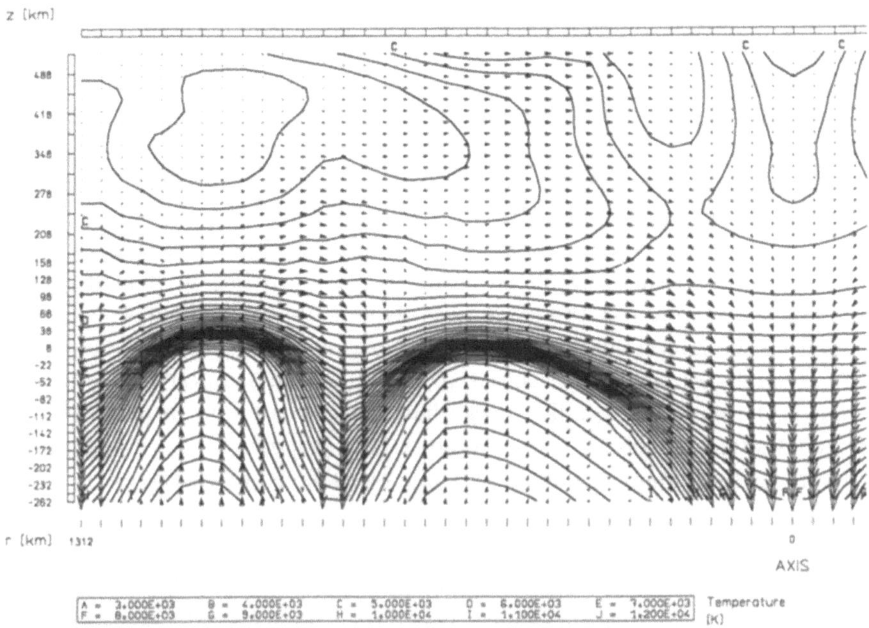

Fig. 2. Snapshot from a numerical simulation of instationary photospheric convection in a cell measuring 2625 km in diameter. This simulation uses 40 vertical grid points (spacing $\Delta z = 15$ km in the lower and $\Delta z = 35$ km in the upper part) and 36 equidistant horizontal grid points ($\Delta r = 37.5$ km). $z = 0$ corresponds to $\tau_{ross} \approx 1$. Temperature contours are shown in steps of 200 K, while arrows indicate the velocity field. Maximum temperature is 11780 K, maximum flow velocity is 8.3 km/s.

flow goes through similar cycles again and again, these are not really periodic in time but occur in irregular intervals, such that the overall behaviour of the convective part of the atmosphere may rather be characterized as chaotic. But superimposed on the convective flow we find quite regular oscillations, probably related to the observed solar 5-minute oscillations. Most prominent periods in the simulations range between 150 and 250 sec.

The important point to be learned from the time-dependent simulations is that the continually changing flow *topology* affects the upper atmosphere much more strongly than a steady state flow does. It seems that in stationary flows the convectively stable layers can adjust to a steady state convective flow in the deeper layers in a way that results in only minor spatial fluctuations of velocity and temperature. This is no longer possible in the time-dependent situation. If, for example, a granule collapses and a downflow develops at a point where a short time before there was an upflow (a situation actually found in the instationary models) or if a granule grows as it rises from below, the overlying layers, which are convectively stable in the classical models, must inevitably undergo substantial temporal variations and exhibit corresponding spatial fluctuations. For this reason we find an "extended overshoot" in the time-dependent case. In the snapshot displayed in Fig. 2 we see much more pronounced velocity fields and temperature inhomogeneities in the photosphere than in the steady state model (Fig. 1). Indeed, the pressure waves excited by the instationary convection carry a substantial amount of mechanical energy into the higher layers. We have estimated the corresponding acoustic flux to be of the order of some $10^7 \mathrm{erg/cm^2/s}$ (Steffen et al. [11]).

Sometimes even shocks develop, preferentially associated with strong downdrafts, as in the example situation displayed in Fig. 3, showing two different shock fronts at the same time. One is almost vertical, travelling away from the intermediate downdraft in horizontal direction for some time until it finally dissipates, the other front extends horizontally and moves upwards. Remarkably, similar shocks have also been found in more idealized convection models by Cattaneo et al. [12], [13] who used a completely different hydrodynamics code and assumed a different geometry.

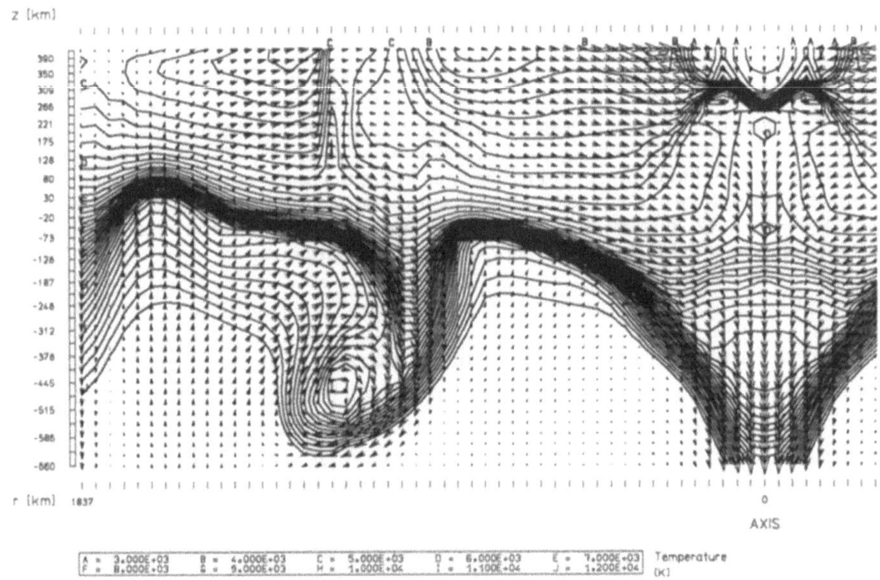

Fig. 3. Snapshot from another instationary simulation of solar granular convection, using 40 vertical (Δz between 40 km in the lower and 20 km in the upper part) and 50 equidistant horizontal grid points (Δr = 37.5 km). This model is somewhat deeper than the one shown in Fig. 2 and has a larger diameter (D = 3675 km). The almost vertical shock front at $r \approx 1160$ km in the upper atmosphere is travelling to the left, while the front at the axis of symmetry moves upwards. Maximum temperature is 13720 K, maximum flow velocity is 11.4 km/s.

4. Observed and Synthetic Spectra

In order to confront the hydrodynamical granulation models described in the preceeding sections with reality, we have compared spectroscopic observations of the solar granulation with synthetic spectra calculated from the simulated convective atmospheres. To obtain spectra corresponding to observations at solar *disk-center*, we integrate the transfer equation along a set of vertical rays, taking into account the depth dependence of temperature, pressure and velocity in the calculation of the source function and the monochromatic continuum and line opacities. These spectrum synthesis calculations, which are independent of the more simplified treatment of radiative transfer of the hydrodynamical simulations, yield the emergent intensity in the continuum at any prescribed wavelength as well as the profiles of arbitrary spectrum lines as a function of horizontal position within the model.

In the following sections we discuss the calculated continuum intensity contrast as well as spatially resolved and unresolved synthetic line profiles with respect to corresponding observations.

4.1 Continuum Intensity Contrast

In the continuum we see horizontal intensity variations corresponding to temperature fluctuations near $\tau = 1$. The rms contrast of a given intensity pattern is defined as $\delta I_{rms} = \sqrt{\sum_i f_i \left(\frac{I_i}{I_0} - 1\right)^2}$, where f_i is the fractional area having intensity I_i, and $I_0 = \sum_i f_i I_i$ is the average intensity.

For the *steady state models*, δI_{rms} is found to increase with cell size, saturating towards the largest diameters (Steffen et al. [9]). An upper limit of $\delta I_{rms} \approx 14\%$ at λ 5000 Å is approached for the largest steady state models ($D \approx 1500$ km).

Evaluation of the rms intensity contrast becomes more complicated for the *time-dependent models*. There are now different possibilities to define δI_{rms}. For each instant of time we can calculate the rms contrast of the intensity pattern, $\delta I_{rms}(t_n)$, as in the steady state case. The *average* rms contrast of the intensity pattern is then calculated as $\overline{\delta I_{rms}} = \sqrt{\frac{1}{N}\sum_n^N (\delta I_{rms}(t_n))^2}$. This quantity does not include possible time variations of the horizontally averaged intensity $I_0(t_n)$. To take this additional contribution into account we define the *total* rms intensity contrast, $\langle \delta I_{rms} \rangle$, including spatial as well as temporal intensity variations.

Depending on the size of the model, the treatment of radiative transfer (grey or frequency dependent), and on other numerical details, the instationary models yield $\overline{\delta I_{rms}}$ values between 16% and 20%, corresponding to $\langle \delta I_{rms} \rangle$ values between 18% and 22%.

These results may be compared with those from Nordlund's compressible 3-D models of solar granulation, as reported by Lites et al. [14]. Depending on details of their models, the rms intensity contrast at λ 6303 Å lies between 14.4% and 17.5%, which can be translated to corresponding values at λ 5000 Å between 18.2% and 22.1%. We note that the 3-D simulations give essentially the same answer as our instationary 2-D models.

Published data for observed δI_{rms} values, corrected for degradations by atmospheric seeing and instrumental effects, range between 8% and 19% at λ 5000 Å (see e.g. Bray et al. [15]). Comparing observed and calculated intensity contrasts, one should keep in mind that observations of the solar granulation contrast are severely affected by the earth's atmosphere and by instrumental effects, making large corrections necessary to derive the true contrast from the observed one. In view of the substantial observational uncertainties, reflected by the large range of δI_{rms} values derived for the solar granulation contrast by different observers, this quantity can hardly be used as a stringent constraint for theoretical granulation models.

4.2 Spatially averaged Line Profiles

The shape of the spectral line profiles is more suitable for a quantitative comparison. In usual solar spectroscopy the light of many granlues enters the entrance aperture of the spectrograph and the resulting line profiles refer to a spatial average. Corresponding line profiles are obtained from granulation models by horizontally averaging the emergent intensity. Although the results presented in this section refer to one particular spectral line, they may be regarded as representative in the sense that lines of different ions or of different strengths behave qualitatively similar.

4.2.1 Steady State Models. As an arbitrary example we compare in Fig. 4a synthetic and observed line profile of the medium-strong Fe II line at λ 5197 Å. The synthetic profile has been computed from the steady state model displayed in Fig. 1 without introducing any additional broadening like classical micro- and macroturbulence. The non-thermal broadening is provided exclusively by the simulated velocity field. Clearly, the calculated line profile is too narrow and too deep. The obvious explanation is that the simulated steady state velocity field is not sufficiently vehement to account for the observed line broadening.

On the other hand, looking at the line asymmetry we can notify an excellent agreement between the shape of calculated and observed bisectors (Fig. 4b), despite the differences in the profiles themselves. Finally it is worth noting that the line core is essentially unshifted in our example, in contrast to observational evidence calling for a convective blueshift of several hundred m/s (cf. Dravins et al. [16]).

4.2.2 Instationary Models. In the case of the time-dependent models the effort needed to compute average spectra is much greater than in the steady

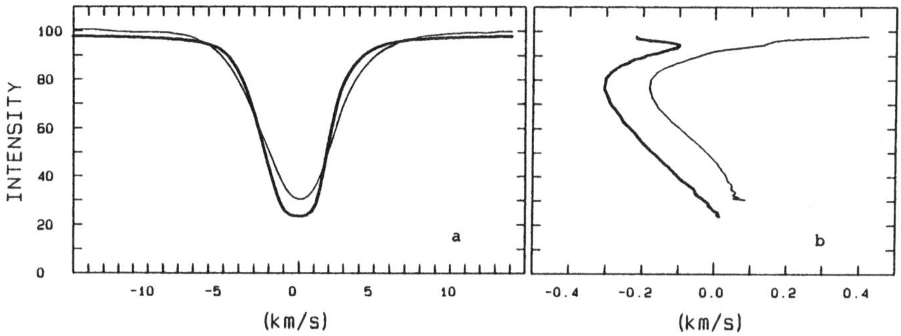

Fig. 4. (a) Comparison of calculated (thick line) and observed (thin line) profile of Fe II, λ 5197.6 Å at solar disk-center. The synthetic line profile has been calculated from the *steady state model* shown in Fig. 1. (b) Comparison of the corresponding line bisectors. The observed bisector (right) is arbitrarily displaced relative to the calculated bisector (left), which is shown on an absolute velocity scale in the solar reference frame. Differences near the continuum are due to a weak blend in the observed spectrum. (FTS observation courtesy of W. Livingston, National Solar Observatory, Tucson, U.S.A.)

state case, because the spatially resolved spectrum synthesis has to be performed for a time series of many models. The synthetic profile for Fe II, λ 5197 Å shown in Fig. 5a has been obtained as a spatial and temporal average of 2450 individual profiles, calculated from a non-stationary simulation covering almost 2 hours of real time (requiring 7 hours of CPU time on a CRAY X-MP 216). For each of 70 snapshots taken at intervals of 100 sec, 35 spatially resolved line profiles contribute to the total average.

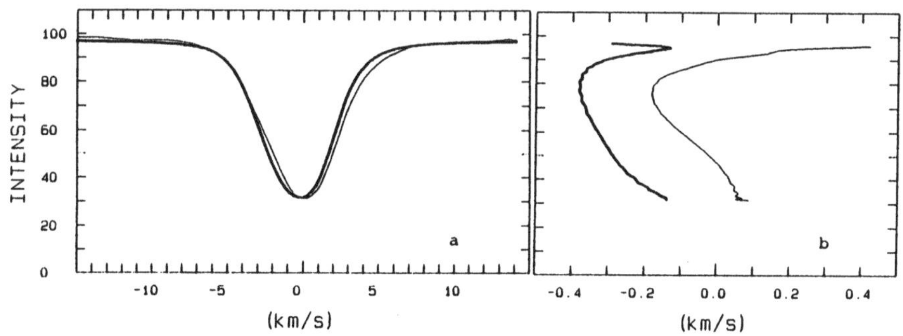

Fig. 5. Same kind of comparison as in Fig. 4, but for a synthetic line profile based on a *non-stationary simulation* in a cell measuring 2100 km in diameter. The calculated profile was obtained as the combined spatial and temporal average of 2450 individual profiles. Again the observed bisector (right) is arbitrarily displaced relative to the calculated bisector (left).

Agreement with observation is much better than for the steady state profile, implying that the velocity field due to the simulated convective and oscillatory motions is sufficient to provide the observed line broadening. A remarkable result! It means that we can explain why classical spectroscopic analyses have to introduce the ad hoc parameters micro- and macroturbulence: these quantities obviously represent the combined effects of the granular velocity field and the associated oscillations. A similar suggestion has been made by Holweger et al. [17]. This explanation was substantiated numerically by Nordlund [18] from 3-D anelastic models and by Lites et al. [14] from compressible 3-D simulations.

As in the case of the steady state model, the shape of observed and synthetic line bisector looks very similar (Fig. 5b). Moreover, now a net blueshift is predicted for the line core, resulting in an improved agreement between calculated and observed line shift.

4.3 Spatially resolved Line Profiles

4.3.1 Observed Spectra. Observations of spatially resolved spectra of the solar granulation were carried out by one of the authors (M.S.) in November 1989 at the German Vacuum-tower-telescope (VTT) at Izaña, Tenerife. The 70 cm telescope gives a scale of 4.6″/mm at the focal plane where the entrance slit of the Echelle spectrograph is located. Choosing a slit width of 100μ allows a maximum spatial resolution of 0.5″ and corresponds to a spectral resolving power of ≈ 500000. In the focal plane of the spectrograph the scale is 4.7″/mm and the dispersion at λ 4912 Å is about 83 mÅ/mm. The detector used was a RCA CCD camera with 512 by 320 pixels, with a pixel size of 30μ by 30μ. This chip covers a total of about 72″, or roughly 50 granules, along its larger dimension (0.14″/pixel) and approximately 800 mÅ in the direction of dispersion (2.5 mÅ/pixel).

Spectrograms at wavelenghts between 4909 Å and 4913 Å were taken in quiet regions near the center of the solar disk. Exposure times were kept as short as possible to minimize the degrading influence of atmospheric seeing. For most of the frames we used 0.5 sec, the minimum exposure time still giving a reasonable signal to noise ratio. Atmospheric conditions were such that spectrograms taken with longer exposure times suffered a distinct deterioration of contrast and spatial resolution.

The basic reductions were carried out on a SUN workstation at the Kiepenheuer-Institut in Freiburg, using the IDL graphics package to run the standard reduction procedures, developed and made available to visiting observers by the Kiepenheuer-Institut. Final reductions and analysis were performed at Kiel with software written by Freytag [19].

Fig. 6a shows one of the best spectra of Fe I, λ 4909.4 Å, after having gone through the whole series of reduction steps. The granular structure is clearly visible. We estimate the spatial resolution of this spectrogram to be somewhere between 0.5″ and 0.8″. The measured intensity contrast in the continuum is 6.5% and the Doppler shift of the line core indicates peak to peak velocity differences of almost 2 km/s. The corresponding rms velocity fluctuation is about 400 m/s in this example. A strong correlation

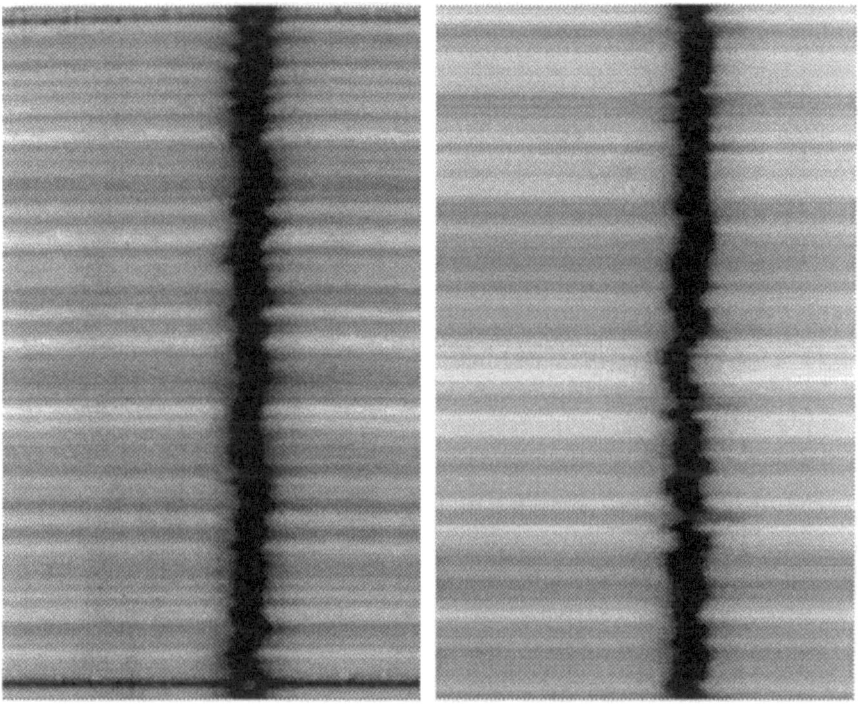

Fig. 6. Left-hand panel (a): spatially resolved spectrogram of the solar granulation, centered on Fe I, λ 4909.4 Å as taken with the German vacuum-tower-telescope at Izaña on November 7, 1989, using the RCA CCD detector covering 72″ in the vertical (spatial) direction and 800 mÅ in the horizontal direction (wavelength increasing to the right). Right-hand panel (b): spatially resolved spectrum as calculated from numerical simulations of the solar granulation for the same spectral line, reproduced to be directly comparable to the observed spectrum (a). Degrading by atmospheric seeing and intrumental effects was modelled in a schematic way. For details see text.

between continuum intensity and line shift is evident, clearly indicating hot rising and cool sinking material. Similar results have been reported by other observers (e.g. Wiehr and Kneer [20]).

Fig. 7 displays the variations of the spatially resolved line profiles and their corresponding bisectors for a small but typical section of the spectrogram shown in Fig. 6a covering 2.8″. Evidently, granulation is associated with considerable spatial variations of the line profiles. In particular, the Doppler velocity of the line core changes by more than 1.5 km/s over granular scales, directly demonstrating that substantial velocity fields must exist even in the upper photosphere where the line core is formed.

4.3.2 Simulated Spectra. The first step for the simulation of spatially resolved spectra is to generate a representative granulation pattern from a time series of instationary hydrodynamical models, assuming it is permitted to replace temporal for spatial variations. To circumvent the problem

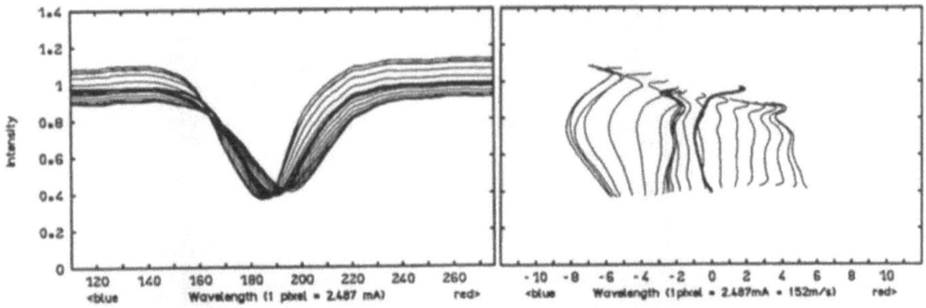

Fig. 7. Left-hand panel (a): Spatially resolved line profiles of Fe I, λ 4909.4 Å, obtained by scanning the *observed* spectrogram shown in Fig. 6a over a small subsection extending 2.8″. Right-hand panel (b): Bisectors of the line profiles shown in (a), plotted on an extended scale. Heavy line: bisector of the line profile averaged over the complete spectrogram (72″).

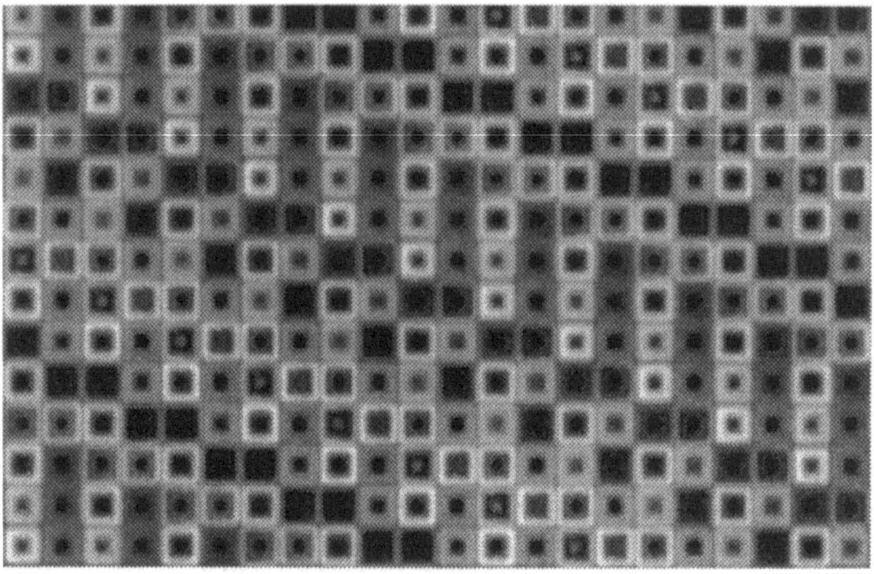

Fig. 8. Artificial granulation pattern, generated from an instationary simulation in a cell measuring 2625 km in diameter, as seen at disk-center in the continuum at λ 4909 Å, taking into account instrumental and atmospheric smearing. The total area covers 52000 km by 32500 km (72″ by 45″).

of uniformly covering a surface with models having a cylindrical cross section, we transform the cross sections into squares of equal area. In this way the fractional areas of dark and bright surface elements are retained and we can easily cover a rectangular surface section without having a problem with interstices in between the actual models. We arrange a sequence of 48 snapshots from an instationary simulation in an essentially random way to cover an area of 52000 km by 32500 km (72″ by 45″) as illustrated in Fig. 8, showing the artificial granulation pattern in continuum light at λ 4909 Å. To simulate seeing and instrumental effects, the original 2-dimensional intensity pattern has been convolved with an axially symmetric point spread function, given by $PSF(r) = (1-p)L(r,a) + pL(r,b)$, where

$$L(r,a) = \frac{a}{2\pi(a^2+r^2)^{3/2}}$$

is the *inverse Abel transform* of the Lorentz-Function. Following Nordlund [21] we used $p = 0.4$, $a = 180$ km and $b = 1800$ km.

Obviously the resulting artificial granulation pattern is topologically different from the real granulation an looks more like an assembly of "exploding granules". Nevertheless we use it as the basis for the calculation of spatially resolved synthetic spectra, presuming that it is primarily the *vertical* structure of individual granulation elements that determines the characteristics of the resolved spectra.

To generate a simulated spectrum, a numerical slit is put onto the artificial granulation pattern, oriented in such a way as to intersect the "granules" in a largely random, non-repetitive way. Scanning the slit in steps of 0.14 $''$ (to match the pixel structure of the observations), we calculate the line profile corresponding to the respective position within the artificial granulation pattern in steps of 2.5 mÅ (= 1 pixel in the observed spectrogram), applying a smearing due to seeing and instrumental effects according to PSF(r) for every wavelength point of the profile and in addition acounting for the finite slit width of 100μ. The resulting simulated spectrogram is shown in Fig. 6b.

The first impression is that observed and simulated spectra look very similar, at least qualitatively. There are some quantitative differences, though. The simulated intensity contrast in the continuum is 8.2%, while the value measured from the observed spectrogram is 6.5%. Similarly, the simulated rms velocity fluctuation is 575 m/s, somewhat higher than the observed value of 400 m/s. These systematic differences suggest that the adopted smearing probably underestimates the effects of seeing adequate for the actual observing conditions. It is unlikely that the reason lies in the hydrodynamical granulation model, because the simulated velocity field has been found to reproduce the observed broadening of spatially averaged line profiles correctly (see 4.2.2).

Looking for a section in the simulated spectrogram that is similar to the observed detail shown in Fig. 7, we came across the example displayed in Fig. 9. Variation of continuum intensity and Doppler shift as well as the arrangement of the line *profiles* is strikingly similar in obervation and simulation. Although there are differences in the shape of the resolved line bisectors, we note a good agreement between the average bisectors.

If the simulated spectrogram shown in Fig. 6b had *not* been degraded to account for seeing and instrumental effects, the resulting rms intensity

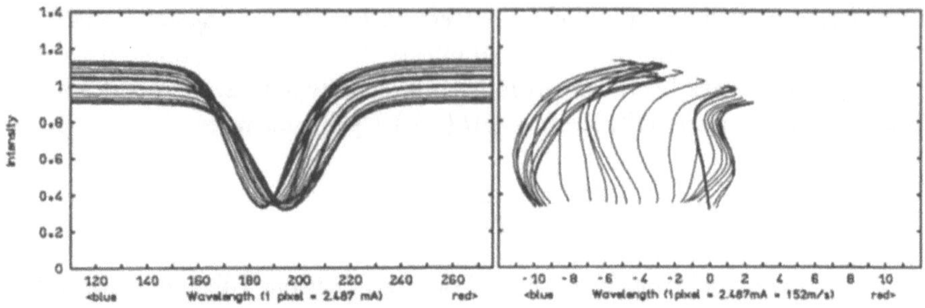

Fig. 9. Left-hand panel (a): Same as Fig. 7a, but obtained by scanning a similar subsection of the *simulated* spectrogram shown in Fig. 6b extending over 3.5″. Right-hand panel (b): Corresponding line bisectors. Heavy line: spatially averaged bisector.

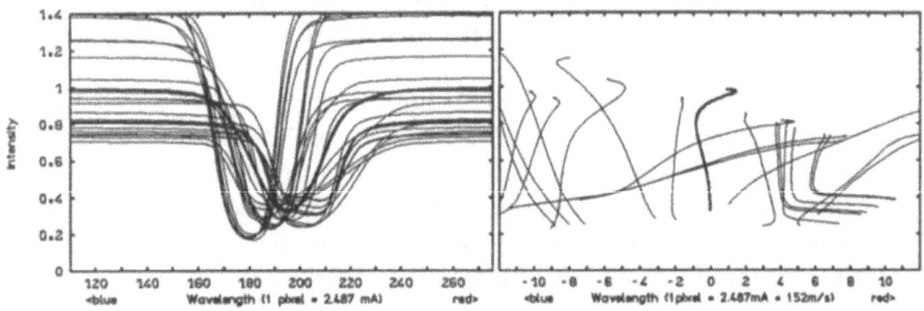

Fig. 10. Same as Fig. 9, except that the simulated spectrogram was *not* artificially degraded by atmospheric seeing and instrumental effects. Per definitionem, the spatially averaged bisector (heavy line) is unaffected. Note that some bisectors fall completely outside the plotted range.

contrast in the continuum would be as high as 21.7%, and the rms velocity fluctuation in the line core would amount to 1.1 km/s. Fig. 10 illutrates how the scans displayed in Fig. 9 would appear in this case. Our models predict that such enormous fluctuations should actually be seen if all observational limitations could be overcome. From the striking differences between Fig. 9 and Fig 10 we conclude that present-day spectroscopic observations are far from really resolving the solar granulation!

5. Conclusions

Numerical simulations of 2-dimensional, axially symmetric compressible convection under conditions appropriate for the solar photosphere, including a detailed treatment of non-local radiative transfer, result in two different kinds of solutions, depending on the adopted size of the convection cell. For small model diameters, the convective flow develops towards a *steady state solution*, whereas for the larger models we obtain *time-dependent convection* that may be characterized as chaotic.

Comparison with various spectroscopic observations strongly suggests that the non-stationary granulation models are a more realistic approximation to the real solar photosphere than the steady state models. Accepting the time-dependent solutions as representative for the hydrodynamical conditions prevailing in the solar atmosphere, we conclude that the solar photosphere up to the level of the temperature minimum is neither static nor plane-parallel nor in radiative equilibrium. Disturbances generated by the continually changing topology of the flow pattern in the underlying convective layers carry a substantial amount of acoustic flux and have considerable effects on the higher layers. This is at variance with the classical models which consider the solar photosphere as static and in radiative equilibrium.

Observational evidence in fact indicates the presence of substantial velocity fields in the higher photospheric layers. It is well known that non-thermal velocities of the order of 2 km/s are needed to explain the observed broadening of spatially averaged line profiles, the reason for introducing classical micro- and macroturbulence (without understanding their physical meanings). In spatially resolved spectra, even the cores of stronger lines show Doppler shift variations of at least 1 km/s on granular scales, directly illustrating the dynamical state of the solar photosphere. We have shown that our simulations are able to reproduce such spectroscopic observations, replacing both micro- and macroturbulence by the effect of convective granular flows overshooting into the higher photosphere.

From the favourable comparison between observed and synthetic spectra we may conclude that many basic properties of photospheric (stellar) convection can be modelled even quantitatively by 2-dimensional numerical simulations. Moreover, we found basic agreement between the results obtained from 2-dimensional and 3-dimensional granulation models, respectively, suggesting that for many purposes 2-dimensional simulations can be a reasonable alternative, requiring considerably less computational effort than fully 3-dimensional calculations. In view of the extensive work that has yet to be done in the field of stellar convection, this is an important aspect.

Acknowledgements. We would like to thank H. Wöhl and H. Schleicher from the Kiepenheuer-Institut for their kind assistance during the obervations at the VTT and the reductions of the observed granulation spectra. We are grateful to H. Holweger for many stimulating discussions and for encouraging this work. The hydrodynamical simulations were carried out at the CRAY X-MP 216 of the Rechenzentrum der Universität Kiel. Financial support by the Deutsche Forschungsgemeinschaft under grants Ho 596/21-2 and Ho 596/27-1 is gratefully acknowledged.

References

1. A. Unsöld: Z. Astrophysik **1** 138 (1930)
2. R.L. Kurucz: Astrophys. J. Suppl. **40** 1 (1979)
3. G. Shaviv, E.E. Salpeter: Astrophys. J. **184** 191 (1973)
4. I.W. Roxburgh: Astron. Astrophys. **65** 281 (1978)
5. J.S. Smagorinsky: Month. Weather Rev. **91** 99 (1963)
6. J.W. Deardorff: J. Fluid Mech. **41** 453 (1970)
7. J.W. Deardorff: J. Comput. Phys. **7** 120 (1971)
8. K.L. Chan, S. Sofia: Astrophys. J. **307** 222 (1986)
9. M. Steffen, H.-G. Ludwig, A. Krüß: Astron. Astrophys. **213** 371 (1989)
10. M. Steffen: in *Stellar Atmospheres: Beyond Classical Models*, eds. L. Crivellari et al., NATO ASI Series (in press)
11. M. Steffen, A. Krüß, H. Holweger: in *Mechanisms of Chromospheric and Coronal Heating*, ed. P. Ulmschneider, Springer (in press)
12. F. Cattaneo, N.E. Hurlburt, J. Toomre: in *Solar and Stellar Granulation*, eds. R.J. Rutten, G. Severino, Kluwer Academic Publishers, p. 415 (1989)
13. F. Cattaneo, N.E. Hurlburt, J. Toomre: Astrophys. J. **349** L63 (1990)
14. B.W. Lites, Å. Nordlund, G.B. Scharmer: in *Solar and Stellar Granulation*, eds. R.J. Rutten, G. Severino, Kluwer Academic Publishers, p. 349 (1989)
15. R.J. Bray, R.E Loughhead, C.J. Durrant: *The Solar Granulation*, Cambridge University Press, 2nd ed., p. 70 (1984)
16. D. Dravins, B. Larsson, Å. Nordlund: Astron. Astrophys. **158** 83 (1986)
17. H. Holweger, M. Gehlsen, F. Ruland: Astron. Astrophys. **70** 537 (1978)
18. Å. Nordlund: in *Small-Scale Dynamical Processes in Quiet Stellar Atmospheres*, ed. S.L. Keil, Sunspot, NM 88349. p. 181 (1984)
19. B. Freytag: *Spektroskopie solarer Granulationszellen und Vergleich mit numerischen Simulationen*, Diplomarbeit, Kiel (1991)
20. E. Wiehr, F. Kneer: Astron. Astrophys. **195** 310 (1988)
21. Å. Nordlund: in *Small-Scale Dynamical Processes in Quiet Stellar Atmospheres*, ed. S.L. Keil, Sunspot, NM 88349. p. 174 (1984)

Solar Spectroscopy with a 100 × 100 Diode Array

A.D. Wittmann

Universitäts-Sternwarte Göttingen,
Geismarlandstr. 11, W-3400 Göttingen, Fed.Rep.of Germany

1. Introduction

Among the standard postfocus devices of the Gregory-Coudé solar telescope (GCT) at Observatorio del Teide/Tenerife (Kneer et al., 1987) is a 100 x 100 self-scanned Silicon photodiode array manufactured by EG&G RETICON. This device was formerly used at the Locarno solar observatory (see, e.g., Küveler and Wöhl, 1983; Scholiers and Wiehr, 1985), and, after its adaption to the GRECOS operating system of the GCT (Wittmann and Kroll, 1990), has been in use at Tenerife since 1986 (e.g. Hanslmeier et al., 1990; Balthasar, 1990).

In July 1989 the Echelle grating of the GCT spectrograph (which had 300 grooves/mm) was replaced: Our new grating has 316 grooves/mm blazed at 63.4 degrees, and a ruled area of 186 x 366 mm. When used in combination with the predisperser spectrograph, exposure times are typically reduced by a factor of two. Early in 1990, efforts were also made to improve the glass window and the electronic readout of the Reticon diode array. This resulted in a significant improvement of performance, in particular a twofold increase in the dynamic range, and a greatly reduced amplitude of the interference fringes caused by the window. In 1990 the array was supplemented by a much more sensitive, high-resolution, Nitrogen-cooled CCD system manufactured by Wright Instruments Ltd, which can be used at both the GCT and the VTT (cf. Soltau, 1990).

The Reticon array is Peltier-cooled and has a binsize (pixel centre-to-centre distance) of 60.0 microns. The 10000 individual square pixels each have a size of 46 x 46 microns, and the total light-sensitive area is 5.99 x 5.99 mm. When placed in the exit focal plane of the spectrograph (where the spatial scale is 8.3 "/mm), 1 bin corresponds to 0.5" or about 370 km on the sun. At a wavelength of 618 nm the dispersion is 10.7 mÅ/pixel, and the exposure time in the photosphere is about 3 seconds. Using the RETICON command of the GRECOS operating system, both single exposures and equally-spaced time series may be taken, with the image data (10000 integer values per frame, optionally in FITS format) being written to standard 1/2-inch magnetic tape.

In 1989 and 1990 the 100 x 100 diode array was used to obtain high-resolution spectrograms of sunspots, pores, and the quiet photosphere. In particular, extended time series of the solar granulation at the disc centre were taken in two different spectral lines during very good - but still variable - seeing conditions: The purpose of this investigation was to measure granular line shifts (i.e. velocities) as a function of seeing on a statistical basis, and to try to extrapolate the measured quantities to ideal conditions, i.e. to the case of almost diffraction-limited observations. The present paper reports about the first results from 4 long time series that were taken on 1 May 1989 and on 6-8 July 1990, while results from the sunspot observations will be published elsewhere.

2. Observations

Two iron lines were chosen for the present investigation, viz. the magnetically sensitive line Fe I 617.334 nm (henceforth called the 6173 line), and the magnetically much less sensitive line Fe I 618.022 nm (henceforth called the 6180 line): The first line was selected because it is one of the very few widely-split Zeeman triplets in the red part of the spectrum (where the Silicon array is most sensitive). The second line was selected with regard to the granulation observations (because it is almost insensitive to granular magnetic fields), and because of its vicinity to the former line, which helps to minimize differences in predisperser setup, exposure time, readout noise, interference patterns, etc. The calculated depth of formation of both lines (using the method of Wittmann, 1974, Section 10) in the photosphere is $\log\tau \simeq -0.8$ (with the 6180 line formed just slightly deeper).

Figure 1 shows the two lines in a sunpot umbra which is split by a light bridge: It is clearly seen that the magnetic field (of about 2600 Gauss in the umbra) practically disappears inside the light bridge, and that the splitting of the 6180 line - which is due to 21 overlapping Zeeman components - is only about one-fourth of that of the 6173 line. Data on the granular time series are given in Table 1. Made entirely under computer control, the exposures were triggered at regular intervals of about one frame every 9.7 seconds. Positional control was achieved through our ZEISS guiding system which also made the slight displacements required to compensate solar rotation.

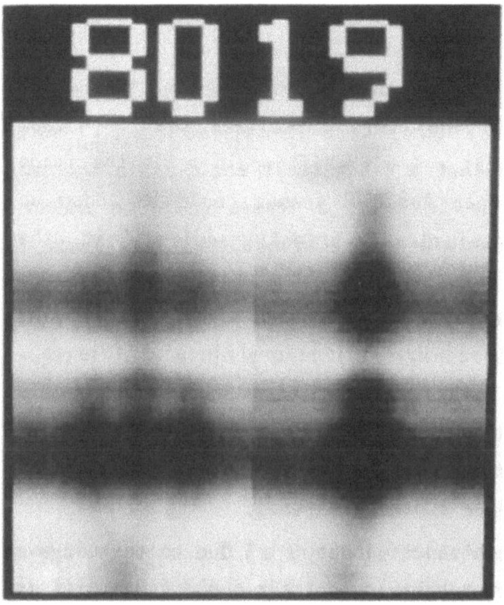

Fig. 1: The iron lines used for the present investigation in a sunspot umbra with a light bridge (image composed from two frames taken almost simultaneously, with wavelength increasing towards the right). Taken on 1 July 1990 at 9:47 UT at x=+28.4" West, y=+95.5" North, with the slit almost perpendicular to the light bridge (Position angle +117.1°). Exposure time 8 sec; 6173 is at left, 6180 at right.

Table 1: Reticon 100 x 100 time series

Date	Start(UT)	End(UT)	Line	Frames
1 May 1989	07:59:43	10:02:47	6173	762
6 July 1990	08:41:31	10:47:18	6180	777
7 July 1990	08:06:36	10:05:32	6180	740
8 July 1990	08:05:13	12:03:27	6180	1480

3. Image Reduction

The reductions were made with a Computer Automation (CA) NM-4/30 computer equipped with a Dawnstoke color graphics controller. The data were stored on Maxtor XT-1085 harddisks and backed up on magtape as needed. First, the dark current frames exposed before and after each series were averaged (to reduce shot noise) and subtracted from the spectral images. Next, the flatfields taken in the nearby continuum (with the solar image drifted in order to smoothe the granular brightness fluctuations) were corrected for the dark current and averaged, then a pixelwise division of each frame by its corresponding flatfield was made. Note that the theory of flatfielding requires that <u>all</u> pixels must have a positive (significantly non-zero) sensitivity. During flatfielding the first 3 and the last 2 rows (diode lines) of each frame were stripped off because they contained too many defective pixels; this leaves us with a total of 95 significant rows (or 9500 significant pixels) per frame.

Then the continuum intensities (averaged over a suitable area to the left and to the right of the spectral line) were normalized to I_{cont}=3136. The latter value was chosen such that a 4-bit shift would scale the image data to the graphics range, while at least 8 frames could be added up without overflow, even if they contained very bright granules. Next, the interference fringe pattern (which overlays each frame and has a typical amplitude of about ± 2 %) was extracted: This was done by first adding up all frames to get an 'average (non-wiggly) line profile plus fringes'. Then, by adding up all 95 rows of the latter frame and by subsequently re-expanding into 95 rows, one gets an 'average line profile without fringes'. Then the difference between these two frames is the 'fringe pattern' frame which is to be subtracted from all other frames.

Then a very slight residual modulation of about ± 1 % due to the odd-even effect (which is caused by different behaviour of the two ADC and amplifier circuits for odd and even rows) was corrected by an algorithm based on the assumption that the average continuum contrast between odd and even rows (i.e. I(n-1)/I(n) - 1, where n = 2,4,6,...) is zero, while a value significantly different from zero would be due to the odd-even effect. Finally, some standard digital image processing techniques were applied to a small number of selected frames in order to quantitatively improve their quality and to gain additional insight into the residual noise statistics. In some cases a minor improvement of the S/N ratio could be achieved by adding up a small number (N = 2-6) of adjacent images whose high degree of structural congruency had been verified before on the basis of their difference frames. In this way it was found that the individual frames typically have a residual noise rms of about ± 2 %. Figure 2 shows two examples from the time series as they have emerged from the reduction process (note that, contrary to Figure 1, the two halves of this figure show different areas on the sun).

The average continuum intensity ($I_{mean} \approx 3136$) and the root mean-square brightness fluctuation of the continuum (I-rms) were calculated using N = 40 continuum pixels to the left and to the right of the iron line in each row (N = 3800 per frame). Apart from its modifications due to instrumental broadening and noise, I-rms may serve as a quantitative measure of the seeing quality: When the seeing is very bad, I-rms is essentially zero, it increases monotonically with improving seeing conditions, and it finally reaches an asymptotic value: From many recent investigations of granular brightness fluctuations (e.g. Schmidt et al., 1981; Wittmann, 1981; Durrant et al., 1983; Collados and Vázquez, 1987; Muller, 1989;

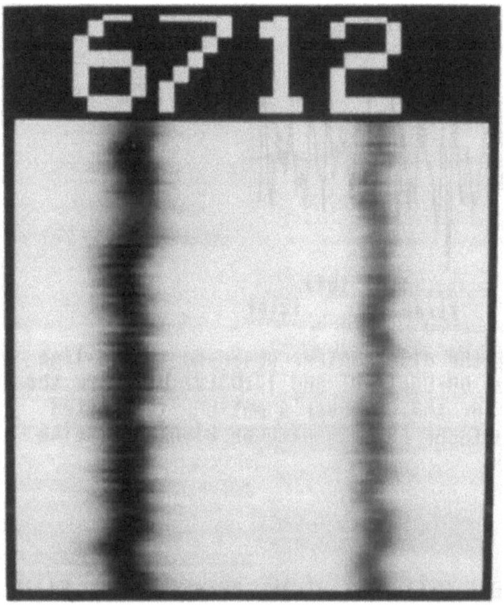

Fig. 2: The two Fe lines in two frames from the time series after reduction: The 6173 line is shown in the left half of the figure, which is an average of 3 single frames taken on 1 May 1989 at 9:18 UT. The section shown (50 pixels) corresponds to 430 mA. The 6180 line is shown in the right half of the figure, which is an average of 6 single frames taken on 6 July 1990 at 9:59 UT. The section shown (50 pixels) corresponds to 536 mA. As in Figure 1, wavelength is increasing towards the right. Granular Doppler shifts are clearly visible

Rodriguez Hidalgo et al., 1990) it has been quite firmly established that the true I-rms at 620 nm (i.e., the value reached for idealized, diffraction-limited seeing conditions) is about ± 12.1 %. Hence we may plot the observed granular Doppler shifts versus I-rms and extrapolate - as good as possible - to the asymptotic value, to get what might be called the 'true' Doppler shift. This method, which is probably applied here for the first time, obviously requires a very large number of spectrograms. The directly observed, average I-rms value (based on 357105 rows) decreases from ± 3.3 % at 8:00 UT to ± 2.7 % at 10:00 UT, reflecting the well-known fact that the best solar seeing at Observatorio del Teide usually occurs during the early morning hours.

The local Doppler shifts of the iron line were determined in the usual way: First, a 9-point parabola centered on the intensity minimum was fitted to the line core intensities of each row. This gave N = 95 line positions per frame, from which the mean line position of the frame was found by averaging: These mean line positions (averaged over a distance of roughly 35000 km on the sun) represent the oscillatory component due to the 5-min oscillations and serve as the reference wavelengths for the local Doppler displacements. As an example, Figure 3 shows the oscillatory component measured at the disc centre (with solar rotation compensated, but diurnal image rotation included) on 8 July 1990: Here the shifts have already been converted into 'true' solar radial velocities by subtracting

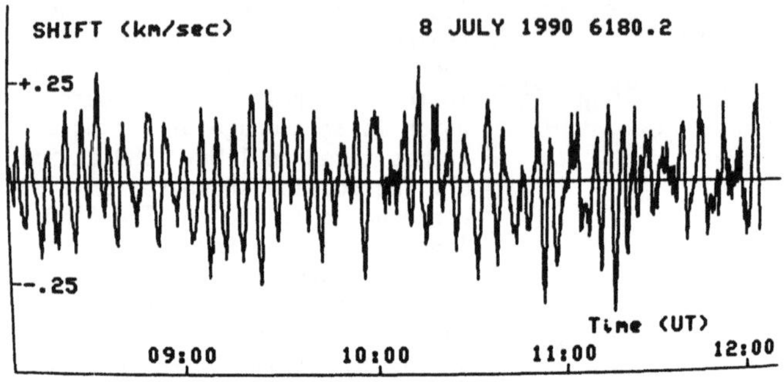

Fig. 3: Five-minute oscillations at the disc centre, observed in the line Fe I 6180.2 A on 8 July 1990 between 08:05:13 UT and 12:03:27 UT. Here the 'true' Doppler velocity, corrected for the observer's motion, is plotted vs. time (1 bin = 9.665 sec, for instance the conspicuous blueshift spike near 10 UT occurred at 09:55:43 UT)

the trend parabola due to the proper motion(s) of the observer (cf. Pallé et al., 1989) and a possible slow temperature drift of the spectrograph.

The mean amplitude of the oscillatory component determined from a phase-adjusted overlay of the 35 best individual oscillations is ± 114 m/sec. The mean (centroid) frequency of the p-modes seen by the diode array is (3.264 ± 0.009) mHz from the overlay, and (3.26 ± 0.06) mHz from a maximum entropy power spectrum analysis. Note that this is slightly larger than the centroid frequency of (3.08 ± 0.08) mHz found by Pallé et al. (1989, their Fig. 5) about three years earlier. The directly observed, average root mean-square value of the granular velocity fluctuation V-rms (based on N = 357105 rows, and corrected for the oscillatory component) decreases from ± 0.30 km/s at 8:00 UT to ± 0.26 km/s at 10:00 UT, again reflecting statistically the worsening of seeing conditions during the late morning hours.

As the spectrograph entrance slit (width roughly 1") would rarely cut right across a bright granule or a dark intergranular lane, most of the rows of any given image will represent a mixture of Doppler (and brightness) signals from both kind of features. Therefore, the maximum redshifts and blueshifts of each frame were used in the subsequent analysis: Figure 4 shows N = 2997 values of V-rms observed in the 6180 line and plotted as a function of I-rms: By extrapolating to I-rms = 12 % we obtain a true V-rms of ±0.5 (± 0.1) km/s. A similar plot for the 6173 line gives a true V-rms of ± 0.7 km/s; both values are in the upper range of results from previous investigations (see, e.g., the review by Wittmann, 1978) and compare very well with more recent measurements based on highly-resolved photographic spectrograms (e.g. Wiehr and Kneer, 1988).

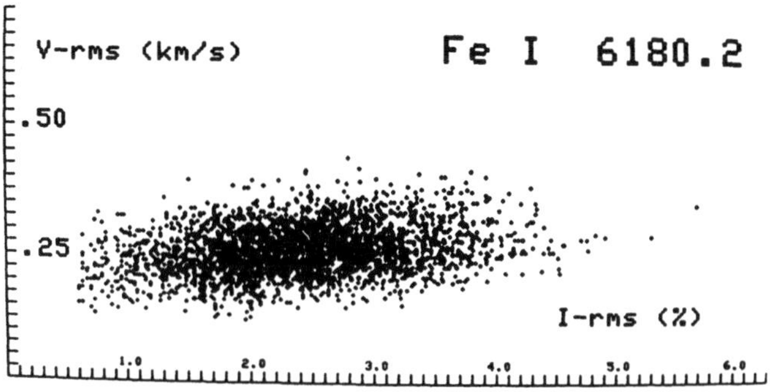

Fig. 4: The root mean-square Doppler velocity fluctuation observed in the line Fe I 6180.2 A: The values derived from N = 2997 individual spectrograms are plotted vs. the root mean-square continuum intensity fluctuation I-rms (for details see text)

Separate plots (scatter diagrams) of the directly measured red- and blueshifts vs. I-rms show that with improving seeing conditions both shifts behave symmetrically, i.e. there is no magnitude difference between downdraft and uprising velocities on resolution scales of the order of 0.3" (the Airy limit of the telescope) to 0.5". Hence, according to the measurements presented here the one-to-one correlation holds down to much smaller scales than found by Wiehr and Kneer (1988). Asymmetries or anticorrelations at still smaller scales can, of course, not be excluded. Extrapolation to I-rms = 12 % leads to peak velocities of ± 1.5 km/s and ± 1.3 km/s for the 6173 and the 6180 line, respectively. Again, this compares very well with the results of Wiehr and Kneer (1988).

Multi-color plots of the magnitude of the measured red- and blueshifts, multiplied by the continuum brightness excess (i.e. the intensity difference including sign, and expressed in %) of the same row, with features brighter than average plotted upwards (in, say, blue), and features darker than average plotted downwards (in, say, red) were shown during the conference; they are omitted here because a reproduction in black and white would be inapt. Obviously the mentioned product is a good measure of the correlation between the brightness and the velocity fluctuations inside the granulation, and the plots clearly demonstrate that the correlation is fairly bad under bad seeing conditions, but greatly improves with increasing resolution until finally all anti-correlations disappear. Thus, at scales of 0.3-0.5", and at a geometric altitude of about 150 km, bright elements are always associated with uprising velocities, and dark elements are always associated with downdraft velocities.

References:

Balthasar,H.: 1990, Solar Phys. **127**, 289

Collados,M., Vázquez,M.: 1987, Astron. Astrophys. **180**, 223

Durrant,C.J., Mattig,W., Nesis,A.: 1983, Astron. Astrophys. **123**, 319

Hanslmeier,A., Wittmann,A.D., Mattig,W.: 1990, Astrophys. Space Sci. **170**, 47

Kneer,F., Schmidt,W., Wiehr,E., Wittmann,A.D.: 1987, Mitt. Astron. Ges. **68**, 181

Küveler,G., Wöhl,H.: 1983, Astron. Astrophys. **122**, 69

Pallé,P.L., Pérez Hernandez,F., Roca Cortés,T., Isaak,G.R.: 1989, Astron. Astrophys. **216**, 253

Rodriguez Hidalgo,I., Collados,M., Vázquez,M.: 1990, Astrophys. Space Sci. **170**, 23

Schmidt,W., Knölker,M., Schröter,E.H.: 1981, Solar Phys. **73**, 217

Scholiers,W., Wiehr,E.: 1985, Solar Phys. **99**, 349

Soltau,D.: 1990, 'The German Solar Telescopes on Tenerife', in: Proc. XXVIII-th COSPAR Plenary Meeting 1990 (in press)

Wiehr,E., Kneer,F.: 1988, Astron. Astrophys. **195**, 310

Wittmann,A.: 1974, Solar Phys. **35**, 11

Wittmann,A.: 1978, in: Stix,M. (ed.), Proc. Kiepenheuer Colloq. on Small Scale Velocity Field in the Sun, p. 29, Kiepenheuer-Institut Freiburg

Wittmann,A.: 1981, Astron. Astrophys. **99**, 90

Wittmann,A.D., Kroll,R.: 1990, Astrophys. Space Sci. **171**, 357

Solar Research at Potsdam: Papers on the Structure and Dynamics of Sunspots

J. Staude

Zentralinstitut für Astrophysik, Sonnenobservatorium Einsteinturm, O-1561 Potsdam, Fed. Rep. of Germany

Abstract. The investigation of sunspots at Potsdam has a tradition of many decades. After some historical remarks the present review considers efforts to determine the magnetic field structure in solar active regions, especially in sunspots, and the modelling of the thermodynamic structure and of the oscillatory regime of umbrae.

The double vector magnetograph of the Potsdam Solar Observatory 'Einsteinturm' enables us to measure the full Stokes vector in the profiles of 2 independently chosen lines. The instrumental calibration and elimination of signal distortions provide the necessary conditions for the theoretical calibration in terms of a transfer theory of polarized radiation. By means of the latter the spatial distribution of the magnetic field vector and related quantities such as electric currents can be derived. Modelling efforts include the derivation of the height dependence of thermodynamic quantities in umbrae from the subphotosphere up to the corona, moreover, the interpretation of umbral oscillations by a model of resonant transmission of magneto-atmospheric waves through umbral models.

Future observations with high spatial resolution from ground-based (Cooperative Observatory at Tenerife, LEST) and space-borne stations (SOHO, OSL) will provide favourable prospects for testing our model predictions, but also for applying our experiences of Stokes polarimetry.

1 Introduction

The interaction between the turbulent solar plasma and the magnetic field plays a key role in the formation of fine structures, in the excitation of dynamic and active processes, and in the heating of the upper solar atmosphere. Sunspots are the strongest concentrations of magnetic flux in the solar mantle and the kernels of active regions, moreover, they provide a unique example of a magnetized stellar atmosphere where up-to-date observational techniques enable us to resolve many details of the fine structure and of the dynamics of various processes.

At Potsdam the investigation of sunspots has attracted the interest of astronomers for more than a century: Solar research and sunspot observations in particular stimulated the foundation of the institute 'Astrophysikalisches Observatorium Potsdam' (AOP) in 1874, and excellent drawings of sunspot fine structure have been obtained from visual observations already before the buildings of the AOP had been completed (Lohse, 1883; see Figure 1). Half a century later, in 1924 the building and the equipment of the 'Sonnenobservatorium Einsteinturm' (SOE) were ready for solar observations, providing one of the

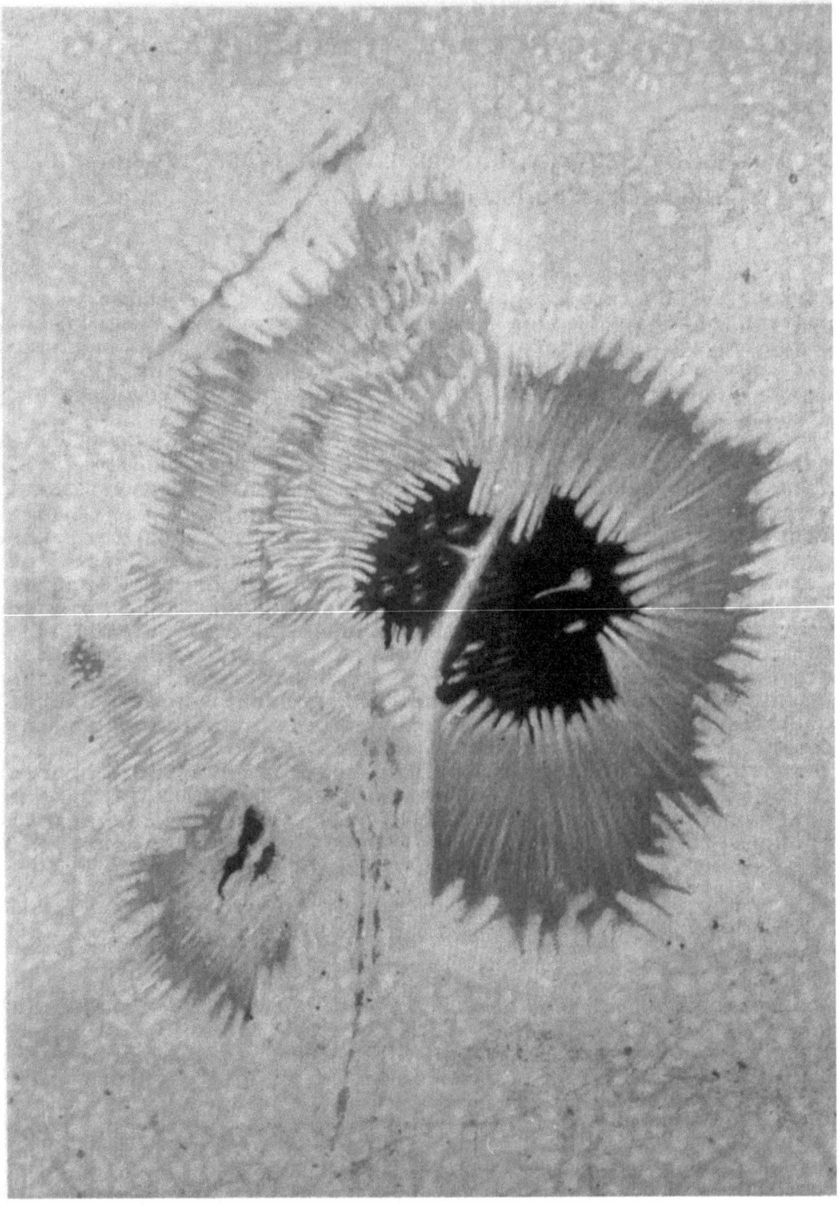

Fig. 1. Sunspot drawing after a visual observation of Lohse (1883), September 25, 1873.

greatest solar tower telescopes and spectrographs at that time and stimulating research in different fields of solar physics, among them also sunspot investigations. In the present paper I intend to review some of these topics, namely measurements of sunspot magnetic fields (Section 2), models of the (subphotospheric and atmospheric) spatial structure of thermodynamic quantities in sunspot umbrae (Section 3), and the interpretation of waves and oscillations observed in umbrae (Section 4). The results are due to many solar physicists working at Potsdam in the past and present time, but the selection of the topics is determined by the individual interests and the point of view of the author who is also responsible for shortcomings in the presentation.

2 Magnetic Field Measurements

2.1 Photographic Zeeman Spectrograms

After the pioneering work of Hale (1908) the SOE at Potsdam was the second observatory in the world where the magnetic fields of sunspots have been measured. Von Klüber (1947) developed and applied since 1942 a well operating photographic technique with a circular analyzer which has subsequently been improved by Grotrian et al. (1953, 1956) and used to measure sunspot magnetic fields over several decades. The measurements have not been limited to the determination of the maximum B_m of the magnetic field strength B of sunspots: The distribution of B across a sunspot in several directions has been measured resulting in maps of B(x,y) (Schröter, 1953; x and y are the Cartesian cooordinates in the image plane), the measured radial dependence of B has been fitted to an analytic expression (Mattig, 1953), and the time dependence $B_m(t)$ of many sunspots has been determined as well (e.g., Künzel, 1964). Investigations of the flare activity of different magnetic field structures in sunspots groups led to the proposal of an extension of the magnetic classification: the term 'δ-group' has been introduced by Künzel (1960) to describe the occurrence of two umbrae with opposite polarity within a common penumbra.

The measurements of magnetic splittings from photographic Zeeman spectra taken through a circular analyzer are less sensitive to signal distortions than photoelectric measurements (see Section 3.2), but photoelectric data are more sensitive to smaller values of B. Both methods complete each other, and the photographic spectra provide useful information to improve the calibration of photoelectric measurements, especially in sunspots. Therefore photographic Zeeman spectra have still been taken when the first photoelectric magnetograph at Potsdam started its work 20 years ago.

2.2 Photoelectric Stokes Polarimetry

In 1970 photoelectric measurements of Stokes parameters started at Potsdam in close cooperation with colleagues from the SibIZMIR institute at Irkutsk which used the same type of vector magnetograph in the Sayan observatory. The equipment, methodical experiences, and first results obtained at Potsdam have been described in a report by Bachmann et al. (1975).

Since then our magnetograph as well as the data handling have been considerably improved. In the present section we intend to outline the actual state of our instrumentation, our procedure of instrumental calibration including methods for reducing signal distortions, the theoretical calibration for deriving the magnetic field components from the measured Stokes parameters, and finally examples of resulting magnetograms and of applications of such data in further investigations of solar active regions are given.

2.2.1 Instrumentation

The Potsdam double line vector magnetograph is working in connection with the solar tower telescope 'Einsteinturm'. The two coelostat mirrors with diameters of 60 cm reflect the light into the vertical direction. The objective lens has a diameter of 60 cm and a focal lenght of 14 m, but for polarization measurements the aperture is usually restricted to 45 cm. By means of an optical enlarging system an equivalent focal length of 28.4 m can be achieved. A 45° plane mirror deflects the light from the vertical into the horizontal direction of the spectrograph installation. Photoelectric guiding and scanning systems are used to enable a scanning of a selected region on the Sun with an extent up to 6'x7' and a spatial resolution of 3"x3".

The double spectrograph consists of 2 independent auto-collimating grating spectrographs, either of them with an objective lens with a diameter of 35 cm and a focal length of 12 m, moreover, with a plane grating with an area of 20x30 cm^2 or 20x25 cm^2 and 600 grooves per mm. Both spectrographs obtain their light through a common slit and polarizing modulator with subsequent beam splitting.

A rectangular modulation of the state of polarization is achieved by means of two step modulated Pockels cells which are working with frequencies of 900 Hz and 300 Hz, followed by a quarter-wave plate and a linear polarizer. The modulation scheme is similar to that described by Stepanov et al. (1975): In 6 subsequent phases signals are obtained which are proportional to $I \pm V$, $I \pm Q$, and $I \pm U$. In either of the two spectrographs photometer slits select 3 ranges (line core: ± 40 mÅ; both wings: $\pm(40\text{-}160)$ mÅ from the line center) in the profile of the respective Fraunhofer line and fed the light to photomultipliers. That means, all 4 Stokes parameters can be measured quasi-simultaneously in 3 ranges of either of 2 independently chosen line profiles. Additional channels measure the intensity in a continuum band, I_c (used for normalization) and in the core of the H-alpha line (used for co-alignment with filtergrams of high spatial resolution). Moreover, the Doppler compensators are working independently from each other in either of the two lines in order to center the lines onto the photometer slits and to measure the line-of-sight velocities v. A sampling rate of 16 Hz is used in each of the 28 channels, the data are then stored on a magnetic tape.

The rms noise corresponds to degrees of polarization of 0.0005. Applying the calibration procedures of the following sections, threshold values of 6 G for the longitudinal component and 120 G for the transverse components of the solar magnetic field are obtained, if the line FeI 5253.5 Å is used.

Since 1982 measurements of the 4 Stokes parameters in one wing of a magnetically sensitive line, that is to say with a reduced variant of the magnetograph described above have been performed; a short description and first results have been published by Bachmann et al. (1983). In 1988 we started test measurements in two lines using the complete instrumentation.

2.2.2 Instrumental Calibration and Elimination of Signal Distortions

A reliable measurement of Stokes profiles requires a careful calibration of the instrumental response of the whole equipment (telescope and modulator) to incident polarized light. This calibration must include an elimination or at least a strong reduction of signal distortions originating in the atmosphere of the Earth and in the telescope.

Temporal variations of the transparency of the atmosphere are eliminated by normalizing all Stokes parameters to I_c. Unfortunately, such a simple procedure cannot exclude distortions by parasitic stray light, that is true scattering of light, for instance at dust particles, and image blurring due to air turbulence. Measurements of stray light have been performed at Potsdam for various seeing conditions, usually by scanning I_c across the solar limb and outside the limb; the results have been compared with other estimates of stray light (intensity contrasts in sunspot umbrae, measurements of Zeeman splittings in umbrae from photographic spectra and their comparison with magnetograph data). These data have shown that blurring can usually be neglected for our moderate spatial resolution of 3". A determination and correction for (unpolarised) scattered light from the quiet photosphere, however, cannot be avoided and has been applied to all measurements. Otherwise errors of factors of 2-3 can easily be achieved under moderate seeing conditions. The earlier discussed discrepancies between 'theoretical' and 'empirical' calibration curves (Severny, 1967, 1977) could largely be due to such simple origins.

Instrumental polarization is mainly due to the oblique reflections at the plane mirrors of the solar telescope. The results are depolarization and crosstalks among the different Stokes parameters, but also a miscentering of the used line within the photometer slits is possible. The latter effect was particularly dangerous for older magnetographs which made use of the symmetry properties of the ideal Stokes profiles with respect to the line center. In earlier papers (Jäger, 1972, 1974; Bachmann et al., 1975) different methods have been discussed for eliminating such distortions. These methods require the knowledge of the Mueller (instrumental response) matrix of the telescope which can be measured but also calculated by means of Fresnel's formulae, if the mirror constants and the reflection angles at the telescope mirrors are known. In each case the elements of the Mueller matrix will depend on time.

At present we are including the determination of instrumental polarization into the calibration of the whole instrumental response. For this purpose a large linear polarizer (diameter 45 cm) or a circular polarizer (diameter 30 cm) can be placed in front of the first coelostat mirror and rotated by definite angles. In this way the Mueller matrix of the whole equipment is measured before and after each scanning of a region on the Sun; interpolation is then sufficient for determining the matrix for each moment during the observation. Small residual errors can still remain due to imperfect properties of the calibration optics. Their removal is only partly possible and requires more sophisticated methods of signal corrections.

As far as no polarization-free telescope is available, each method for reducing the instrumental polarization is useful. For instance, a great part of linear polarization can be compensated by a tilted glas plate with appropriate orientation of its two axes, moreover, one can look for optimum angles of the coelostat mirrors for which instrumental polarization is diminished by partial mutual compensation (Bachmann, 1984).

2.2.3 Theoretical Calibration

In principle the term 'magnetograph' is misleading because such a device measures signals which are proportional to the Stokes line profiles ($I_i=(I,Q,U,V)$) of light emitted from a selected region on the Sun. Such a 'Stokesmeter' becomes a magnetograph only by the theoretical calibration, that is by calculating the relation between the Stokes line profiles and the magnetic field vector **B**. **B** is determined by its components $B_\parallel = B \cos\gamma$, $B_{\perp 1} = B_\perp \cos 2\varphi$, $B_{\perp 2} = B_\perp \sin 2\varphi$ with $B_\perp = B \sin\gamma$, where $B = |\mathbf{B}|$, $B_\parallel$ and $B_\perp$ are the longitudinal and transverse field components, respectively, γ is the angle between the line-of--sight and **B**, and φ is the azimuth of **B**. The problem of theoretical calibration have been reviewed in several papers by SOE authors (Bachmann et al., 1975; Staude, 1980; Staude and Hofmann, 1988), here we will only outline those procedures which are nowadays used at Potsdam.

For calculating the required relation between I_i and **B** a suitable theory of line formation in the presence of a magnetic field is needed. In each case such a theory makes use of a given model atmosphere, that is of the spatial distribution (usually the height dependence) of thermodynamic quantities. This implies that the theoretical calibration becomes model dependent. Additional information on the used line (more generally, on the used atom or ion) such as quantum numbers, excitation potentials, and oscillator strengths, must be provided as well.

In our observatory a computer code BHM has been developed for performing the required calculations. A detailed description of an earlier version of this code has been given in a report (Staude, 1982). This code uses the equations of transfer for the Stokes vector I_i in a form as simple as in LTE. That means, a system of 4 coupled linear ordinary differential equations with variable coefficients can be numerically integrated by a Runge-Kutta technique with automatic choice of step width. Non-LTE effects can be considered by applying coefficients bi which describe the deviations from the LTE populations of the upper and lower energy levels of the considered transition. The bi are obtained from a usual multi-level non-LTE calculation neglecting the magnetic splitting; the LINEAR code of Auer et al. (1972) and the MULTI code of Carlsson (1986) are available for this purpose. Calculations by Rees (1969) and by Domke and Staude (1973) have shown that such a 'zero field approximation' is a good approach for practical purposes. The codes LINEAR and MULTI also contain useful procedures for deriving semi-empirical model atmospheres;for more details of such modelling see Section 3. BHM considers the height dependence of all thermodynamic quantities and hence of all line parameters such as line absorption and emission, damping, Doppler widths, continuum absorption, but also **B** and v can vary. Anomalous dispersion is taken into account. Usually Russell-Saunders coupling and electric dipole transitions are assumed, but other splitting patterns can also be considered. If hyperfine structure splitting is important, the splittings and strengths of subcomponents can be calculated by the HYPER code of Landi Degl'Innocenti (1978). Blends of several magnetically split lines can be calculated as well.

The described code is used to calculate synthetic Stokes line profiles and theoretical calibration curves (integrations of the profiles over the photometer slit widths) for given values of (B, γ, φ) and selected model atmospheres. Comparisons between such data sets and the measured Stokes parameters yield the desired information on **B**. Our calculations show that the results are sensitive with respect to the proper choice of the model atmosphere for the observed point on the Sun, moreover, simple approaches such as the 'weak field approximation' can result in large errors.

Simultaneous measurements in 2 lines can provide information on unresolved fine structures in the observed pixel. Procedures for applying the line-ratio method to vector magnetograph measurements with our equipment have been developed by Bachmann (1990) applying the code described above.

The mean depths of formation of the different Stokes profiles and the range of depths which essentially contribute to the line formation are important for the interpretation of polarimetric measurements. In order to calculate these quantities a method has been proposed which is based on the probability interpretation of a scalar product of Stokes vectors (Staude, 1972).

2.2.4 Examples of Magnetograms and Other Derived Quantities

A first example of a set of vector magnetograms is given in Figure 2. The position of the observed region with sunspots was in the northern hemisphere of the Sun, at a distance of 60° from the central meridian. Therefore projection effects result in a strong asymmetry of the transverse field, we are looking into the spots from a SW direction. The simple pattern of $B_{\perp}$ azimuths of the isolated spot in the western part is only obtained after carefully performing the calibration and correction procedures described in Subsections 2 and 3. In the example of Figure 3 the vector field components and the geometrical mapping of the off-center vector magnetogram were transformed from the image plane into the heliographic plane. The resulting transverse field shows a strong twist, contrary to the example of the spots in Figure 2. The transverse field magnetograms can be used to calculate the vertical current density, j_z. The result is also shown in Figure 3. It is interesting to note that one ribbon of a flare in this active region was concentrated in the umbra, just at the position of the maximum of downflowing j_z in our magnetogram, while the other ribbon was located in a weak field region where it was stretched and nearly parallel to the neutral line (Hofmann et al., 1987; Hofman and Kalman, 1990).

Our magnetograms have also been used to calculate the force-free magnetic field structure above the photospheric region where the measurement has been performed. A procedure for the theoretical extrapolation from a $B_{\parallel}$ magnetogram assuming a linear force-free field ($\nabla \times \mathbf{B} = \alpha \mathbf{B}$ with α = constant) has been developed by Seehafer (1978). The free parameter α is determined by fitting the calculated lines of force of **B** to observed fine structures in Hα filtergrams; $B_{\perp}$ magnetograms can also be used to compare the observed field with the theoretical prediction from the extrapolation procedure. In a first test of the method we used SacPeak magnetograms of the complex active region of August 1972 (Seehafer and Staude, 1979). Interesting results have been obtained for a small active region with a large p-spot of 4 July, 1973 (Seehafer and Staude, 1980): A moving magnetic feature pressed together opposite magnetic fluxes to form an X-type neutral sheet which could be identified in the extrapolated field; the configuration was the cause of chromospheric activity such as subflares and ejections. The same magnetogram has also been

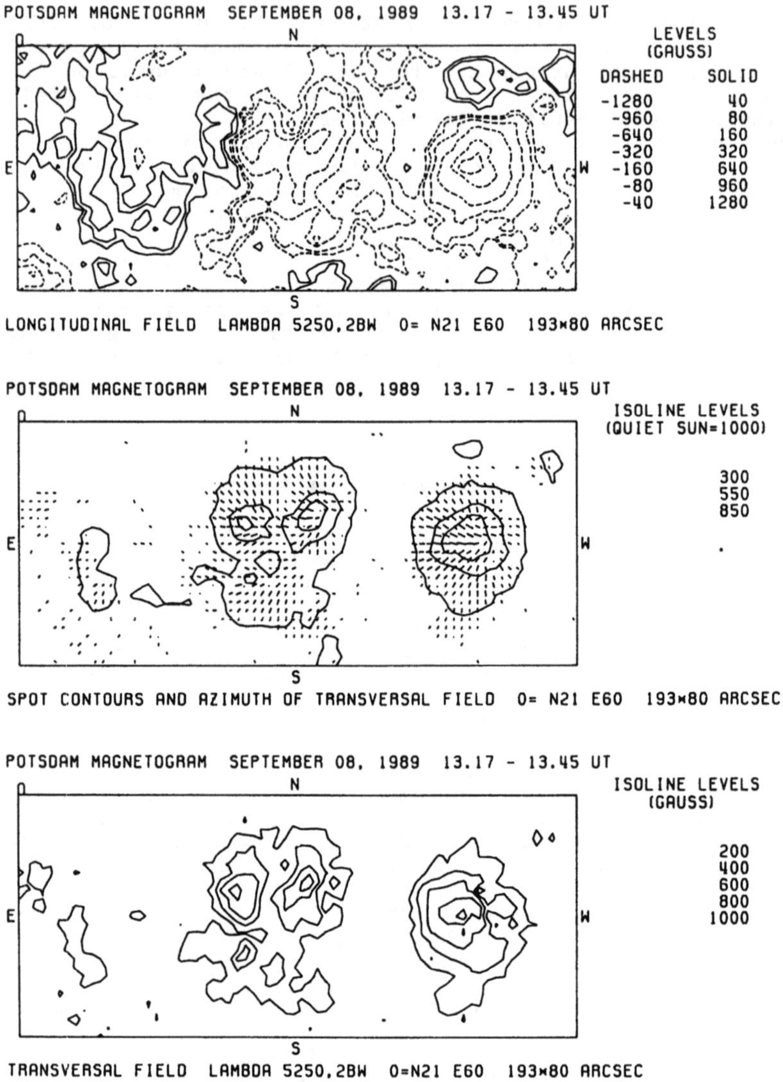

Fig. 2. Vector magnetogram derived from measurements in the blue wing of the line FeI 5250.2 Å.

used to compare the results from our extrapolation procedure with those of other authors such as Nakagawa and Raadu (1972) and Schmidt (1964), the latter only for the special case of a potential field, that is $\alpha = 0$ (Seehafer, 1982; Seehafer and Staude, 1983). The extrapolated magnetic field structures have also been used in model calculations of microwave radio emission (e.g., Hildebrandt et al., 1984) which could be compared with the measurements from radio telescopes such as VLA, WSRT, or RATAN with large spatial resolution. Figures 4 show the $B_{\parallel}$ magnetogram of 4 July, 1973 and the overview of the calculated magnetic lines of force, that is their projection onto the image plane. The maps of radio emission for 4.9 GHz (6

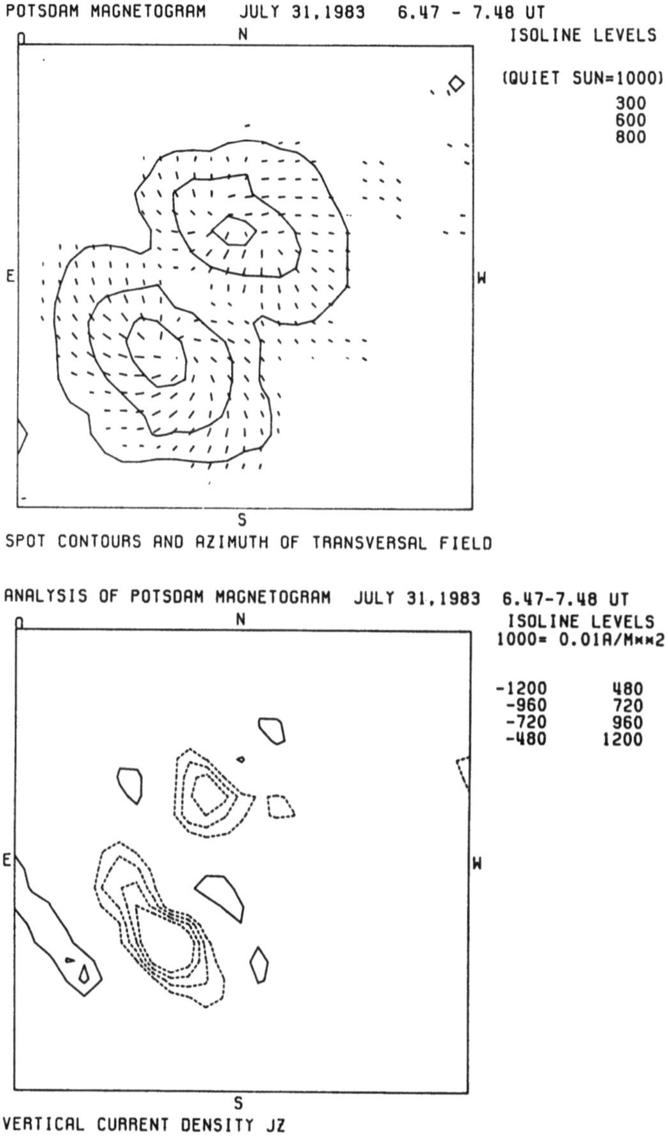

Fig. 3. Magnetic field azimuth and vertical current density derived from a vector magnetogram of the main (double) spot of AR NOAA 4263.

cm) in Figures 5 (brightness temperature T_b and degree of circular polarization p) are calculated considering the extrapolated force-free magnetic field.

The Potsdam magnetographs have been included into several international programs for comparing the results from quasi-simultaneous measurements of the same active region. A comparison of $B_\parallel$ magnetograms has been organized by Pflug (1980) who also proposed methods for finding a uniform zero point and scaling of the different $B_\parallel$ maps. Recently a

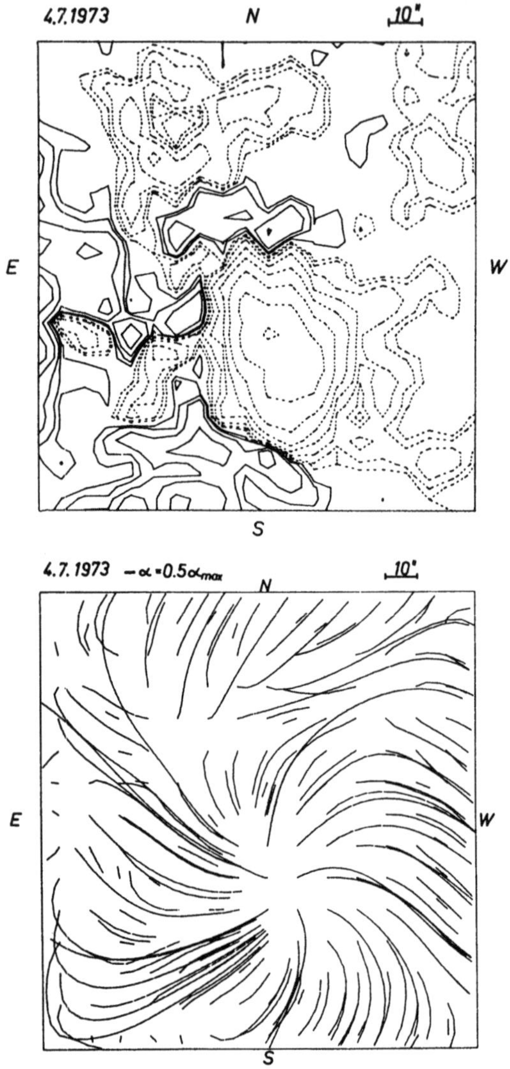

Fig. 4. Longitudinal magnetogram ($B_{\parallel}$; above) and overview of calculated lines of force (below). The $B_{\parallel}$ contour levels are 20, 40, 80, 160, 320, 640, 1280, and 2560 G. The measurements have been obtained at Potsdam on 4 July 1973, 08:30 UT.

cooperative program included vector magnetograms from Potsdam and Sayan and $B_{\parallel}$ data from Ondrejov; the comparison included methods for eliminating signal distortions and for calculating current densities (Hofmann et al., 1988a,b) as well as vertical field gradients (Hofmann and Rendtel, 1989). After applying all possible calibration and correction procedures the data show generally a good agreement, but some differences in details due to different seeing and instrumental conditions and to the use of different lines cannot completely be removed.

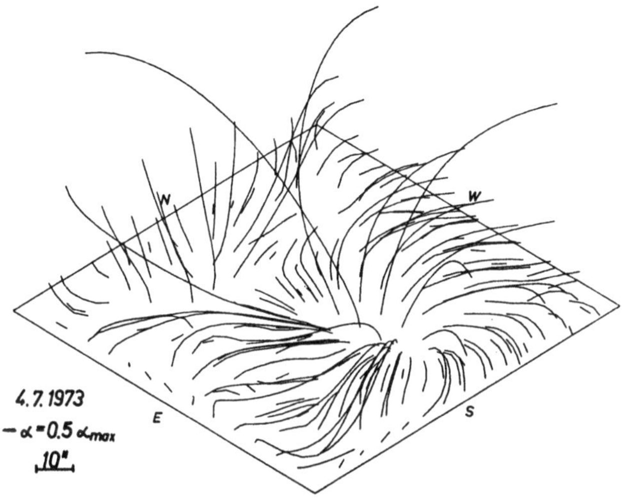

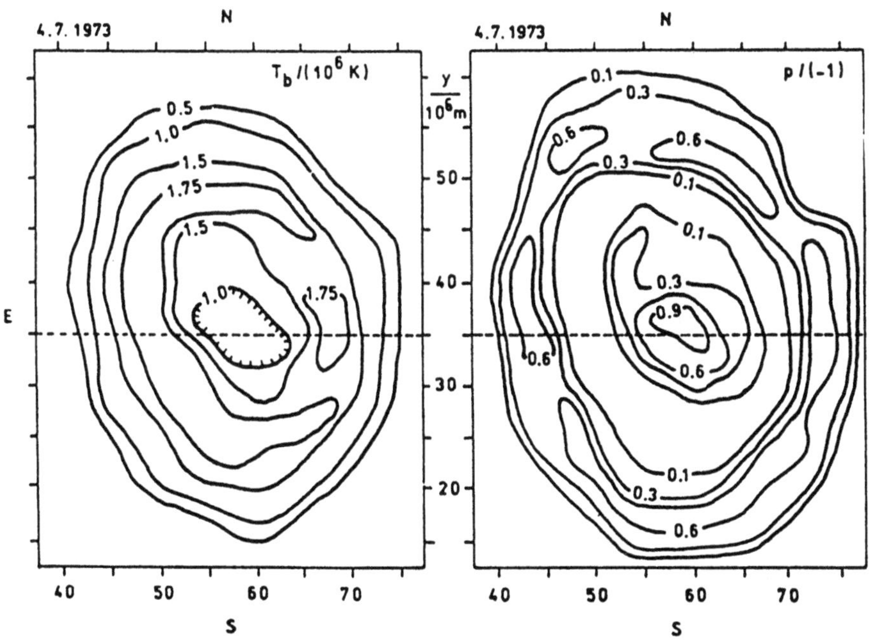

Fig. 5. Above: Perspective view (isometric projection) of the same lines of force as in Figure 4 below. Below: Radio emission for 4.9 GHz (6 cm) calculated for the magnetic field structure above the large spot in Figure 4 (brightness temperature T_b and degree of circular polarization p).

3 Atmospheric Models of Sunspot Umbrae

The theoretical calibration of the data of a Stokes polarimeter in terms of the magnetic field parameters (B, γ, φ) requires the use of atmospheric models, at least for some basic solar structures such as sunspot umbrae, facular flux tubes, and the quiet Sun. Moreover, for the investigation of various physical processes it is necessary to know the spatial distribution of thermodynamic quantities such as temperature T, gas and electron pressures P_G and P_e, or mass density ϱ in the considered region. For instance, in order to investigate the vertical propagation of magneto-atmospheric waves in sunspot umbrae (see Section 4 below), the relevant quantities must be known for a rather large extent of heights. In the present section some approaches to the construction of such sunspot umbral models will be outlined.

Sunspot modelling at Potsdam has a tradition of many decades. Ten Bruggencate and von Klüber (1939, 1944) derived average values of T, P_e, and P_G in the line-forming region of sunspots in comparison with the corresponding values in the surrounding quiet Sun. Contrary to earlier papers by authors from the USA who used simple estimates of line strengths, exact measurements of equivalent widths have been performed by the Potsdam group. Moreover, the concentration on selected large spots strongly reduced the disturbing influence of parasitic stray light in the umbral spectra. Consequently the resulting values of T = 3800-3900 K and P_e = 0.055 Pa are much closer to presently known values of these quantities than those of other authors of that time. Mattig (1958) derived a first umbral model in hydrostatic equilibrium form $T(\tau)$ data by Michard (1953; τ is the optical depth), while Michard obtained a completely different, strongly subhydrostatic model from the same data by using additionally the older spectra from ten Bruggencate and von Klüber to determine P_e, P_G, and ϱ. For many years both models have been used as extreme cases. Nowadays we know that the reality is closer to the hydrostatic model of Mattig.

The present author made an attempt to derive a unified working model for the umbra of a large stable sunspot and the mean undisturbed environment for reference (Staude, 1981). These models extended on a common geometrical height scale from subphotospheric layers up to the base of the chromosphere-corona transition region (CCTR); they were based on uniform and self-consistent physical assumptions for all layers. Earlier published models usually covered only a narrow range of heights and were based on different physical assumptions; this rendered it difficult to combine several models relating to separate regions.Our modelling of subphotospheric layers was based on a generalized version of Öpik's mixing length theory of cellular convection (Staude, 1976) which yielded results in agreement with standard models of the convective zone for the quiet Sun, but also allowed modifications to describe in a semi-empirical way the influence of the strong umbral magnetic field on convection (Staude, 1978). For the photospheric umbral layers the model by Stellmacher and Wiehr (1975) has been adapted, in the lower chromosphere the model is similar to that by Teplitskaya et al. (1978). For the chromosphere and CCTR a new umbral model has been proposed which was able to explain the basis features of many observations at different wavelength, in particular the space-borne data of high spatial resolution in EUV lines which became available just at that time. The resulting T(z) relation of the models is shown in Figure 6 as

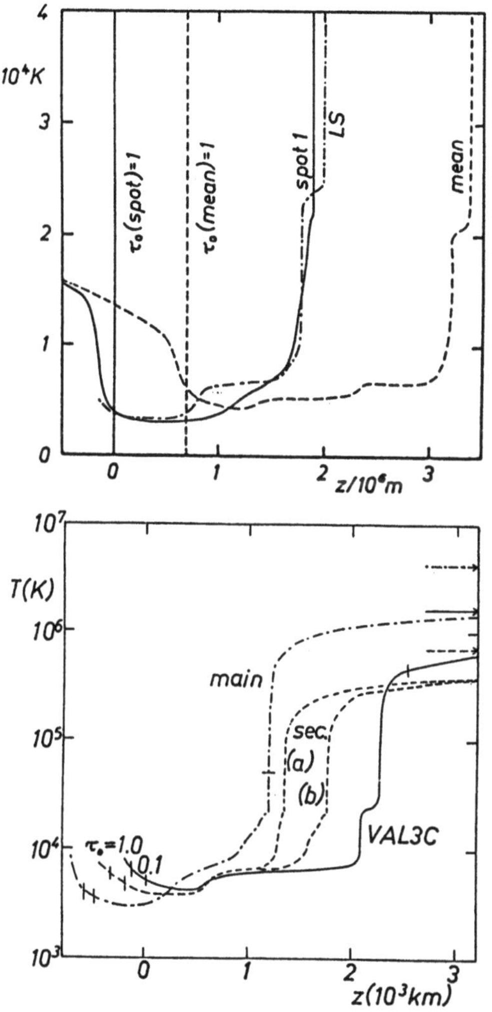

Figs. 6-7. Temperature T versus geometrical height z for the umbral model of Staude (1981; 'spot 1' in Fig. 6 above; models 'LS' and 'mean' undisturbed atmosphere for comparison) and for the components of the model of Obridko and Staude (1988; Fig. 7 below). The arrows indicate coronal values which are asymtotically approached for $z \rightarrow \infty$. VAL3C is the mean quiet Sun model by Vernazza et al. (1981).

an example, z is the geometrical height. The figure also shows the model by Lites and Skumanich (1982) for comparison; this model (LS) was based on new observed data from OSO-8 which were not available to Staude (1981). Both models are similar to each other, but LS introduced two plateaus of T at 7000 K and 23000 K while the model by Staude (1981) shows a steady gradient of T(z) in the chromosphere. Both types of models are able to explain some features of spectroscopic diagnostics, but it is difficult to favour one of the models; seismological diagnostics (see the following section) seem to support the gradient-type model (Staude et al., 1985).

In a subsequent cooperative work the umbral model by Staude (1981) has been extended to the CCTR and lower corona by a two-component atmosphere; the resulting 'Wroclaw-Ondrejov sunspot model' is able to explain high-resolution observations in different spectral ranges, including X-ray, EUV, and radio data (Staude et al., 1983). Finally a two-component structure has also been taken into account in the photosphere and low chromosphere to consider the bright umbral dots (Obridko and Staude, 1988). The main component consists (in agreement with the model by Staude, 1981) of cold matter at photospheric levels, a thin chromosphere, a shallow transition region, and a deep-set corona. The secondary component is assumed to have a volume filling factor of 5 to 10% at all heights and to consist of thin fine structure elements which are elongated parallel to the magnetic field; this matter should be hot at photospheric levels but cooler than the quiet Sun, it should have a more extended chromosphere and relatively cold elements at higher levels, that is an amply extended, almost isothermal transition region at heights where the main component already reaches coronal temperatures. Figure 7 shows T(z) for these models, including two different variants for the chromospheric part of the secondary component. In spite of the good agreement of the model predictions with basic features of sunspot observations it would be highly desirable to test the model by further measurements with high spatial resolution, including simultaneous observations at different wavelengths (X-rays up to radio).

4 The Interpretation of Sunspot Umbral Oscillations

Sunspot oscillations have attracted the attention of astronomers for many years: Magneto-atmospheric waves (that is, waves under the influence of the 3 restoring forces gravity (buoyancy), pressure (compressibility), and magnetic forces) are likely candidates for energy transport and heating in stellar atmospheres. In sunspot observations it is possible to resolve many details of such periodic disturbances and of the basic atmospheric structure as well, thus models of both waves and atmospheres can be tested.

Oscillations of velocity and intensity (that is, of T, ϱ, etc.) have been measured in spectral lines formed at different height levels in sunspot umbrae. The data show closely packed, sharp resonance peaks of the power spectra in two distinct period bands around 3 min and 5 min. A third band seems to exist at periods $\gtrsim$ 20 min. At photospheric heights of the umbra the 5 min oscillations appear to exist similar to the p-modes in the quiet Sun, but with reduced energy. Their amplitudes decrease with increasing height, while the amplitudes of the 3 min oscillations increase upward. In the CCTR the 3 min oscillations are only visible above umbrae; in the chromosphere 3 min oscillations have also been measured in thin magnetic flux tubes. The 3 min and 5 min bands show longitudinal oscillations, that means, the displacements are aligned to the vertical magnetic field. Most observers found horizontal coherence of the oscillations across the whole umbra, others a subdivision of an umbra into several (2-4) oscillating elements. The periods of the 20 min oscillations are close to the lifetimes of bright umbral dots, indicating a possible physical connection. References to original papers and more details on the observed data can be found in reviews by Žugžda et al. (1987), Staude et al. (1987), Kneer (1990), and Staude (1990). These reviews also describe our modelling efforts and proposals for an interpretation of the umbral oscillations which will now be outlined.

For a realistic treatment of the umbral atmospheric structure we used semi-empirical models as explained in Section 3 above. The procedures for constructing such models have been described in more detail by Staude (1981, 1982, 1985) and summarized in earlier work on umbral oscillations (Zugzda et al., 1983, 1984, 1987). The model atmospheres are based on self-consistent solutions of the equations of state, of non-LTE radiative transfer for hydrogen, and of hydrostatic equilibrium. Each model describes the basic state of a considered atmosphere (corresponding quantities are indicated by a subscript o) which has subsequently been subjected to periodic disturbances by magneto-atmospheric waves. The behaviour of longitudinal waves in such a model has been investigated by studying the resonant transmission or filtering of a broadband flux of wave energy F_o incident from deeper convective layers. That means, we calculated the coefficient D of the transmission of F_o from the convective zone up to the corona above the umbra, taking into consideration the partial reflections at all intermediate heights depending on the thermodynamic structure. The resonance frequencies $\omega_i = 2\pi/P_i$ (P_i are the resonance periods) correspond to the maxima of $D(\omega)$. Moreover, the method provides information on the real positions of the boundaries of each resonator and on the height dependence of amplitudes and phases of oscillations of velocity u, of vertical plasma displacement h, and of T′ and P′, the fluctuating parts of $T = T_o + T'$ and $P_G = P_{Go} + P'$, respectively. Such calculations have been carried out for various models of the umbral atmosphere (different values of the temperature minimum, T_{min}, and of the chromospheric extent); in this way we looked for possibilites of using observations of umbral oscillations for sounding the atmospheric structure by seismology. Interactions between longitudinal and transverse modes of magneto-atmospheric waves have been investigated by applying the formulae for mutual linear transformation.

The results can be summarized as follows: The umbra seems to consist of a system of cavities or resonators which are more or less coupled and where various modes of magneto-atmospheric waves are partly trapped and partly transmitted. Immediately below the photosphere (for an optical depth of $\tau_o > 1$ in the continuum at 5000 Å) we have a plasma parameter (ratio of sound speed and Alfvén speed) $\beta \approx 1$ and a region of linear transformation of different waves modes. At larger depths and heights we meet with values of $\beta \gg 1$ and $\beta \ll 1$, respectively, the different modes are decoupled and can be considered separately. In the photosphere and chromosphere we have a resonator for longitudinal, slow waves which are similar to pure sound waves but propagate only parallel to the strong magnetic field. The predicted properties agree with those which are observed in the 3 min band. On closer consideration the umbral atmosphere comprises a system of two resonators, a photospheric-chromospheric resonator with strong reflections from the CCTR and the base of the photosphere and a chromospheric resonator with additional reflections at T_{min}, where the latter resonator forms a subsystem of the other. Quasi-transverse, fast waves exist only as evanescent waves at atmospheric heights of the umbra. At subphotospheric depths two independent resonators are acting. A resonator for slow, quasi-transverse waves can explain the oscillations with periods $\gtrsim$ 20

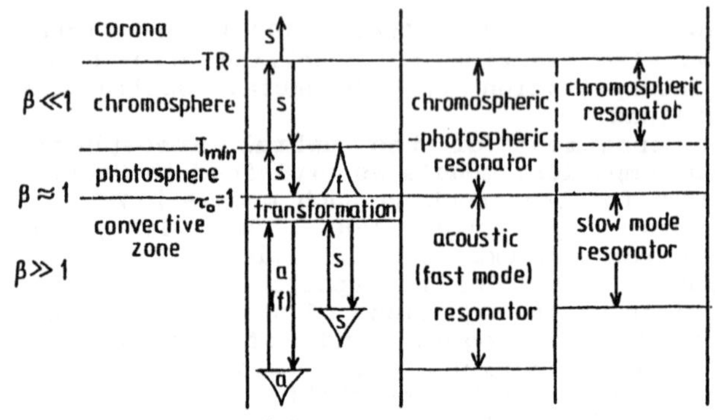

Fig. 8. Scheme of the resonators in an umbra. The letters s, a, and f indicate slow mode, acoustic and fast mode waves, respectively. (L) means quasi-longitudinal waves, (T) quasi--transverse waves. TR is the CCTR. Vertical arrows show the direction of wave progation, triangles mean evanescent waves.

min, while a resonator for fast (acoustic), quasi-longitudinal waves could result in the umbral 5 min oscillations. The subphotospheric acoustic resonator strongly couples with the slow-mode resonator at photospheric and chromospheric heights. The slow, quasi-transverse resonator, on the contrary, is largely isolated from the other resonators; its oscillations appear as oscillator convection (overstability) in the umbral magnetic field. The total conception is based on simplifying assumptions of the applied theory (neglect of nonlinear and nonadiabatic effects) and leaves some problems undecided. Nevertheless, theoretical predictions bear a great resemblance to many observed features, and the good agreement points out that our models seem to reflect essential features of the physical nature of the structure and dynamics of sunspots.

The system of resonators is schematically outlined in Figures 8 and 9. Figure 9 considers two umbral models with different extents (that is, different temperature gradients) of the chromosphere. The curve of transmission $D(\omega)$ for the umbral model of Figure 2 is shown in Figure 10, and the height dependence of kinetic energy for different harmonics (resonance periods P_i) of longitudinal waves in the same model is drawn in Figure 11.

In the modelling details of the predicted oscillations depend critically on the assumed umbral structure. In this way we obtain a useful method of probing the basic umbral model by seismology, thus completing customary spectroscopic diagnostics. For instance,

- the extent between the temperature minimum (T_{min}) and the chromosphere-corona transition region (TR) determines the spectrum of resonance frequencies, thus providing a method to determine the temperature gradient in the umbral chromosphere (Žugžda et al., 1984);
- the values of T_{min} in umbral fine structures can be estimated by investigating the spectrum and the quality of their resonators (Locans et al., 1988);
- observations of UV lines provide information on the umbral TR, where the observed oscillations are probably concentrated in cold fine structure elements which occupy only a few percent of the volume (Staude et al., 1985).

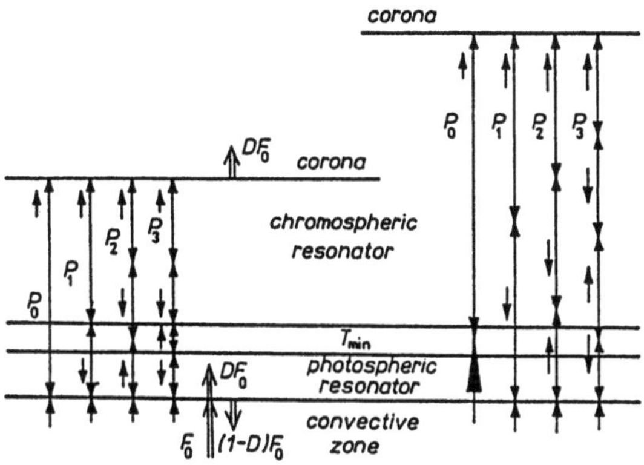

Fig. 9. Scheme of the photospheric-chromospheric resonator in two different models of the umbral atmosphere: (a) "thin" (shallow) chromosphere, (b) "thick" (extended) chromosphere. Short arrows indicate the direction of velocity in a given phase for each mode P_i, encounters of triangles represent nodes, double arrows describe the energy flux.

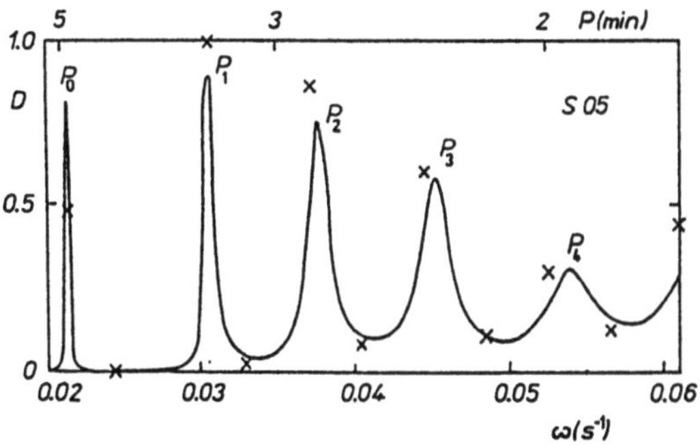

Fig. 10. Transmission D of the basic umbral model of Fig. 7 for longitudinal waves versus frequency or period P. The full curves take into account the convective zone, the crosses indicate maxima and minima from a computation without consideration of the convective zone.

Apart from the Sun the rapidly oscillating Ap (roAp) stars represent the only main-sequence stars to which the methods of asteroseismology can so far be applied. The oscillations of the roAp stars are concentrated in the magnetic poles or large spots of these stars, therefore it is tempting to look for analogies with sunspot oscillations. First steps in this direction have already been done (Staude, 1990a, b), but better observed data and detailed model calculations are required before final conclusions are possible.

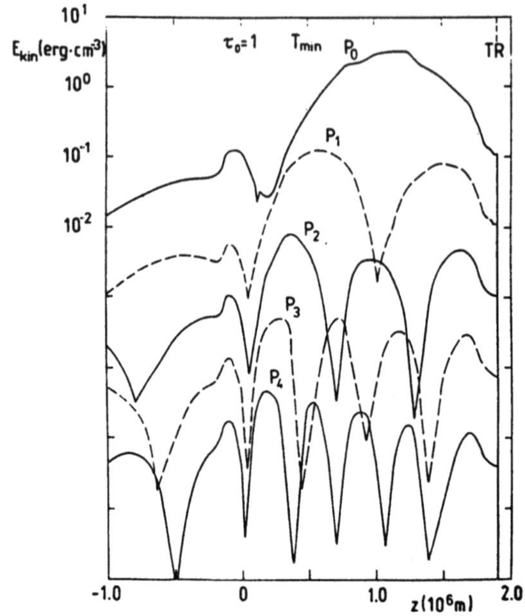

Fig. 11. Kinetic energy density E_{kin} of longitudinal waves versus height z for the umbral model of Fig. 7. The ordinate scale relates to the basic mode P_o, while the curves for each following harmonic P_i have been shifted downwards by one order of magnitude of E_{kin}.

5 Conclusions

In the present paper we intended to describe the instrumentation of our double vector magnetograph as well as our procedures of instrumental and theoretical calibrations, including the methods of diminishing the systematic errors from signal distortions and model assumptions. The examples show that the resulting magnetograms provide reliable and useful data for further investigations of physical processes and structures in solar active region. Of course,analogous to other instruments of this kind our magnetograph is a compromise between different demands, determined by our scientific aims and limited by our instrumental and atmospheric conditions. For instance, we intended to scan complete active regions with moderate spatial resolution (3" is the limit determined by our seeing conditions), but to have also sufficient spectral information to reduce uncertainties from model assumptions and unresolved fine structures. Suggestions for an improvement are straightforward: Better detectors such as CCD cameras could provide more complete information in much shorter time, e.g., more (at least 3 or 4) lines should be used to measure simultaneously the magnetic field vector at different height levels, and more measuring points in each Stokes profile would improve the theoretical calibration.

Comparisons between recent observations of umbrae and our detailed numerical model calculations achieved a substantial agreement. This fact points out that the hitherto developed models for the structure and dynamics of sunspot umbrae seem to reflect essential features of the nature of sunspots, notwithstanding the various simplifying assumptions. Recent analytic calculations for simplified umbral models also confirmed our model of a system of umbral resonators (Campos, 1989; Evans and Roberts, 1990); Campos (1989) only argued in favour of an additional resonator for pure Alfvén waves. The model provides an important tool for sounding subphotospheric and atmospheric layers of sunspots and possibly of other magnetized stellar atmospheres. Further observations of oscillations should try to improve the spatial resolution on the Sun and the covering of the data in space and time. Model calculations should abandon the restricting assumptions by considering nonlinear and nonadiabatic processes such as radiative dissipation of waves, moreover, self-consistent calculations of wave propagation and non-LTE line formation are needed to improve the diagnostics.

A close cooperation with colleagues from eastern countries (Soviet Union, Poland, Czechoslovakia, Hungary) was important for achieving many of the results reported in this paper. The new possibilities of cooperation with the west will provide favourable prospects for common work in the future. For example, observations with high spatial resolution from ground-based observatories (Cooperative Observatory at Tenerife; LEST) as well as from space-borne stations (SOHO; OSL) will provide useful possibilities for testing our model predictions. That means, our modelling can contribute to an interpretation of the data, and our experiences of Stokes polarimetry could be applied to the new observing facilities as well.

References

Auer, L.H., Heasley, J.M., and Milkey, R.W.: 1972, Kitt Peak Nat. Observ. Contr. No. **555**.

Bachmann, G.: 1984, Bull. Astron. Inst. Czechosl. **35**, 308.

Bachmann, G.: 1990, Astron. Nachr. **311**, 63.

Bachmann, G., Hofmann, A., and Staude, J.: 1983, Publ. Debrecen Heliophys. Observ. **5**, 369.

Bachmann, G., Jäger, F.W., Künzel, H., Pflug, K., and Staude, J.: 1975, HHI-STP Report (Berlin) No. **4**, 38pp.

Campos, L.M.B.C.: 1989, Mon. Not. R. astr. Soc. **241**, 215.

Carlsson, M.: 1986, Uppsala Astron. Observ. Report No. **33**.

Domke, H., and Staude, J.: 1973, Solar Phys. **31**, 279.

Evans, D.J., and Roberts, B.: 1990, Astrophys. J. **348**, 346.

Grotrian, W., Künzel, H., Mattig, W., and Schröter, E.H.: 1953, Publ. Astrophys. Observ. Potsdam No. **97**; 1956, Publ. Astrophys. Observ. Potsdam No. **99**.

Hale, G.E.: 1908, Ap. J. **28**, 315.

Hildebrandt, J., Seehafer, N., and Krüger, A.: 1984, Astron. Astrophys. **134**, 185.

Hofmann, A., Rendtel, J., Aurass, H., and Kalman,B.:1987, Solar Phys. **108**, 151.

Hofmann, A., Grigorjev, V.M., Selivanov, V.L., and Klvana, M.: 1988a, Astron. Nachr. **309**, 331. 1988b, Astron. Nachr. **309**, 373.

Hofman, A., and Kalman, B.: 1990, Astron. Astrophys.(in press).

Hofmann, A., and Rendtel, J.: 1989, Astron. Nachr. **310**, 61.

Jäger, F.W.: 1972, Solar Phys. **27**, 481.

Jäger, F.W.: 1974, Solar Phys. **39**, 499.

Kneer, F.: 1990, Proc. 6th Europ. Meeting on Solar Physics, Publ. Debrecen Heliophys. Observ., Hungary (in press).

Künzel, H.: 1964, Astron. Nachr. **288**, 177.
Künzel, H.: 1965, Astron. Nachr. **289**, 233.
Landi Degl'Innocenti, E.: 1978, Astron. Astrophys. Suppl. **33**, 157.
Lites, B.W., and Skumanich, A.: 1982, Astrophys. J. Suppl. **49**, 293.
Locans, V., Škerse, D., Staude, J., and Žugžda, Y.D.: 1988, Astron. Astrophys. **204**, 263.
Lohse, O.: 1883, Publ. Astrophys. Observ. Potsdam Nr. **13** (Bd. 3), 293.
Mattig, W.: 1953, Z. Astrophys. **31**, 273.
Mattig, W.: 1958, Z. Astrophys. **44**, 280.
Michard, R.: 1953, Ann. d'Astrophys. **16**, 217.
Nakagawa, Y., and Raadu, M.A.: 1972, Solar Phys. **25**, 127.
Obridko, V.N., and Staude, J.: 1988, Astron. Astrophys. **189**, 232.
Pflug, K.: 1980, Phys. Solariterr. (Potsdam) **13**, 5.
Rees, D.E.: 1969, Solar Phys. **10**, 268.
Schmidt, H.U.: 1964, NASA-SP **50**, 107.
Schröter, E.H.: 1953, Z. Astrophys. **33**, 20.
Seehafer, N.: 1978, Solar Phys. **58**, 215.
Seehafer, N.: 1982, Solar Phys. **81**, 69.
Seehafer, N., and Staude, J.: 1979, Astron. Nachr. **300**, 151.
Seehafer, N., and Staude, J.: 1980, Solar Phys. **67**, 121.
Seehafer, N., and Staude, J.: 1983, Phys. Solariterr. (Potsdam) **22**, 5.
Severny, A.B.: 1967, Izv. Krymsk. Astrofiz. Observ. **36**, 22.
Severny, A.B.: 1977, Izv. Krymsk. Astrofiz. Observ. **56**, 142.
Staude, J.: 1972, Solar Phys. **24**, 255.
Staude, J.: 1976, Bull. Astron. Inst. Czechosl. **27**, 365.
Staude, J.: 1978, Bull. Astron. Inst. Czechosl. **29**, 71.
Staude, J.: 1980, Phys. Solariterr. (Potsdam) **14**, 58.
Staude, J.: 1981, Astron. Astrophys. **100**, 284.
Staude, J.: 1982, HHI-STP Report (Berlin) No. **14**, 24.
Staude, J.: 1990a, in: G. Scholz (ed.),'Hot Chemically Peculiar and Magnetic Stars', Proc. Meeting Potsdam, Nov. 1989. Mitteil. Karl-Schwarzschild-Observ. **125**, 157.
Staude, J.: 1990b, in: P.B. Byrne and D.J. Mullan (eds.), 'Surface Inhomogeneities on Late-type Stars', Proc. Colloquium Armagh, July 1990. Lecture Notes in Physics, Springer (in press).
Staude, J., and Hofmann, A.: 1988, ESA SP-**285** Vol. II, 123.
Staude, J. et al.: 1983, Acta Astron. **33**, 441.
Staude, J., Žugžda, Y.D., and Locans, V.: 1985, Solar Phys. **95**, 37.
Staude, J., Žugžda, Y.D., and Locans, V.: 1987, in: L. Hejna and M. Sobotka (eds.), 'The Sun'. Proc. 10th ERAM, Vol. 1; Publ. Astron. Inst. Czechosl. Acad. of Sci. **66**, 161.
Stellmacher, G., and Wiehr, E.: 1975, Astron. Astrophys.**45**, 69.
Stepanov, V.E., Grigorjev, V.M., Kobanov, N.I., and Osak, B.F.: 1975, Issled. po Geomagn., Aeron. i Fiz. Solntsa (Irkutsk) **37**, 147.
ten Bruggencate, P., and von Klüber, H.: 1939, Z. Astrophys. **18**, 284.
ten Bruggencate, P., and von Klüber, H.: 1944, Veröff. Univ.-Sternwarte Göttingen **78**.
Teplitskaya, R.B., Grigoryeva, S.A., and Skochilov, V.G.: 1977, Issl. po Geomagn., Aeron. i Fizike Solntsa (Irkutsk) **42**, 48.
Vernazza, J.E., Avrett, E.H., and Loeser, R.: 1981, Astrophys. J. Suppl. **45**, 635.
von Klüber, H.: 1947, Z. Astrophys. **24**, 121.

Žugžda, Y.D., Locans, V., and Staude, J.: 1983, Solar Phys. **82**, 369.
Žugžda, Y.D., Locans, V., and Staude, J.: 1985, Astron. Astrophys. **143**, 201.
Žugžda, Y.D., Locans, V., and Staude, J.: 1987, Astron. Nachr. **308**, 257.
Žugžda, Y.D., Staude, J., and Locans, V.: 1984, Solar Phys. **91**, 219.

Time-Resolved Stokes V Polarimetry of Small-Scale Magnetic Structures on the Sun

B. Fleck

Institut für Astronomie und Astrophysik,
Am Hubland, W-8700 Würzburg, Fed. Rep. of Germany

Abstract

We address problems of the dynamics of small-scale magnetic fluxtube concentrations by analyzing high spatial resolution time series of spectra of selected lines taken simultaneously in left-hand and right-hand circularly polarized light. This approach is complementary to the one of the Zürich group. Our observations were obtained with the Vacuum Tower Telescope and the Echelle spectrograph at Izaña, Tenerife. The spectral range covered the lines of Fe 6302.5Å and Fe 6301.5Å. All parameters extracted from the Stokes V and Stokes I profiles (except the Stokes V line ratio) reveal significant spatio-temporal variations. The dominant oscillatory signal in the velocity data is due to the 5-min oscillations. Of special interest are the strong spatially and temporally varying differences between the (I+V) and (I–V) spectra in the magnetic structures. They cause large differences between the velocities deduced from the Stokes I profile and the Stokes V zero-crossing wavelength (up to 1 km/s), whereas their spatio-temporal averages are almost identical. We observe also an anticorrelation between the amplitude and asymmetry of the Stokes V profile.

1. Introduction

Small-scale magnetic elements play a fundamental rôle for the structure of the solar atmosphere. In recent years, they have therefore attracted much interest among solar physicists, both theoreticians and observers (for reviews see e.g. Schüssler, 1987, 1990; Solanki, 1987, 1990; Spruit, 1981a; Spruit and Roberts, 1983; Stenflo, 1989). One of the key problems in the understanding of the structure of the solar atmosphere is still the unsettled question of the chromospheric/coronal heating. There is broad agreement that genuine sound waves which dissipate their energy after steepening to shock waves (for a review see Ulmschneider, 1991) provide only a basal flux (Schrijver, 1987, 1991), and that the more important agent for the chromospheric and coronal heating are the small-scale magnetic field concentrations. The two scenarios proposed involve either various kinds of MHD-waves (see e.g. Spruit, 1981b; Ulmschneider and Muchmore, 1986; or the review of Hammer, 1987, and references therein) or the direct dissipation of magnetic energy by continuous magnetic field rearrangement and neutral point reconnection (Parker, 1983, 1991). Which one of these models describes the real nature appropriately remains to be shown. Here, we wish to underline the importance of investigating the *dynamics* of fine-scale magnetic field concentrations in addition to their static properties.

On the theoretical side numerous investigations have been published addressing the propagation characteristics of all kinds of MHD-modes (longitudinal sausage modes, transversal kink modes, torsional Alfvén modes, fast and slow body and surface waves) in various geometries of the field configuration and in different approximations as for instance the thin tube approximation (Abdelatif, 1988; Cram and Wilson, 1975; Defouw, 1976; Edwin and Roberts, 1982, 1983; Evans and Roberts, 1990; Roberts, 1981a,b, 1983; Roberts and Webb, 1978, 1979; Ryutova, 1990; Spruit, 1981b, 1982; Webb and Roberts, 1980a, 1980b, Wentzel, 1979; Wilson, 1979). On the observational side, studies concerning the dynamics and waves in small-scale magnetic field concentrations are very scarce. The observations of Giovanelli et al. (1978) and those of Wiehr (1985) certainly need to be mentioned here. These authors measured the periodic shifts of the Stokes V zero-crossing sampled from an area of several arcsec square. Hence, it has to be assumed that information from many magnetic elements was averaged. Furthermore, the spectral resolution was only moderate and they did not obtain the complete Stokes V profile. From the observational point of view, we think, one can state that waves in magnetic fluxtube concentrations are almost new territory. This discrepancy in number between a few observational attempts and numerous theoretical investigations addressing the dynamical behaviour of waves in magnetic fluxtube concentrations reflects the substantial observational difficulties encountered in these fine-structure studies.

With the new telescopes now being fully operational on the Canary Islands — a site where on can expect extended intervals of excellent seeing conditions — we have just started a new project aiming at time-dependent high resolution polarimetry. Our approach is complemetary to the one of the Zürich group who made excellent use of FTS data obtained with the Mc Math telescope at Kitt Peak (Stenflo et al., 1984). While Stenflo, Solanki and coworkers have analyzed temporally and spatially averaged Stokes V and I spectra of hundreds of Fe I and II lines laying particular emphasis on the internal structure and properties of fluxtubes (e.g. Stenflo and Harvey, 1985; Solanki, 1986; Pantellini et al., 1988; Zayer et al., 1989), we study the *dynamical behaviour* of fluxtube concentrations by investigating *time series* of right-hand (I+V) and left-hand (I–V) circularly polarized spectra of several selected Fraunhofer lines taken strictly simultaneously with the *highest spatial* and *spectral resolution* achievable, i.e. we sacrifice the information of the wide wavelength range of the FTS in order to gain high temporal and spatial resolution. The high spectral resolution is maintained. Preliminary results of this project were presented by Fleck and Deubner (1990, 1991).

2. Observations and Data Reduction

The observations were obtained in October 1989 with the Vacuum Tower Telescope and the Echelle spectrograph at Izaña, Tenerife, by F.-L. Deubner and me. We used the RCA CCD-camera with 320×512 pixels. The polarisation optics consisted of an achromatic $\lambda/4$–plate and two calcites at ±45° placed immediately behind the entrance slit of the spectrograph. We consider the employment of two calcites rotated by 90° with respect to each other highly advantageous because the optical path difference between the ordinary and extraordinary ray of a single calcite element causes different focus positions for the two polarisation states. Subtraction of the two images (I+V) and (I–V) would then lead to a spurious Stokes V signal even in the absence of polarisation. As a by-product of using a split calcite element the linear polarization of the two beams leaving the Stokes analyser is under ±45° to the grooves of the strongly polarizing echelle grating, so that both polarisation states suffer the same attenuation. The lines chosen are the two Fe I lines at 6302.5Å (g=2.5) and 6301.5Å (g_{eff}=5/3). The projected slit width and length were 0.5 and 19.6 arcsec, respectively. The time series we present in this paper was obtained in a very active plage near disk center ($\mu \approx 0.98$). It consists of 520 frames taken in intervals of 18.1 s with an exposure time of 1.75 s. The spatial and spectral resolution are 0.14 arcsec/pixel and 3.2 mÅ/pixel, respectively. Two other series one of which was obtained at $\cos\Theta = \mu \approx 0.8$ will be discussed in a later paper.

The spectra were carefully corrected for dark current and time dependent flat-field effects. Then the oppositely circularly polarized spectra were aligned in the spatial and dispersion direction applying crosscorrelation and Fourier shift techniques. We want to stress this very important step in the reduction procedure, which requires great exactness and conscientiousness. Already minute residual shifts of small fractions of a pixel (even with this highly oversampling spatial and spectral resolution) between the two polarization states cause spurious Stokes V signals. After normalization to equal continuum intensity the two polarization states (I+V) and (I–V) were added and subtracted to obtain Stokes I and Stokes V, respectively. From these data several special diagnostic parameters were extracted: the zero-crossing wavelength V_0 of Stokes V, the wavelength and amplitude of the blue and red V wing maxima (V_b, V_r, a_b, a_r) the area of the blue and red V wings (A_b and A_r). From these quantities we derived the relative and absolute amplitude asymmetry $\delta a_r = \frac{a_b - a_r}{a_b + a_r}, \delta a_a = \frac{a_b}{a_r}$ (and corresponding area asymmetries), the wavelength and intensity of the minima as well as the center of gravity (Rees and Semel, 1979) of Stokes I, (I+V) and (I–V). Finally, random excursions of the image due to seeing and guiding errors in the slit direction were eliminated by crosscorrelation techniques.

3. Results and Discussion

In Fig.1 we have displayed the spatio-temporal variations of the Stokes V zero-crossing wavelength. The spectrograph slit covered two flux tube concentrations, centered near the x-positions 5" and 11", with Stokes V amplitudes sufficient (> 1.5%) for a reliable interpretation. In the grey background no significant Stokes V signal has been detected. Note that, in this paper, with the name of "fluxtube concentrations" we actually refer to clusters of fluxtubes or "magnetic knots" (Beckers and Schröter, 1968), rather than to isolated concentrated structures ("solitary" fluxtubes).

The dominant oscillatory signal of the Stokes zero-crossing V_0 (as well as in all the other velocity data sets) are the 5-min oscillations (see Fig.1). The signature of the 5-min oscillations in the zero-crossing wavelength has been reported earlier by Giovanelli et al. (1978) and Wiehr (1985). Their observations, however, sampled an area of a few arcsec square. Here, we also present the spatial structure of these oscillations (see Fig.1): The two magnetic concentrations do not oscillate in phase, and, more importantly, the left and the right edges of each structure reveal very often significant phase differences. This leads to high spatial velocity gradients of up to 1 km/s across the magnetic fluxtube concentrations, which demonstrate the presence of physically and dynamically independent substructures within them. These spatial gradients certainly cannot be accounted for by seeing effects.

The RMS of the Stokes I velocity and of V_0 amounts to approximately 225 and 175 m/s, respectively, being more than a factor of 2 smaller than the RMS velocity derived from the quiet sun. These small RMS values pose a serious problem with respect to the width of spatially and temporally averaged I_V line profiles (integrated Stokes V profiles; see Solanki, 1986). These are much broader than the corresponding quiet sun line profiles, and

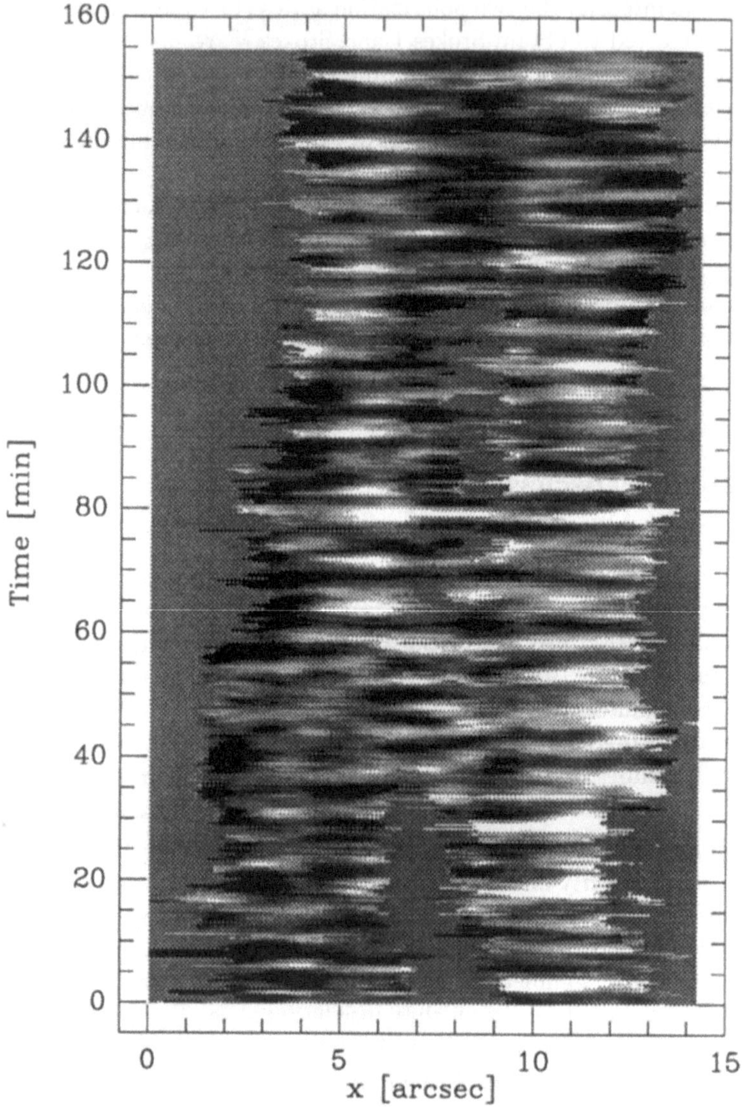

Fig. 1. Spatio-temporal variations of the Stokes V zero-crossing wavelength V_0. The dominant signal is due to the 5-min oscillations. The limits of the grey scale mapping are -600 m/s and +200 m/s, respectively. Black/white shading indicates down(−)/upflowing(+) phase.

can only be explained by small-scale non-stationary mass motions with amplitudes of about 2 to 3 km/s (Solanki, 1986). Since we cannot detect any indications of such large amplitudes (not even in our best individual spectra) we conclude that either these mass motions take place on extremly small spatial scales ($\ll$250 km) or at rather high frequencies where the modulation

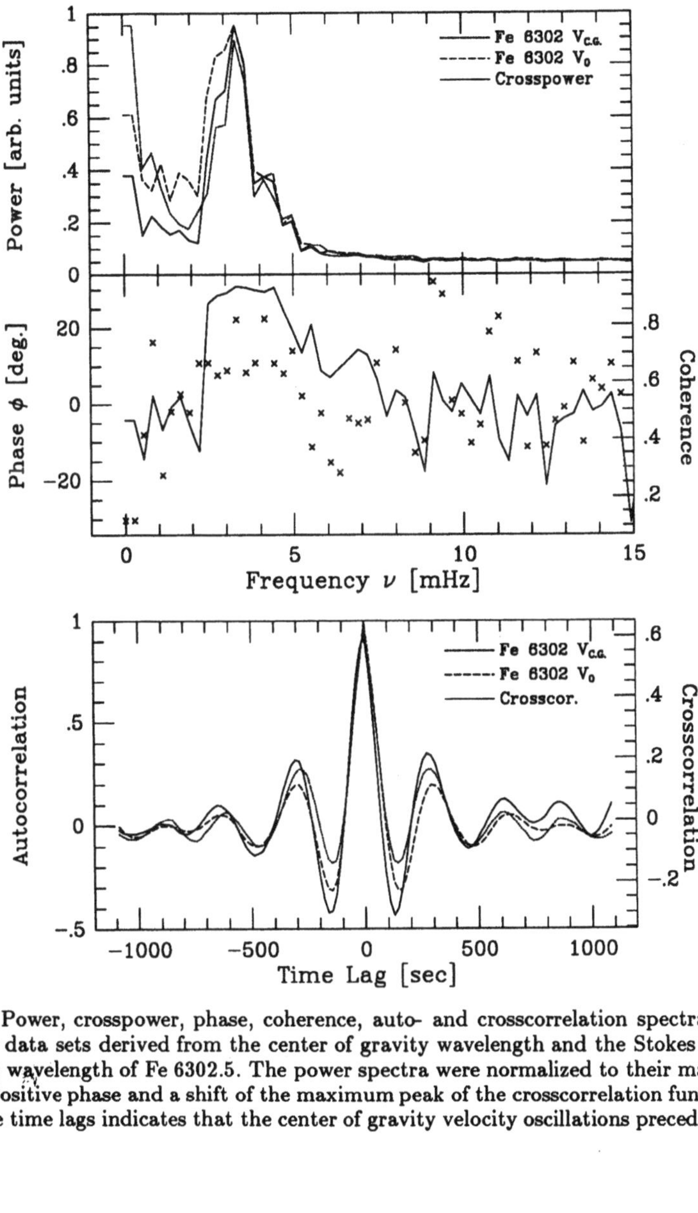

Fig. 2. Power, crosspower, phase, coherence, auto- and crosscorrelation spectra of the velocity data sets derived from the center of gravity wavelength and the Stokes V zero-crossing wavelength of Fe 6302.5. The power spectra were normalized to their maximum value. Positive phase and a shift of the maximum peak of the crosscorrelation function to negative time lags indicates that the center of gravity velocity oscillations precedes those of V_0.

transfer function of the velocity response function has dropped to almost 0. The present power spectra reach the noise level already at about 6 mHz (see Fig.2). Therefore, we are not able not detect a real high frequency component as observed e.g. in the quiet sun as running acoustic waves (see e.g. Fleck and Deubner, 1989).

Fig.2 shows the power, crosspower, phase, coherence, auto- and cross-correlation spectra of the center of gravity velocity and V_0 of Fe 6302.5.

As is obvious from the power spectra, we were able to detect a significant oscillatory signal only in the frequency range from 2.8 to about 5 mHz. In this frequency regime, longitudinal sausage modes are basically still evanescent (see Roberts, 1983). However, due to the elasticity of the tubes and because of radiation damping effects, their propagation characteristics differ considerably from genuine hydrodynamic evanescent "sound" waves. For photospheric conditions, the latter have a very high phase speed (about 60 to more than 100 km/s), i.e. they are "mainly evanescent", whereas tube waves propagate with a phase speed of about 10 km/s only, i.e. they are "mainly progressive" (see Roberts, 1983). This slower propagation speed which was already indicated by the observations of Giovanelli et al.(1978) is apparent in the phase spectrum displayed in Fig.2: The 5-min Stokes V zero-crossing oscillations lag the Stokes I center of gravity oscillations by more than 10° (more than 9 s for 300 s oscillations). This phase lag is truly remarkable, taking into account that due to the Wilson depression the Stokes V signal — which originates exclusively in the magnetic structures — is believed to be formed geometrically deeper in the atmosphere than the velocity derived from Stokes I which is affected by "straylight" from the non-magnetic surroundings. However, not all of our phase spectra are in agreement with a model requiring a slower phase speed of the tube waves: The phase difference between the Stokes I velocity of the two Fe lines the line forming heights of which differ by about 100 km is small ($\approx 3°$) but significantly different from zero. The Stokes V zero-crossing wavelengths of Fe 6302 and Fe 6301, on the other hand, oscillate almost perfectly in phase. Only further observations with higher S/N ratio and better statistical stability can clarify this issue. A very important point would be to extend the useful frequency range into the regime of truely propagating waves beyond 5 mHz. In our observations, at these frequencies, the calculated phase values are meaningless and scatter randomly around 0° (note also the drop of coherence at 4.5 mHz). We hesitate to interpret, but note the slightly longer average period of the Stokes V_0 oscillations compared with the Stokes I oscillations (see the corresponding autocorrelation spectra and the enhanced power of V_0 at the low frequency side of the 5-min peak in Fig.2).

A surprising observational result is the strong variance in core brightness we have found between the (I+V) and (I–V) spectra in the magnetic structures (Fig.3). The differences fluctuate and have a distinct 300 s periodicity, leading to a significant variable offset between the Stokes I velocity and the Stokes V zero-crossing wavelength of up to $\pm$ 1 km/s by causing highly asymmetric Stokes I profiles (see Fig.4). The spatio-temporal averages of the Stokes I velocity and V_0, on the other hand, are almost identical: for Fe 6302.5Å we obtain a red shift of V_0 of approximately -150 m/s, and for Fe 6301.5Å only -10 m/s. Connected with the occurrence of strong differences among (I+V) and (I–V) spectra — a phenomenon which is also apparent in Fig.8 of Scholiers and Wiehr (1985), but not discussed further

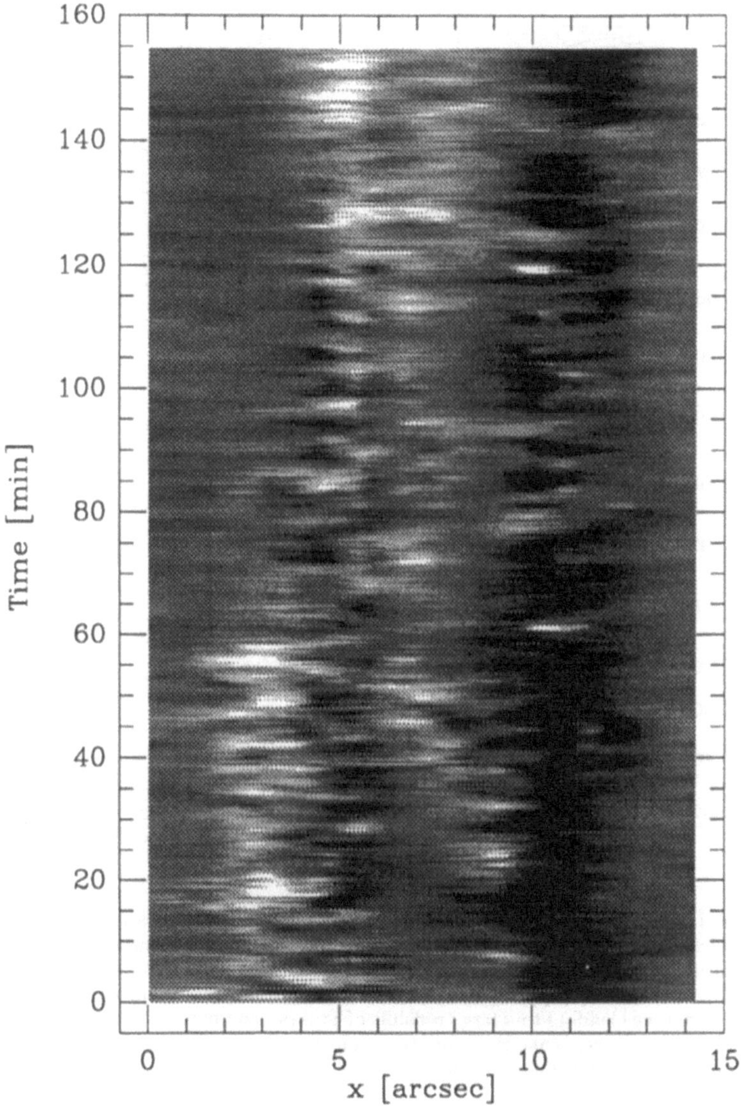

Fig. 3. Spatio-temporal behaviour of the core intensity difference between the two circularly polarized spectra (I+V) and (I–V) of Fe 6302. The min/max values of the grey scale mapping are ±6%.

there — are considerable differences in the C-shapes of both the (I+V), (I–V) and I spectra. The C-shapes even change their sign. Since the corresponding Stokes V profiles reveal always the same (positive) amplitude and area asymmetry (the blue Stokes V wing being stronger), one can conclude that the time dependent asymmetry between (I+V) and (I–V) is solely due to the unpolarized light (Stokes I) and is therefore not related to the

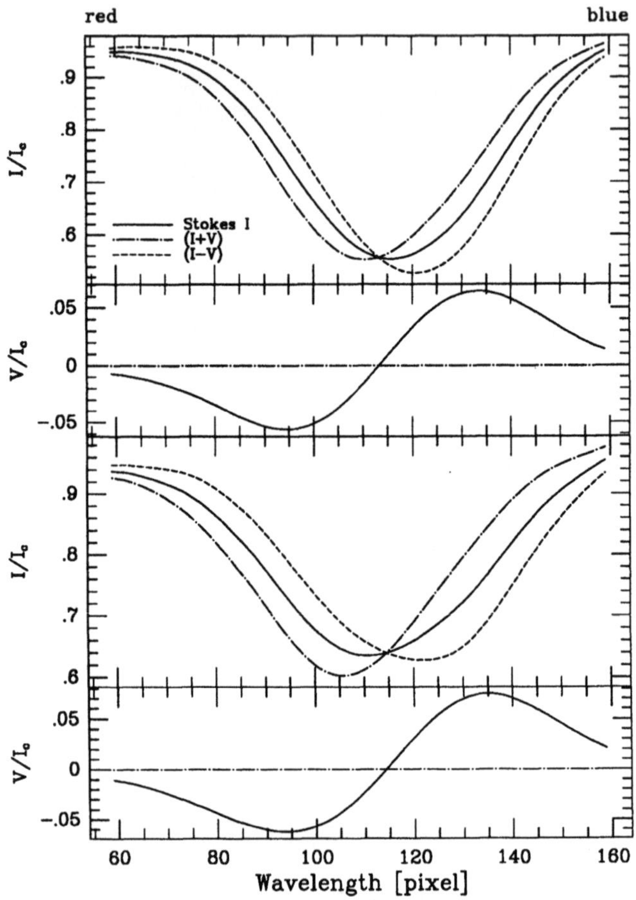

Fig. 4. Stokes (I+V), (I−V), I and V spectra of two regions (two rows on the CCD camera) selected out of the same frame. One region reveals a strong positive difference between the core intensity of (I+V) and (I−V) causing a large blue shift of the Stokes I velocity compared with V_0 (upper panel). The other region exhibits a strong negative (I+V) − (I−V) difference leading to a large redshift of Stokes I compared with V_0 (second lowest panel). Note, however, that the amplitude and the area asymmetry of Stokes V remains the same, in particular also the sign!

magnetic structure itself. One could for instance speculate that the temperature structure and the flow pattern of the abnormal granulation (Dunn and Zirker, 1973; Muller, 1989) cause this difference between the (I+V) and (I−V) spectra by changing the line asymmetry. However, the spatio-temporal behaviour of the (I+V)–(I−V) core intensity difference with its distinct 300 s periodicity displayed in Fig.3 does not resemble very much granular patterns but rather oscillatory motions which are known to cause asymmetric line profiles as well (see e.g. Roddier, 1967; Snider, 1970; Roca-Cortes et al., 1983; Cavallini et al., 1986). An important parameter as regards the shape of line profiles is the phase difference between velocity oscillations and the perturbations of the thermodynamic quantities (cf. Mihalas and Mihalas, 1984). This phase difference is significantly affected by radiation damping. In the neighbourhood of fluxtube concentrations horizontal radiative transfer plays an important rôle. One can therefore assume that there the dynamics of waves and oscillations and hence the line asymmetries are considerably changed compared to the quiet sun. In order to interpret and understand these important effects properly we need detailed theoretical model calculations as well as further observations with higher spatial and temporal resolution.

Another important feature we have found are the strong spatial (and temporal) variations of the relative amplitude and area asymmetry δa_r and δA_r. In most spectra, the amplitude asymmetry typically increases from about 3 to 5% in the center of the magnetic structure to more than 15% towards the edges.

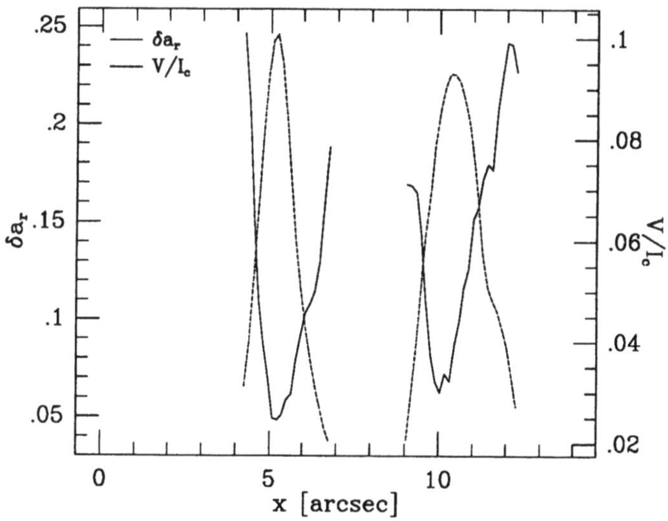

Fig. 5. Relative amplitude asymmetry and Stokes V amplitude of a selected frame.

Fig.5 represents a particularly nice example of this effect. The general trend of the anticorrelation between the Stokes V amplitude and the amplitude asymmetry is displayed in Fig.6. This finding is in good agreement with theoretical calculations of Solanki (1989) and confirms a stationary flow of cool material outside of the fan-shaped magnetic fluxtubes as the cause of the observed red/blue asymmetry of the Stokes V profiles (Grossmann-Doerth et al., 1988, 1989). We want to note, however, that the observed line ratio between the two Fe lines is constant across the magnetic structures and

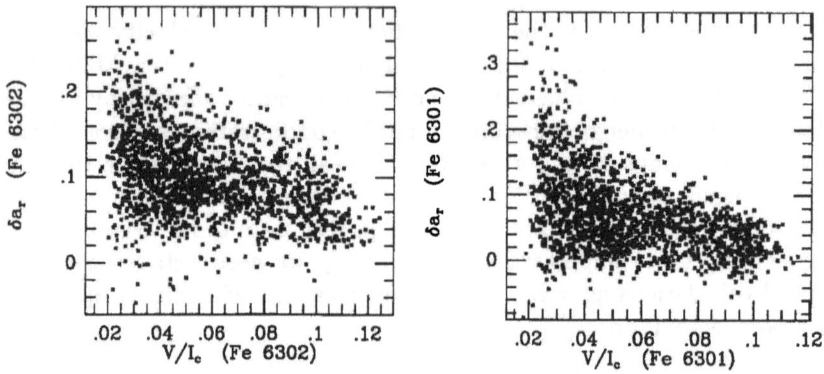

Fig. 6. Scatter plot of the relative amplitude asymmetry of Fe 6302 (left diagram) and Fe 6301 (right diagram) vs. the Stokes V amplitude. The data were taken from the 35 best spectra of our time series.

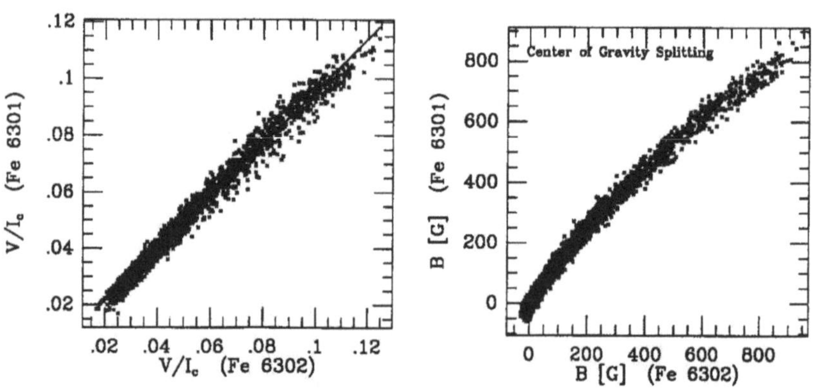

Fig. 7. Scatter plot of the Stokes V amplitude of Fe 6301 vs. the Stokes V amplitude of Fe 6302 (left diagram), and scatter plot of the center of gravity splitting of Fe 6301 vs. Fe 6302 (right diagram). The data were taken from the 35 best spectra of our time series.

therefore does not agree with this intepretation which requires rays passing through both non-magnetic and magnetic structures.

In Fig.7 we display scatter plots of the Stokes V amplitude and center of gravity splitting of Fe 6301 vs. Fe 6302. It is interesting to note that — in contradiction to theoretical arguments proposed by Rees and Semel (1979) — the center of gravity method seems not to be entirely free of saturation effects (at least not for the two Fe lines chosen). For weak apparent flux densities, the Fe 6301 line with moderate splitting (g_{eff}=5/3) indicates a higher field strength than the Fe 6302 line with large splitting (g=2.5). For high apparent flux densities, on the other hand, the field strength derived from the center of gravity splitting of Fe 6301 increases more slowly than the field strength derived from Fe 6302. This effect is even more pronounced if one displays the core splitting of the σ-components (instead of the center of gravity splitting). Some hint of this effect is also present in the scatter plot of the Stokes V amplitudes (left diagram in Fig.7). However, the effect has just the opposite sign of what would be expected if there were real weak fields. For more detailed interpretations one has to take into account that the two Fe lines at 6302Å have different *gf*-values and therefore do not meet the conditions for the line ratio analysis as for instance the Fe line pair at 5250/5247Å.

At the end of this section a few remarks concerning the instrumental polarization seem to be appropriate. We are mainly interested in time-dependent dynamical phenomena of magnetic structures. Typical time scales range from about 50 s to 10 minutes. For these time intervals the instrumental polarization can be considered constant. Absolute measurements like those of the relative amplitude asymmetry, on the other hand, have to be interpreted with some care. We want to note that, in our spectra, we did not detect a conspiciuous π-component as observed for instance in spectra taken in a sunspot penumbra. One can therefore assume that the linear polarization within the flux elements was small, and likewise also any V signal introduced by "crosstalk" due to the retardation of the telescope.

4. Conclusions and Future

We have presented first results of a novel project which addresses the dynamics of small-scale magnetic fluxtube concentrations in the solar atmosphere. This project aims at the identification of the MHD-modes present in the magnetic structures, a proper understanding of their propagation characteristics, and their contribution to the heating of the chromosphere and corona. From our results we conclude that it is necessary to measure the complete (I+V) and (I–V) profiles with the highest spatial and temporal resolution achievable in order to disentangle the effects responsible e.g. for the various asymmetries, otherwise essential information is lost. To have the complete Stokes V profile available instead of applying the common magnetogram technique (e.g. Tarbell et al., 1979; Topka et al., 1986; Keller et al., 1990) is even more valuable as theoreticians have now the tools ready to provide not only the temperature, density and magnetic field structure of fluxtubes but also to calculate the emergent Stokes profiles (see e.g. Grossmann-Doerth et al., 1989).

Several new questions have arisen from our preliminary results, but the quality of our data appears very promising. The situation is somewhat comparable to the mid to late seventies when several investigations concerning the propagation of acoustic-gravity waves in the quiet solar atmosphere were initiated (e.g. Deubner, 1974; Schmieder, 1976; Lites and Chipman, 1979).

Two of the most valuable improvements in the future will be the possibility of running several large CCD cameras in parallel while the solar image will be stabilized by a correlation tracker (von der Lühe et al., 1989; Rimmele and von der Lühe, 1990). We will then be able to measure simultaneously various lines which respond differently to magnetic field and temperature fluctuations, i.e. we will be able to carry out a spatio-temporal analysis of the magnetic and thermal line ratio. To study the dispersion relation of the waves in magnetic elements such observations should also include lines formed higher in the atmosphere. We hope that we will be able to make a big step forward already in the near future with the instrumentation presently available. Ultimately, the LEST and OSL projects should enable us to observe and understand the MHD-processes of the magnetic flux elements which are of crucial importance for the physics of the solar atmosphere.

Acknowledgements: I am very grateful to F.-L. Deubner for encouraging advice. I would also like to thank Sami Solanki for helpful discussions. Part of this work was supported by the Deutsche Forschungsgemeinschaft (grant De 226/9–1).

References

Abdelatif, T.E.: 1988, *Ap. J.* **333**, 395
Beckers, J.M., Schröter, E.H.: 1968, *Solar Phys.* **4**, 142
Cavallini, F., Ceppatelli, G., Righini, A., Alamanni, N.: 1986, *Astron. Astrophys.* **156**, 310
Cram , L.E., Wilson, P.R.: 1975, *Solar Phys.* **41**, 313
Defouw, R.J.: 1976, *Astrophys. J.* **209**, 266
Dunn, R.B., Zirker, J.B.: 1973, *Solar Phys.* **33**, 281
Edwin, P.M., Roberts, B.: 1982, *Solar Phys.* **76**, 239
Edwin, P.M., Roberts, B.: 1983, *Solar Phys.* **88**, 179
Evans, D.J., Roberts, B.: 1990, *Ap. J.* **348**, 346
Deubner, F.-L.: 1974, *Solar Phys.* **39**, 31
Fleck, B., Deubner, F.-L.: 1989, *Astron. Astrophys.* **224**, 245
Fleck, B., Deubner, F.-L.: 1990, in *The Dynamic Sun*, Proc. of the 6th European Meeting on Solar Physics, *Publ. Debrecen Obs.* **7**, 306
Fleck, B., Deubner, F.-L.: 1991, in *Mechanisms of Chromospheric and Coronal Heating*, ed. P. Ulmschneider, Springer, in print
Giovanelli, R.G., Livingston, W.C., Harvey, J.W.: 1978, *Solar Phys.* **59**, 49
Grossmann-Doerth, U., Knölker, M., Schüssler, M., Weisshaar, E.: 1989, in *Solar and Stellar Granulation*, eds. R.J. Rutten and G. Severino; NATO ASI Series **C 263**, p. 481
Grossmann-Doerth, U., Schüssler, M., Solanki, S.K.: 1988, *Astron. Astrophys.* **206**, L37
Grossmann-Doerth, U., Schüssler, M., Solanki, S.K.: 1989, *Astron. Astrophys.* **221**, 341
Hammer, R.: 1987, in *The Role of Fine-Scale Magnetic Fields on the Structure of the Solar Atmosphere*, eds. E.H. Schröter, M. Vazques, and A.A. Wyller, Cambridge University Press, p. 255
Keller, C.U., Solanki, S.K., Tarbell, T.D., Title, A.M., Stenflo, J.O.: 1990, *Astron. Astrophys.* **236**, 250
Lites, B.W., Chipman, W.: 1979, *Ap. J.* **231**, 570
Mihalas, D., Mihalas, B.W.: 1984, *Foundations of Radiation Hydrodynamics*, Oxford Univ. Press, New York

Muller, R.: 1989, in *Solar and Stellar Granulation*, eds. R.J. Rutten and G. Severino; NATO ASI Series **C 263**, p. 101
Pantellini, F.G.E., Solanki, S., Stenflo, J.O.: 1988, *Astron. Astrophys.* **189**, 263
Parker, E.N.: 1983, *Ap. J.* **264**, 642
Parker, E.N.: 1991, in *Mechanisms of Chromospheric and Coronal Heating*, ed. P. Ulmschneider, Springer, in print
Rees, D.E., Semel, M.: 1979, *Astron. Astrophys.* **74**, 1
Rimmele, Th., von der Lühe, O.: 1990, *Rev. Mod. Astron.***2**, 105
Roberts, B.: 1981a, *Solar Phys.* **69**, 27
Roberts, B.: 1981b, *Solar Phys.* **69**, 39
Roberts, B.: 1983, *Solar Phys.* **87**, 77
Roberts, B., Webb, A.R.: 1978, *Solar Phys.* **56**, 5
Roberts, B., Webb, A.R.: 1979, *Solar Phys.* **77**, 5
Roca-Cortes, T., Vazquez, M., Wöhl, H.: 1983, *Solar Phys.* **88**, 1
Roddier, F.: 1967, *Ap. J.* **147**, 1113
Ryutova, M.P.: 1990, in *Solar Photosphere: Structure, Convection and Magnetic Fields*, ed. J.O. Stenflo, IAU-Symp. No. 138, Kiev, p. 229
Schmieder, B.: 1976, *Solar Phys.* **47**, 435
Scholiers, W., Wiehr, E.: 1985, *Solar Phys.* **99**, 349
Schrijver, C.J.: 1987, *Astron. Astrophys.* **172**, 111
Schrijver, C.J.: 1991, in *Mechanisms of Chromospheric and Coronal Heating*, ed. P. Ulmschneider, Springer, in print
Schüssler, M.: 1987, in *The Role of Fine-Scale Magnetic Fields on the Structure of the Solar Atmosphere*, eds. E.H. Schröter, M. Vazques, and A.A. Wyller, Cambridge University Press, p. 223
Schüssler, M.: 1990, in *Solar Photosphere: Structure, Convection and Magnetic Fields*, ed. J.O. Stenflo, IAU-Symp. No. 138, Kiev, p. 161
Snider, J.L.: 1970, *Solar Phys.* **12**, 352
Solanki, S.K.: 1986, *Astron. Astrophys.* **168**, 311
Solanki, S.K.: 1987, in *The Role of Fine-Scale Magnetic Fields on the Structure of the Solar Atmosphere*, eds. E.H. Schröter, M. Vazques, and A.A. Wyller, Cambridge University Press, p. 67
Solanki, S.K.: 1989, *Astron. Astrophys.* **224**, 225
Solanki, S.K.: 1990, in *Solar Photosphere: Structure, Convection and Magnetic Fields*, ed. J.O. Stenflo, IAU-Symp. No. 138, Kiev, p. 103
Spruit, H.C.: 1981a, in *The Sun as a Star*, ed. S. Jordan, NASA SP-450, p. 385
Spruit, H.C.: 1981b, *Astron. Astrophys.* **98**, 155
Spruit, H.C.: 1982, *Solar Phys.* **75**, 3
Spruit, H.C., Roberts, B.: 1983, *Nature* **304**, 401
Stenflo, J.O.: 1989, *Astron. Astrophys. Rev.* **1**, 3
Stenflo, J.O., Harvey, J.W.: 1985, *Solar Phys.* **95**, 99
Stenflo, J.O., Harvey, J.W., Brault, J.W., Solanki, S.: 1984, *Astron. Astrophys.* **131**, 333
Tarbell, T.D., Title, A.M., Schoolman, S.A.: 1979, *Ap. J.* **229**, 387
Topka, K.P., Tarbell, T.D., Title, A.M.: 1986, *Ap. J.* **306**, 304
Ulmschneider, P.: 1991, in *Mechanisms of Chromospheric and Coronal Heating*, ed. P. Ulmschneider, Springer, in print
Ulmschneider, P., Muchmore, D.: 1986, in *Small Scale Magnetic Flux Concentrations in the Solar Photosphere*, eds. W. Deinzer, M. Kn"olker and H.H. Voigt, Vandenhoeck&Ruprecht, Göttingen, p. 169
von der Lühe, O., Widener, A.L., Rimmele, Th., Spence, G., Dunn, R.B., Wiborg, P.: 1989, *Astron. Astrophys.* **224**, 351
Webb, A.R., Roberts, B.: 1980a, *Solar Phys.* **68**, 71
Webb, A.R., Roberts, B.: 1980b, *Solar Phys.* **68**, 87
Wentzel, D.G.: 1979, *Astron. Astrophys.* **76**, 20
Wiehr, E.: 1985, *Astron. Astrophys.* **149**, 217
Wilson, P.R.: 1979, *Astron. Astrophys.* **71**, 9
Zayer, I., Solanki, S., Stenflo, J.O.: 1989, *Astron. Astrophys.* **211**, 463

Instabilities in Astrophysical Shear Flows

W. Glatzel

Universitäts-Sternwarte Göttingen,
Geismarlandstr. 11, W-3400 Göttingen, Fed. Rep. of Germany

Summary. Shear flows are found in many astrophysical objects such as accretion disks, jets and rotating stars. The study of their stability is particularly important with respect to the occurrence of turbulence in accretion disks and the stability of stellar rotation laws. Properties and mechanisms of shear instabilities relevant in various astrophysical situations are discussed.

1 Introduction

Shear flows are a common phenomenon in astrophysics. All astrophysical disks which cover a wide range from planetary to galactic scales involve shearing motions. On the stellar scale models of thin accretion disks are discussed to explain the observations of cataclysmic variables and X - ray binaries. A basic parameter in these models is the shear viscosity which is thought to cause angular momentum and mass transport which leads to accretion onto the central object. Since molecular viscosity is too small to fit the observations some kind of turbulence is regarded as the origin of the required viscosity. However, necessary for an accretion disk to become turbulent is the instability of the basic Keplerian accretion flow. Although a turbulent viscosity is incorporated in all accretion disk models the crucial question remains, if such an instability exists and how it works, since standard stability criteria guarantee stability. (For instance, according to Rayleigh's criterion a flow is stable if its angular momentum h increases with radius ($\frac{\partial h}{\partial r} \geq 0$) which is satisfied for a Keplerian disk.)

A similar problem exists in the theory of star formation. Here the specific angular momentum of a molecular cloud is too high for single star formation and a mechanism is needed which efficiently redistributes and removes angular momentum in and from the protostellar disk. One possibility to overcome the angular momentum problem is angular momentum transport through viscous friction caused by turbulence (see, e.g. Tscharnuter, 1989). However, again the question about the origin and mechanism of an instability leading to turbulence requires a satisfactory answer.

Galactic disks, where the flow speed can be measured directly, show spiral structures, some of which could possibly be related to shear instabilities, whose study is thus strongly suggested.

Accretion tori orbiting massive black holes are an essential ingredient in models for the central engine of active galactic nuclei. For small deviations from constant specific angular momentum distribution they exhibit steep funnels which makes them particularly attractive for the collimation and generation of jets. However, even centrifugally stable accretion tori ($\frac{\partial h}{\partial r} \geq 0$), in particular those having constant angular momentum distribution were found unstable when subject to nonaxisymmetric perturbations (Papaloizou & Pringle, 1984). In this case it is of interest, whether the violent shear instabilities found do not allow for models of AGN's involving accretion tori or if and how the models can and have to be changed to avoid instabilities. Although the instability had already been found by Drury (1977), it was the discovery by Papaloizou & Pringle (1984) which initiated investigations concerning its mechanism, its dependence on parameters and possible stabilisation (see, e.g., Blaes 1985, 1987, Frank & Robertson 1988, Ghosh & Abramowicz 1991, Goldreich et al. 1986, Hanawa 1987, 1988, Kato 1987, Kojima 1986, 1989, Kojima et al. 1989, Narayan et al. 1987, Papaloizou & Pringle 1985, 1987). These studies turned out to be relevant not only for accretion tori but also for rotating shear flows in general, i.e. for the thin accretion disks, protostellar and galactic disks mentioned above.

Rotating shear flows are not restricted to disk like objects only but are likely to occur in any rotating star. Compared to disks where the shear rate is high and involves supersonic velocity differences, shear rates in rotating stars are low, i.e. the velocities lie in the subsonic regime. In an inviscid analysis the rotation law of a star cannot be determined and is usually treated as a free parameter. However, restrictions on the possible form of rotation laws may be derived from stability considerations. The investigation of shear instabilities in rotating stars can thus provide information on admissible stellar rotation laws.

A second class of astrophysical shear flows is found at the boundary jets, where matter is expelled by some compact source into the interstellar or intergalactic medium respectively. Instabilities of the shear layer at the edge of the jet are studied with respect to the formation of knots and wiggles (Ferrari et al. 1982, Ferrari & Trussoni 1983, Hardee 1979, Ray 1982). Apart from ordinary Kelvin - Helmholtz instabilities a second type of shear instabilities can be identified, which is - except for minor changes due to geometry and rotation - qualitatively identical with the instabilities found in rotating disks and stars (see Glatzel, 1988).

This surprising result is due to a common origin of the instabilities mentioned: The shear motion. However, not only the origin, also the mechanism of the instabilities seems to be similar in very different physical environments. Therefore we shall try to give here a unified description of the origin and mechanism of shear instabilities based on a simple highly idealized model which still allows for qualitatively correct results and - concerning the instabilities - shows all properties of more realistic models (Section 2). The application of different subtypes of shear instabilities to various astrophysical objects will be discussed in Section 3.

2 The mechanism of the instability

2.1 General remarks

The discussion presented here and in the following sections is based on a linear analysis, i.e. momentum and continuity equation are linearized in the dependent variables. The energy equation is replaced by assuming a polytropic relation between pressure and density or by requiring entropy conservation for the perturbations. Under very general conditions it is then possible to derive an evolution equation for the perturbation energy from the perturbation equations which describes the redistribution of energy, in particular the energy transfer from the stationary flow to the perturbations (Glatzel, 1987a). This energy transfer term depends on the perturbation itself and is proportional to the shear rate. It vanishes, if there are no shearing motions, e.g. for a rigidly rotating star. Thus for a perturbation to grow shear is a necessary condition and may be regarded as the origin of instabilities.

Another way to understand the destabilizing tendency of shearing motions consists in comparing the energy of systems with and without shear but satisfying the same integral constraints, e.g. two stellar models with rigid and differential rotation but having the same mass and angular momentum. The minimum energy is always achieved for vanishing shear, i.e. in the example given above for a rigidly rotating star (see, e.g. Lynden - Bell & Pringle, 1974). This is immediately obvious in a plane flow. If there is no shear, its kinetic energy vanishes in the frame of reference comoving with the fluid, while for a shear flow having the same mean velocity the kinetic energy is greater than zero in any frame of reference.

One of the first models for which the violent instabilities considered here have been discovered were twodimensional accretion tori orbiting massive compact objects (Papaloizou & Pringle, 1984). Meanwhile it has become clear that for the instabilities shear is the important quantity while geometry and rotation play a minor role only. In fact, all their basic properties are already found in a simple linear shear layer to which our discussion will be restricted in the following. Furthermore, this simplification allows for a unified treatment of shear layers in different astrophysical situations, e.g. in accretion disks and at the boundary of jets.

The model considered is a shear layer (see also Glatzel, 1988) where the fluid is confined between two infinite parallel planes moving with relative velocity $2V_0$ parallel to the planes. The flow velocity is assumed to vary linearly in space and to match the velocities of the confining boundary planes. For convenience we introduce a coordinate system where the x - axis is taken in flow direction, the z - axis perpendicular to the boundary planes and the origin is placed on the midplane between the boundaries. Lengths are measured in units of half of the thickness of the shear layer, z_0. With these definitions the shear layer extends in z from $z = -1$ to $z = +1$ and the velocity may be written as $\mathbf{V} = (V_0 z, 0, 0)$. In general the system allows for perturbations proportional to $\exp(ik(\omega t + x))$, i.e. for waves travelling in x - direction with wavenumber k and pattern speed ω.

A standard approach to the study of stability is a local analysis, where the coefficients of the perturbation equations are kept constant and boundary conditions are disregarded. The problem then allows for an analytical solution and leads to a dispersion relation describing the dependence of the pattern speed ω on the local wavenumber k. However, none of the instabilities occurring in a correct treatment is found in the local analysis. Accordingly, a global analysis is necessary to study these instabilities and the differential equations describing the perturbation problem have to be integrated together with suitable boundary conditions.

In Sects. 2.2 and 2.4 we shall describe the mechanism of shear instabilities for two physically entirely different situations: A compressible configuration only allowing for sound waves and a nonisentropic incompressible system allowing for internal gravity waves only. The common properties of these two situations will be of particular interest.

2.2 Sound waves

In this section the medium within the shear layer is assumed to be compressible and isentropic with a constant sound speed, a_0. If velocities are measured in units of a_0, the flow is entirely determined by one dimensionless number, the Mach number M, which is defined as the ratio of flow velocity and sound speed at the edge of the shear layer: $M = \frac{V_0}{a_0}$. The properties of the shear layer will be discussed as a function of the Mach number which may also be regarded as the normalized flow velocity at its edges or a measure of the shear strength.

The perturbations in the isentropic compressible configuration considered exhibit the physical properties of sound waves travelling in x - direction (with wavenumber k and pattern speed ω) and having the character of standing sound waves perpendicular to the boundaries. They may be ordered according to n, their number of nodes in z.

For $M = 0$, i.e. for a medium at rest, the perturbation problem is easily solved analytically and the spectrum of these sonic modes is given by:

$$\omega = \pm\left[1 + \left(\frac{\pi}{2}\frac{n}{k}\right)^2\right]^{1/2} \quad ; \quad n = 0, 1, 2, \ldots \tag{1}$$

Due to the symmetries of the configuration we have - for given wavenumber k - a twofold infinite set of discrete neutrally stable sonic modes travelling in positive and negative x - direction respectively.

Considering now the case of finite shear strength we follow the evolution of the sonic spectrum for fixed wavenumber k when M is increased continuously from $M = 0$ to some finite value. For illustration the pattern speed ω for the neutrally stable sonic modes having $n \leq 12$ and $k = 1$ is plotted as a function of M in Fig. 1. Indicated by dashed lines is the range of the flow speed ($-M \leq V_x \leq M$), i.e. the dashed lines correspond to the extrema of the flow speed at the boundaries.

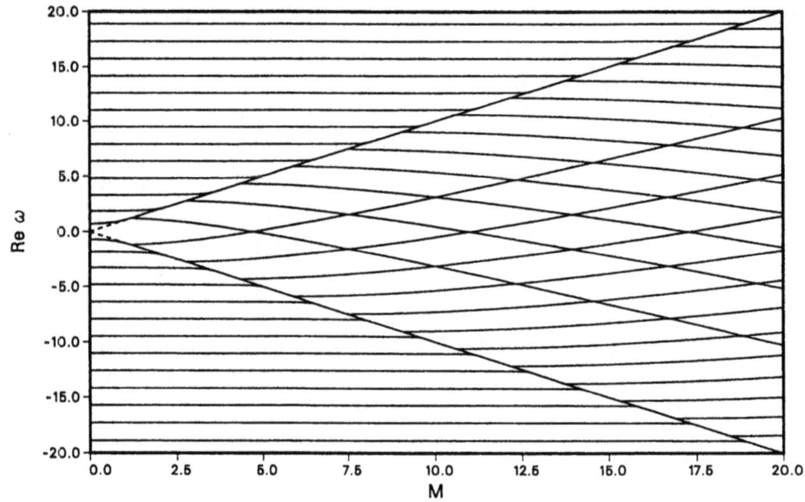

Fig. 1. The pattern speed ω of sonic modes as a function of the Mach number M for the wavenumber $k = 1$. The plot is based on an analytical approximation excluding mode interaction.

In the limit $M \to 0$, i.e. if the pattern speed of a mode is large compared with the flow speed ($|\omega| \gg M$), the spectrum becomes identical with the sonic spectrum of the corresponding medium at rest discussed above. With increasing M, i.e. increasing shear strength, the pattern speed of a mode is distorted if it gets comparable with the flow velocity, where the pattern speed of the mode having $n = 0$ is increased by the shear. Always remaining outside the flow velocity range ($|\omega| > M$) it approaches the flow velocity at the boundaries for large values of M. Contrary to the mode having $n = 0$ the pattern speed of all modes having $n \geq 1$ is decreased by the shear and enters the flow speed range, i.e. the mode gets a critical layer within the flow. For higher Mach numbers the pattern speed even reverses its sign and approaches the velocity of the boundary ($|\omega| < M$) moving in the direction opposite to the mode in the medium at rest. Thus multiple mode crossings are generated, where crossings involving the mode with $n = 0$ occur outside ($|\omega_{crossing}| > M$), all other mode crossings inside ($|\omega_{crossing}| < M$) the flow velocity range.

In order to clarify the situation and to enable a separation of different effects Fig. 1 is not based on the solution of the full problem but on an analytical approximation which does not allow for mode interaction. The discontinuity of modes with $n \geq 1$ appearing at $|\omega| \approx M$ is an artefact of the approximation in which different expansions of the exact dispersion relation are used for $|\omega| < M$ and $|\omega| > M$.

The counterpart of Fig.1 based on the exact dispersion relation is shown in Figs. 2 and 3. Apart from slight modifications due to the approximate treatment used for Fig. 1, Figs. 1 and 2 are identical except for the mode crossings. This is not surprising since mode interaction is essential around resonances or mode crossings and was artificially excluded in the approximation. The unfolding of mode crossings requires a separate treatment - in particular since in some cases it is related to the occurrence of instabilities (see Figs. 2 and 3) - and is discussed in Sect. 2.3.

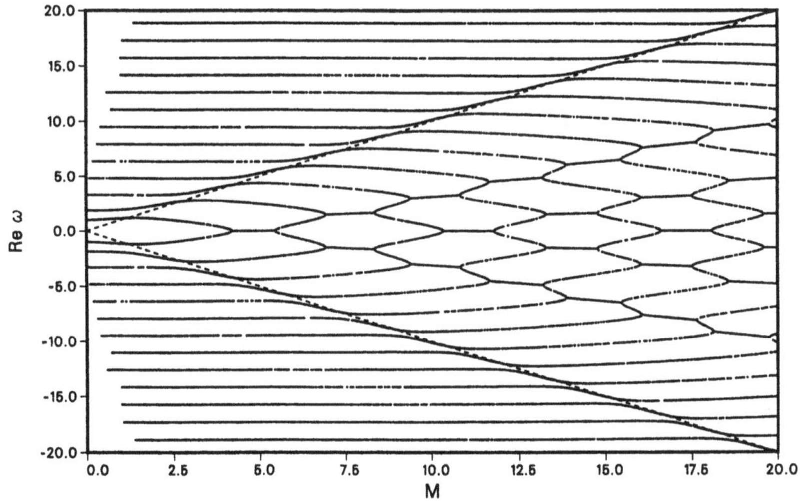

Fig. 2. Same as Fig. 1 but according to the exact dispersion relation. Note the unfolding of mode crossings into avoided crossings and instability bands.

2.3 The unfolding of mode crossings

The unfolding of mode crossings is independent of the physical nature of the crossing modes. We shall briefly discuss here some of its basic features. According to the general theory (see, e.g. Glatzel, 1987b) mode crossings

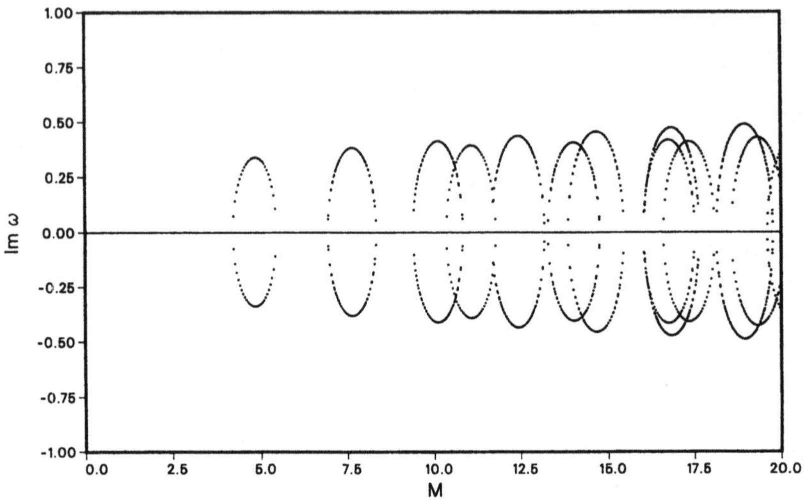

Fig. 3. The imaginary part of the pattern speed for the modes shown in Fig. 2.

of neutrally stable modes A and B unfold - except for the very rare case of vanishing mode interaction - in two different ways. Either they remain neutrally stable and mode A before crossing is connected to mode B after crossing and vice versa. This case is called avoided crossing and occurs in our problem at all crossings involving the mode having $n = 0$ (see Figs. 1 and 2). The second possibility is the formation of a complex conjugate pair of modes - an instability band - around the original crossing by merging of modes A and B close to the crossing. Its pattern speed is the mean of the uncoupled crossing modes. This case is realised here at all crossings where the pattern speed lies within the flow speed range (see Figs. 1, 2 and 3).

Physically the two different results of mode interaction - instability band and avoided crossing - may be interpreted in terms of the energy of a mode (see Glatzel, 1987b). If the energy of the crossing modes differs in sign, the amplitude of both of them can grow without bound by resonant exchange of energy, even if the energy of the entire system is to be conserved. On the other hand, resonant exchange of energy cannot lead to an instability if the energy of the crossing modes has the same sign.

The term "energy" requires a precautionary remark. Here the "energy" of a mode is defined as energy difference of the shear flow with the mode superimposed and the stationary configuration, which is not a scalar gauge invariant quantity. Even its sign depends on the frame of reference chosen. Hence this "energy" is also referred to as pseudoenergy.

The existence of negative energy modes also provides a physical explanation for the instability of sonic modes when energy loss by acoustic radiation is considered: The amplitude of a negative energy mode grows by loss of energy. However, instability by energy loss due to acoustic radiation does not necessarily imply instability by energy loss in general as claimed by several authors. For instance, energy loss by viscous dissipation does not lead to an instability. (Viscous instabilities occurring in the flows considered here have a different physical origin (see Glatzel, 1989).) The reason for this apparent contradiction is that we are dealing with a pseudoenergy and different pseudoenergies, in particular different signs may be appropriate in different physical situations. In the case of energy loss by acoustic radiation a sonic mode is correctly characterised by a negative, for energy loss by viscous dissipation by a positive pseudoenergy.

2.4 Gravity waves

In order to allow for internal gravity waves we introduce an external gravity of the form $g = -g_0 z$ which qualitatively resembles the effective gravity in a slender accretion torus. Convective stability of this configuration requires a parabolic density profile: $\rho \sim (1 - z^2)$. As we are interested in the behaviour of pure gravity waves, sound waves are excluded by assuming the fluid to be incompressible.

Measuring velocities in units of $\sqrt{z_0 g_0}$ the flow is - similar to the case of sound waves - characterised by the dimensionless number $\frac{V_0}{\sqrt{z_0 g_0}} = Ri^{-1/2}$, where Ri can be identified with the Richardson number. Similar to the Mach number M we may use $Ri^{-1/2}$ as a parameter and regard it as the normalized flow velocity at the edges of the shear layer and as a measure of the shear strength. The perturbations correspond to gravity waves travelling in x - direction (with wavenumber k_p and pattern speed ω_p) and having the character of standing waves perpendicular to the boundaries.

In the medium at rest ($Ri^{-1/2} = 0$) the gravity modes form two sets having (with respect to z) symmetric and antisymmetric eigenfunctions respectively (for details see Glatzel, 1990). For $\omega_p \to 0$ their eigenvalues are approximately given by:

$$\omega_p \approx \pm \frac{1}{\sqrt{2}} \left(n\pi + \frac{5\pi}{8}\right)^{-1} \quad ; \quad n = 0, 1, 2, \ldots \quad \text{symmetric} \tag{2}$$

$$\omega_p \approx \pm \frac{1}{\sqrt{2}} \left(n\pi + \frac{7\pi}{8}\right)^{-1} \quad ; \quad n = 0, 1, 2, \ldots \quad \text{antisymmetric} \tag{3}$$

The modes are symmetric with respect to propagation in positive and negative x - direction and n is an order parameter related to the number of nodes of the eigenfunction in z.

Similar to the sonic modes the pattern speed of the neutrally stable gravity modes is distorted by a shear flow. Their behaviour as a function of the shear strength $Ri^{-1/2}$ for fixed wavenumber $k_p = 0$ is shown in Fig. 4, where dashed lines denote the flow speed at the boundaries. (Fig. 4 corresponds to Fig. 1 for the case of sonic modes.)

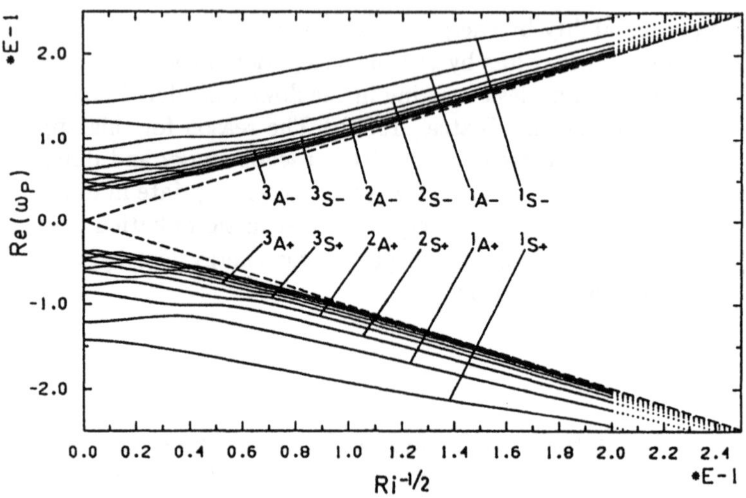

Fig. 4. The pattern speed $Re(\omega_p)$ of neutrally stable gravity modes as a function of the Richardson number Ri for the wavenumber $k_p = 0$. The modes are labelled with the order parameter n, where subscripts indicate the symmetry property and propagation direction. Note the sequences of avoided crossings.

If the pattern speed of a mode is large compared with the flow speed ($|Re(\omega_p)| \gg Ri^{-1/2}$) we obtain the spectrum of the gravity modes in the corresponding medium at rest. Accordingly the modes in Fig. 4 are labelled with the order parameter n, the symmetry property and propagation direction of the gravity mode whose pattern speed is approached for $Ri^{-1/2} \rightarrow 0$. Note that for $Ri^{-1/2} \rightarrow 0$ (in accordance with equations (2) and (3)) the gravity modes occur in pairs which is an immediate consequence of the symmetry of the Brunt - Väisälä frequency profile.

As the shear strength is increased, the pattern speed of the gravity modes attains several minima and maxima and ultimately approaches the flow velocity at the boundaries. For small values of Ri ω_p increases with $Ri^{-1/2}$ always remaining real and outside the flow velocity range. Using the general properties of mode couplings the modal diagram in Fig. 4 may be interpreted by the crossing of two sets of modes, where the pattern speed of one of them increases with the shear strength and decreases for the second. This leads to multiple mode crossings which have unfolded into sequences of avoided crossings and thus explain the minima and maxima in the pattern speed. If this interpretation is correct, gravity modes should exist whose pattern speed lies within the flow speed range and can be found by extrapolating the sequences of avoided crossings into the range $|Re(\omega_p)| < Ri^{-1/2}$.

In fact, these modes have been found and their pattern speed is shown in Fig. 5, where modes with $|Re(\omega_p)| > Ri^{-1/2}$ do not have a physical meaning and are plotted in order to illustrate that these modes for $|Re(\omega_p)| < Ri^{-1/2}$ are related to the sequences of avoided crossings in Fig. 4. Modes with

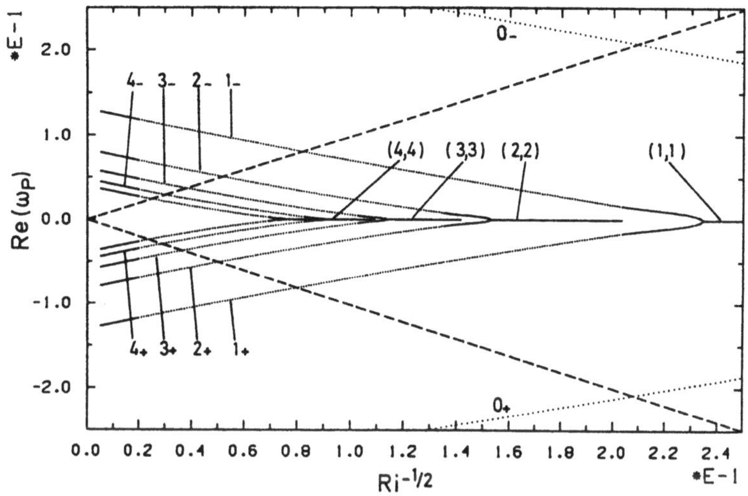

Fig. 5. Same as Fig. 4 for modes having $|Re(\omega_p)| < Ri^{-1/2}$. Points outside the range confined by the dashed lines are shown for illustration and have no physical meaning. Note the mode resonances labelled (n, n), where n is the order parameter of the crossing modes.

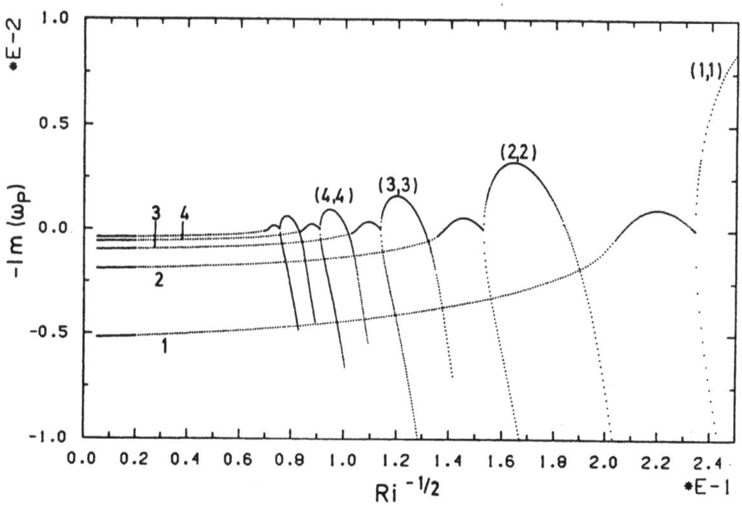

Fig. 6. The imaginary part of the pattern speed $Im(\omega_p)$ for the modes shown in Fig. 5.

$|Re(\omega_p)| < Ri^{-1/2}$ are no longer neutral (the imaginary part of the pattern speed $Im(\omega_p)$ is plotted in Fig. 6) but are damped if their pattern speed is close to the maximum flow speed. Their crossings at $Re(\omega_p) = 0$ have unfolded into instability bands (see Figs. 5 and 6).

Similar to the sonic modes also the pattern speed of a gravity mode is distorted by the shear, if it becomes comparable to the flow speed, i.e.

predominantly in the subsonic regime for the case of gravity modes. The direction of distortion depends on the particular mode and seems to be related to its symmetry properties. If the pattern speed of symmetric gravity modes is assumed to decrease with the shear strength and to increase for antisymmetric modes, the modal diagrams Figs. 4 - 6 can be interpreted in a natural way. The shear induced distortion ultimately causes multiple mode crossings which unfold - as for the sonic modes - into avoided crossings or instability bands.

Both the properties of mode crossings and the mechanism of a resonance instability together with its physical interpretation in terms of the energies of the crossing modes is identical for gravity and sonic modes and seems to be quite well understood. However, since in both cases discussed the direction of distortion of the pattern speed of a mode by the shear depends on the particular mode considered and only an appropriate distortion pattern generates the mode crossings necessary for resonance instabilities to occur, it is not sufficient to explain the properties of mode crossings. For an understanding of the shear instabilities at least equally important is, how and why shear distorts the pattern speed of a mode. So far this question has been entirely ignored.

3 Application

The mechanism of shear instabilities discussed in Section 2 does not only hold for the sound and gravity waves considered but seems to be applicable quite generally. A qualitatively similar behaviour is found for surface waves (see, e.g. Blaes & Glatzel, 1986 or Goldreich et. al., 1986) and magnetosonic modes (Glatzel, 1991) too and even the classical Kelvin - Helmholtz instability may be subsumed in the picture discussed here. Whenever the pattern speed of a particular type of modes becomes comparable to the flow speed, it is distorted by the shear which generates multiple mode crossings and shear instabilities through resonance instability bands. Thus different types of modes are responsible for shear instabilities in different ranges of the shear rate.

Sonic modes and fast magnetosonic modes having supersonic pattern speed are important in supersonic shear flows, which occur in astrophysical disks and jets. Gravity modes have low pattern speeds. Accordingly the shear instabilities associated with them have to be considered for small shear rates, i.e. in differentially rotating stars and possibly in disks. The pattern speed of slow magnetosonic modes which are currently investigated (Glatzel, 1991) strongly depends on the magnetic field. Consequently the range where these modes may lead to shear instabilities is also sensitive to the field strength. Astrophysical application is expected for disks and rotating stars. Mode coupling of surface waves can only lead to instabilities, if the interacting surfaces are sufficiently close, i.e. these instabilities depend on the size of the object. They are astrophysically relevant in accretion tori around massive compact objects. Kelvin - Helmholtz - type modes are due to the velocity profile itself and can always be generated by an appropriate choice of the vorticity distribution. Physically meaningful velocity fields in which Kelvin - Helmholtz instabilities seem to be inevitable are found in astrophysical jets and possibly in rotating stars.

The way of looking at shear instabilities presented here also explains in a natural way why laboratory Couette flow is found to be linearly stable, a fact which has misled many authors to believe that accretion disks are stable thus causing problems to justify the assumption of a turbulent viscosity. Since entropy gradients together with an external gravity are not considered, gravity modes cannot exist in the environment of standard Couette flow. Sound waves are excluded by the requirement of incompressibility, surface waves by rigid boundary conditions and Kelvin - Helmholtz modes by a vanishing vorticity gradient together with rigid boundary conditions. Thus laboratory Couette flow does not allow for any type of standing waves which could produce shear instabilities through mode resonances. In other words, the lack of any discrete mode which is necessary to establish an instability is the reason for the stability of Couette flow.

References

Blaes, O.M., 1985. Mon. Not. R. astr. Soc., **216**, 553.
Blaes, O.M., 1987. Mon. Not. R. astr. Soc., **227**, 975.
Blaes, O.M. & Glatzel, W., 1986. Mon. Not. R. astr. Soc., **220**, 253.
Drury, L.O'C., 1977. PhD thesis, Cambridge University, U.K.
Ferrari, A., Massaglia, S. & Trussoni, E., 1982. Mon. Not. R. astr. Soc., **198**, 1065.
Ferrari, A. & Trussoni, E., 1983. Mon. Not. R. astr. Soc., **205**, 515.
Frank, J. & Robertson, J.A., 1988. Mon. Not. R. astr. Soc., **232**, 1.
Ghosh, P. & Abramowicz, M.A., 1991. Astrophys. J., in press.
Glatzel, W., 1987a. Mon. Not. R. astr. Soc., **225**, 227.
Glatzel, W., 1987b. Mon. Not. R. astr. Soc., **228**, 77.
Glatzel, W., 1988. Mon. Not. R. astr. Soc., **231**, 795.
Glatzel, W., 1989. J. Fluid Mech., **202**, 515.
Glatzel, W., 1990. Mon. Not. R. astr. Soc., **242**, 338.
Glatzel, W., 1991. in preparation.
Goldreich, P., Goodman, J. & Narayan, R., 1986. Mon. Not. R. astr. Soc., **221**, 339.
Hanawa, T., 1987. Astron. Astrophys., **185**, 160.
Hanawa, T., 1988. Astron. Astrophys., **196**, 152.
Hardee, P.E., 1979. Astrophys. J., **234**, 47.
Kato, S., 1987. Publ. Astron. Soc. Japan, **39**, 645.
Kojima, Y., 1986. Progr. Theor. Phys., **75**, 251.
Kojima, Y., 1989. Mon. Not. R. astr. Soc., **236**, 589.

Kojima, Y., Miyama, S.M. & Kubotani, H., 1989. Mon. Not. R. astr. Soc., **238**, 753.
Lynden - Bell, D. & Pringle, J.E., 1974. Mon. Not. R. astr. Soc., **168**, 603.
Narayan, R., Goldreich, P. & Goodman, J., 1987. Mon. Not. R. astr. Soc., **228**, 1.
Papaloizou, J.C.B. & Pringle, J.E., 1984. Mon. Not. R. astr. Soc., **208**, 721.
Papaloizou, J.C.B. & Pringle, J.E., 1985. Mon. Not. R. astr. Soc., **213**, 799.
Papaloizou, J.C.B. & Pringle, J.E., 1987. Mon. Not. R. astr. Soc., **225**, 267.
Ray, T.P., 1982. Mon. Not. R. astr. Soc., **198**, 617.
Tscharnuter, W.M., 1989. in Theory of Accretion Disks, Proc. NATO ARW, eds. Meyer, F. et al., Kluwer, Dordrecht.

Simultaneous Observations with a Tunable Filter and the Echelle Spectrograph of the Vacuum Tower Telescope at Tenerife

W. Schmidt

Kiepenheuer-Institut für Sonnenphysik,
Schöneckstr. 6, W-7800 Freiburg, Fed. Rep. of Germany

1 Summary

The observing project described herein makes use of the great flexibility of the instrumentation of the Vacuum Tower Telescope at Tenerife. The most important characteristics of the telescope are described. The science observing programs were dedicated to the measurements of small scale magnetic structures and the velocity fields in both horizontal coordinates and at different heights in the solar atmosphere. Narrow band filtergrams in left and right handed circular polarized light with a field of view of about 120 arcseconds squared were recorded in order to construct longitudinal magnetograms while the spectrograph was equipped with a Stokes-V polarimeter for the measurement of Stokes-V profiles with high spectral resolution. Slit jaw reference pictures were taken in white light in order to locate the spectrograph slit in the magnetograms.

2 Introduction

The observing program described below is very well suited to illuminate the status of the Vacuum Tower Telescope at Tenerife (VTT), its capabilities and its state of development. The program benefits from quite a number of outstanding characteristics of the VTT and it makes use of many of its resources. In addition, several improvents of the instrument's hardware and software have been initiated and carried out during the preparations for this program.

The Echelle Spectrograph, the main post-focus instrument of the VTT, shows practically the same high spatial resolution as the telescope and its spectral resolution meets the design value of 6 mÅ at a wavelength of 5000 Å.

The short term wavelength stability of the spectrograph ("Spectrograph seeing") has been measured to be less than 0.05 mÅ (rms). Details of the telescope and the spectrograph design are given by Schröter et al. [1] and Soltau [2], [3] .

The coelostat system provides a nonrotating solar image at the telescope's focal plane. The limb guiding servo system uses the commandable secondary mirror of the coelostat to compensate for all types of additional slow motions of the solar image, caused, e.g. by the change of declination and atmospheric refraction. The Sun's rotation at the observed region can also be compensated for, thus leading to a non-moving field of view. This greatly facilitates long term observations.

The ultimate step in guiding, the compensation of fast motions (produced by Image Motion or wind-induced vibrations of the telescope) with a Correlation Tracking System, is presently under construction.

Besides the standard configuration of the VTT with the spectrograph as the main focal plane instrument, there is the possibility to set up additional instruments in a laboratory which then are used instead of the spectrograph. In order to operate an instrument in the laboratory and the spectrogaraph simultaneously, a large format beam splitter is used, which reflects 50 per cent of the incoming beam to the laboratory while the remaining light passes to the spectrograph.

The observing campaign was prepared and carried out jointly with the Lockheed Palo Alto Research Laboratory (LPARL). The observing run lasted from September 1, 1990 through September 26, 1990. The Lockheed group provided their Tunable Filter together with the equipment for polarization analysis and data aquisition. The Tunable Filter is an evaluation model of the corresponding instrument of NASA's *Orbiting Solar Laboratory, OSL* [4]. One of the goals of the program was the test of hardware components and software which is presently being developed for the OSL. The optical setup of the whole experiment is described in the following section. Details of the observing programs and an outline of the data processing techniques are given in Section 4.

3 Optical Setup

3.1 Spectrograph setup

This observing program was the first one which attempted to use both the vertical Echelle spectrograph and another major instrument in the laboratory. To this end a large format (20 cm x 26 cm) beam splitter had to be mounted in the main vertical beam. Within the wavelength range from

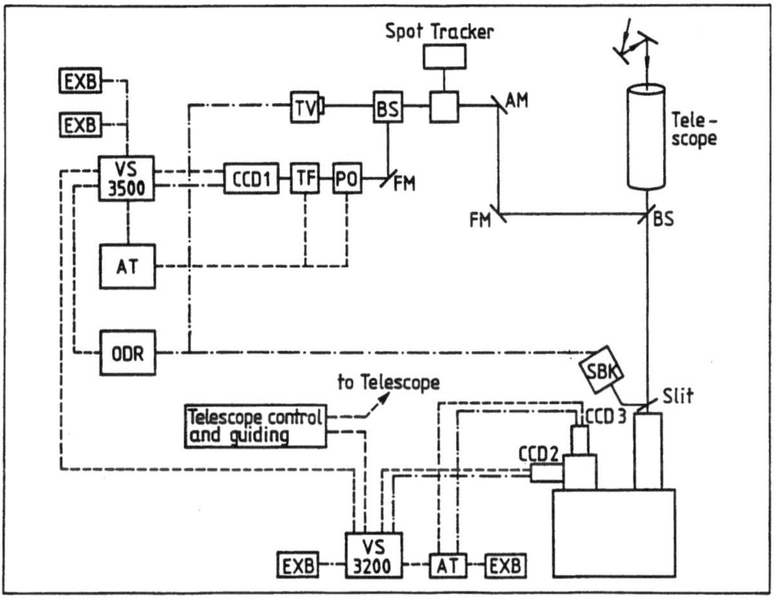

Fig. 1. Scheme of the observational setup at the Vacuum Tower Telescope. Full lines indicate the light paths, dashed lines show control and network links, dash-dotted lines denote the flow of data. BS: beam splitter, AM: active mirror, FM: folding mirror, PO: polarization optics , SBK: Slit Jaw Reference Camera, ODR: Optical disc Recorder

4500 Å to 6500 Å it is designed to reflect 50 % of the beam to the laboratory while the remaining 50 % pass to the spectrograph. The backside of the beam splitter carries an antireflection coating which minimizes the backside reflection. In addition, it has a slight wedge angle which tilts the light reflected from the backside just enough to block the unwanted light with a light baffle at the first image of the entrance pupil in the horizontal beam.

At the spectrograph focal plane we used three different CCD camera to record the wavelength bands under consideration (see below). The limitation to three wavelength bands is threefold: firstly it is the maximum of bands which can -with reasonable effort- be selected with the spectrograph predisperser, secondly all the CCD cameras available at the VTT (cf. Table 1) have rather large sized vacuum tanks or mounting plates which inhibits the use of more than three of these devices at the spectrograph focal plane and thirdly this happens to be the number of cameras which could be operated in a coordinated manner. Two cameras are operated by an IBM-type PC/AT computer (No. 2 and 4 in Table 1), while the third one is essentially identical to the one used at the Tunable Filter (see below) and is operated by a Vaxstation 3200.

The slit jaw reference camera of the spectrograph produces TV images in H_α, Ca K and in a continuum band at 5800 Å. A continuum image

Table 1. Characteristics of the CCD cameras presently available at the Vacuum Tower Telescope.

Unit:	1	2	3	4
Manufacturer:	Photometrics	Wright Instr.	JPL/LPARL	Wright Instr.
Chip:	SID53612x0	EEV P8603	TI 1024	THX 31156
Pixels:	512 x 320	578 x 385	1024 x 1024	1024 x 1024
Pixel size:	30 μ	22 μ	18.3 μ	19 μ
Readout time:	17 s	20 s	1 s	10 s
A/D conversion:	14 bit	14 bit	12 bit	16 bit
Wavelength range:	400 - 800 nm	400 - 800 nm	250 - 1100 nm	250 - 1100 nm

was recorded on Optical Disc Video Recorder simultaneously with each spectrogram.

3.2 Tunable Filter Setup

The first optical element in the light reflected off the beam splitter (see Figure 1) is a collimating lens which produces an image of the telescope's entrance pupil. The plate scale at the secondary image was adjusted to the characteristics of the detector by the proper choice of the reimaging lens. The scale at the prime focus of the telescope is 4.6 arcsec/mm while a value of 6.9 arcsec/mm was chosen for the Tunable Filter setup. The main components of the Tunable Filter setup, which is shown in Figure 2, are:

* a 1024 x 1024 pixel CCD camera, identical to the one used at the spectrograph (see Table 1). The chip operates at a temperature of 0 deg. C.
* the Tunable Filter with a bandwidth of about 80 mÅ at H_α
* a blocking filter wheel with 8 blocking filters (6 - 8 Å bands)
* a polarization analyzer (RCP,LCP, 4 orientations of linear polarization)
* a high speed active mirror, located at the first image of the entrance pupil providing the image stabilisation
* a quadrant cell sunspot tracker which generates the servo commands for the active mirror

The whole Tunable Filter (subsequently abbreviated TF) setup is controlled by an IBM PC/AT. A Vaxstation 3500 controls the CCD camera, collects the data , displays them and writes them to magnetic tape cassettes.

The cycle time of this system is limited by the 250 kbyte/second transfer rate of the Exabyte tape recorders used to store the data. Since a short cycle time is essential especially for the filtergrams used for the magnetograms, only a 512 x 512 pixel subarray of the CCD camera was read out. In some cases the full chip was used with reduced spatial resolution, by summing over 2x2 pixels, which again resulted in a 512 x 512 pixel data array. In the first case the pixel size corresponds to 0.13 arcseconds and in the latter one it is 0.26 arcseconds. The transfer rate of the Exabyte drive (250 kbyte/s), the readout rate of the camera (1 Mbyte/second) and a typical integration time of 0.5 seconds lead to a cycle time of about 2.5 seconds for the 512 x 512 pixel images.

3.3 Coordinated Operation

All the CCD cameras and the corresponding computers used for data aquisition and camera control were basically standalone systems. Only the two Vaxstation operated cameras had been used before in a joint program . Cameras 2 and 4 (cf. Table 1) had the capability of receiving external trigger signals for frame recording. This trigger signal was generated by one of the two Vaxstations and sent to both PC/AT computers and also to the Optical Disc Recorder, thus synchronizing the operation of all cameras operating at the spectrograph. Figure 1 shows the units and the various command lines connecting them. Since the cycle time at the spectrograph and at the Tunable Filter were significantly different, the TF system made 20 exposures while the spectrograph system took six (three in some cases). Therefore the synchronization between the spectrograph and TF systems was achieved via a *handshake* procedure between the computers after each cycle of 20 resp. 6 (3) exposures.

In order to perform raster scans at the solar image (by moving the image across the spectrograph slit), the timing of the whole system had to be made by the telescope control computer. In the scan mode, six computers were running the cameras, the TF and the telescope and all were operating on a common network.

4 Observations and Data Processing

The goal of the observings programs was to obtain simultaneous filtergrams and spectrograms, with the spectrograph field of view being centered in the filtergraph field of view. Slit jaw reference images were also recorded in order to determine the location of the spectrograph slit in the filtergrams by cross-correlating the reference images with continuum filtergrams.

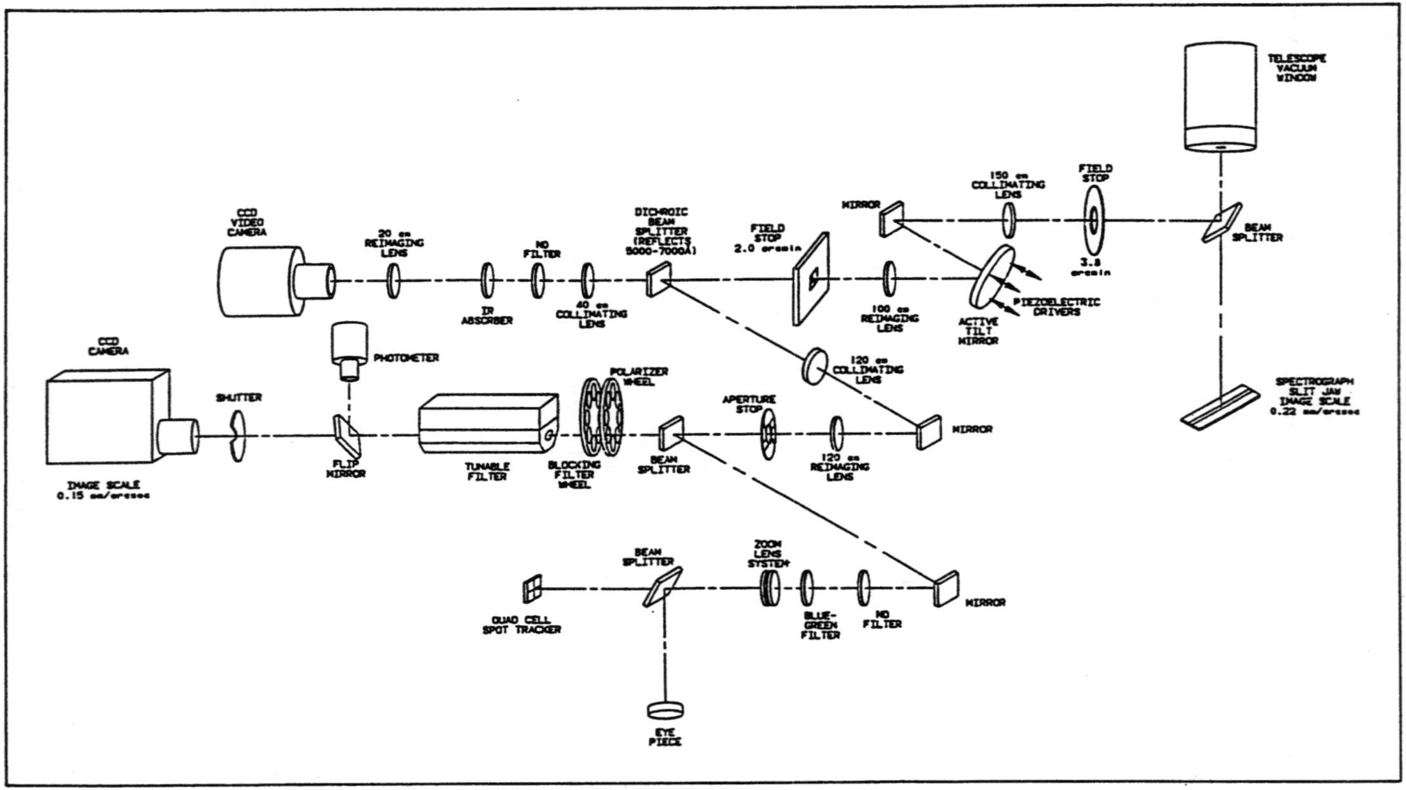

Fig. 2. Optical layout of the Tunable Filter section of the experiment

One of the science programs was the investigation of the dynamics of the penumbrae of sunspots. To this end, a sunspot penumbra was moved in small steps (1 arcsec) across the spectrograph slit and at each position three magnetically insensitive lines with different heights of formation were recorded (Fe I 5123 Å, FE I 5576 Å, Fe I 7090 Å). At the same time the TF collected the following data: four images for Dopplergrams, in Fe I 5576 Å at ±30 mÅ and ±90 mÅ from the line center, an image in clean continuum at 6303 Å and four images for magnetograms, two each in left and right circular polarized light (LCP/RCP) at ±60 mÅ from the center of the Fe I 6302.5 line and an image in H_α for reference. One cycle of ten images at the TF took about 30 seconds, while the cycle time of the spectrograph CCD cameras was about 20 seconds, dictated by the slowest of the three cameras used.

A second program concentrated on the small scale structure and the magnetic field of sunspots. The spectral lines used for this investigation were Fe I 6302.5 Å and Ti I 6303.7 Å (a low temperature line visible only in sunspots) for the spectrograph, while the corresponding TF sequence consisted of 20 filtergrams: taken at ±120, ±90, ±60, ±30 mÅ from the center of the 6303.7 Å line, both in right and left hand circular polarized light, continuum images at 6304.4 Å, near 5200 and 5800 Å, and an image in H_α for reference.

The third part of the campaign dealt with small magnetic structures in and outside of active regions. A Stokes-V polarimeter was mounted behind the entrance slit of the spectrograph, providing the Stokes I+V and I-V signals at the spectrograph focal plane. The V profiles of four neutral Iron lines were measured: 6302.5 Å, 6301.5 Å, 5250.22 Å and 5247.06 Å, which have Landé factors of 2.5, 1.5, 3.0 and 2.0 respectively. For this measurement three CCD cameras were needed at the spectrograph, because of the large linear dispersion (1 Å/cm) of the instrument. The TF took filtergrams in both circular polarizations in the wings of the 6302.5 Å line at ±60 mÅfrom line center , in unpolarized light at four positions in the wings of Fe I 5576 Å(±30 and ±90 mÅ), in clean continuum at 6302 Å and in H_α. For these observations, the targets were selected visually by means of the Ca K slit jaw TV image. For all observations 'white light' slit jaw reference pictures were recorded on an Optical Disc Recorder, synchronized with the spectrograms. Figure 3 shows a filtergram of a sunspot at disc center taken with the TF at a wavelength of 6303.2 Å(continuum). This is an example of a time series of images where a preliminary processing of the data has been carried out, while for the main part of the data image processing has not yet started. The necessary tools and procedures for the data processing have already been developed for the filtergrams and are presently being prepared for the spectrograms. Besides the ususal correction for noise and variable gain of the pixels of the CCD detector the main steps of data processing for the filtergrams are (i) spatial reregistration of time adjacent images and (ii) correction for image distortion (see Topka et al. [5]). These corrections eliminate residual image motion and seeing effects.

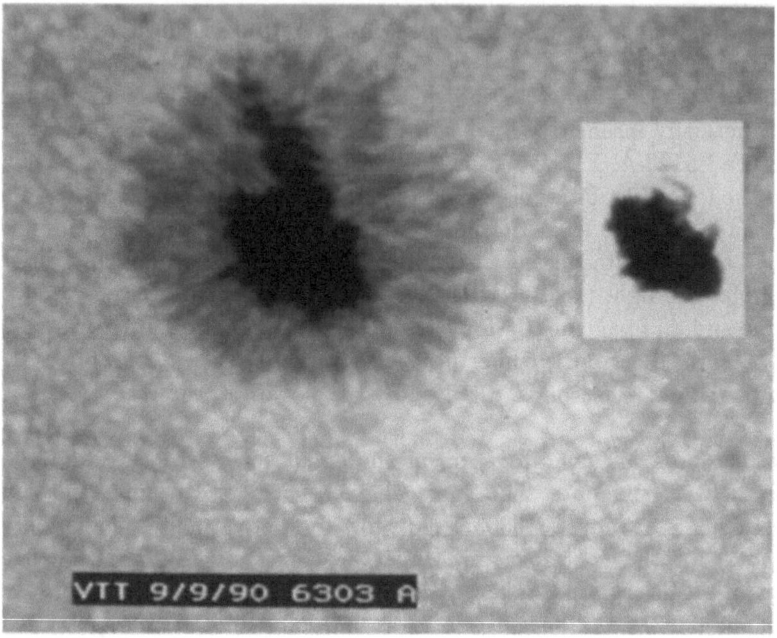

Fig. 3. Filtergram of a sunspot at disc center taken in clean continuum at a wavelength 6303.2 Å. The length of the inserted text bar is 40 arcseconds on the Sun. The inset shows the inner part of the sunspot umbra on a different intensity scale. The picture was taken on September 9, 1990, using a 1024 x 1024 pixel CCD camera, integration time was 0.3 s. (Photographic reproduction of the TV display of the frame, resulting in some degradation and a slight distortion.)

5 Concluding Remarks

The scientific data analysis has still to be made and its treatment is beyond the scope of this contribution. The observations described herein have demonstrated the capablities of the VTT and its focal plane instruments. The usage of a high speed active mirror in the TF part of the experiment has led to a significant improvement in image quality (especially during periods with strong wind) by removing most of the image motion. The horizontal laboratory has proven to be very well suited for high resolution observations.

It was a big effort to run a system with various cameras having different control computers and, due to different data formats, the analysis of the coordinated observations will be complicated and tedious. An integrated system for camera control and data aquisition is highly desireable.

An image stabilization system for the VTT using a correlation tracker is presently under construction [6] and will provide a stabilized image for the main beam which feeds the spectrograph **and** the horizontal laboratory. In contrast to the active mirror system described above, which needs a sunspot to generate the guiding signal, the correlation tracker uses the photospheric granulation for the image stabilization. This provides full flexibility for the observations, independent of the presence (and proximity to the region to be observed) of a sunspot. The correlation tracker system will be installed at the VTT in 1991.

Acknowledgements

The observing program was carried out at the Vacuum Tower Telescope at Tenerife from September 1 through September 26, 1990 jointly by the Lockheed Palo Alto Research Laboratory (LPARL) and the Kiepenheuer- Institut für Sonnenphysik (KIS). K. L. Smith (LPARL) and A.v. Alvensleben (KIS) prepared and managed the shipments of equipment to and from the observing site. G. Slater, T. Tarbell and K. Topka (LPARL), T. Rimmele and D. Soltau (KIS) formed the scientist team who (together with the author) carried out the observations. J. Staiger (KIS) spent various days (and nights) at the telescope computer in order to customize the software for this program. M. Levay and K. Wong (LPARL) made all the important modifications of the camera control and data aquisition software. R. Benedikt (Astron. Institut Würzburg), T. Cruz (LPARL), M. Knobloch (KIS) and D. Torgerson (LPARL) provided the on-site technical support. Parts of this program has been supported by NASA contracts NAS 8-32805 and NAS 5-26813 and by the BMFT, Förder-Nr. 010E86059.

References

1. Schröter, E.H., Soltau, D., Wiehr, E.: 1985 Vistas in Astronomy, Vol. 28, p. 519
2. Soltau, D, 1988, in **High Spatial Resolution Solar Observations**, Proc. 10. Sacramento Peak Summer Workshop, ed. O.v.d. Lühe
3. Soltau, D., 1989, in **Solar and Stellar Granulation**, eds. R.J. Rutten and G. Severino, Kluwer Academic Publishers, Dordrecht, Holland,
4. Title, A.M., Tarbell,T.D., Wolfson,C.J., 1989, in **Solar and stellar Granulation**, eds. R.J. Rutten and G. Severino, Kluwer Academic Publishers, Dordrecht, Holland, p. 25
5. Topka, K.P., Tarbell, T.D., Title, A. M.: 1986, Astrophys. J., 306,p. 304
6. Rimmele, T. and von der Lühe, O.: 1989, Rev. Mod. Astron., 2, 105

Aspects of the Present Heliospheric Research

H.J. Fahr

Institut für Astrophysik und Extraterrestrische Forschung, Universität Bonn, Auf dem Hügel 71, W-5300 Bonn 1, Fed. Rep. of Germany

Abstract

First of all the question is raised here why the sun at all is surrounded by a circumsolar heliosphere, instead by a Strömgren sphere configuration. The fact that counts most in this context is the evident peculiar motion of the sun relative to the ambient interstellar medium. An interesting puzzle thus is how the inner heliospheric plasma properties reflect this fact. Here we briefly will follow the lines by which the interstellar medium influences the heliospheric plasma.

Neutrals and ions are strongly coupled in interstellar/interplanetary space by mutual charge exchange reactions. Over distance scales of the order of charge exchange mean free paths the thermodynamic state of neutral and ionized media thereby is mutually communicated. In the following we study the penetration of neutral interstellar atoms through the perturbed plasma region ahead of the solar system into the inner regions of the heliosphere where their presence has been proven by their typical resonance glows. First we give the main outlines of the theory connected with this problem and then address in detail the question of the neutral hydrogen and oxygen presence in the heliosphere. The conclusion is reached that for interstellar ion densities of $10^{-2} cm^{-3}$ or greater neutral LISM hydrogen or oxygen will be depleted at least by factors of 0.5 or 0.1, respectively. The problem still stands whether or not the anomalous cosmic ray components showing up as He^+, O^+, Ne^+, N^+-ions of about 10 MeV/Nucleon in the inner heliosphere can be ascribed to LISM neutrals entering the heliosphere and becoming ionized and reenergized there.

We follow these claims for the origin of the anomalous cosmic rays and discuss in this context quantitative results connected with theoretical predictions for the LISM penetration into the heliosphere. We reach the conclusion that the ratios between the production rates for He^+, O^+, Ne^+, N^+-ions cannot be brought into accordance with those values needed to explain the anomalous cosmic ray observations, unless element-specific reenergization efficiencies are admitted.

1 Why is there a heliosphere at all?

In principle, each star with an effective radiation in the UV part of the spectrum (i.e at around $\lambda \leq 912$ Å) should generate its own Strömgren sphere which defines a specific circumstellar region where hydrogen ionization rates dominate over corresponding recombination rates. This famous old concept of Strömgren (1939) was applied to the sun as a UV-radiating star by Williams (1965) and Brandt (1964). While Williams, considering the sun as being at rest with respect to the ambient interstellar medium, obtains a Strömgren sphere radius of a few $10^4 AU$, Brandt (1964), admitting a diffusive velocity of the interstellar medium of $1km/s$ relative to the sun, only obtains 10^3 AU. This drastically shows the velocity-sensitivity of such estimates for a circumsolar plasmasphere (for a review see e.g. Fahr, 1980).

In fact, however, the heliosphere with its observationally well confirmed evidence is something conceptually very different. It is considered to be a circumsolar region of space which is dominated by plasma of solar origin clearly separated by its boundary layer, the heliopause, from the surrounding space dominated by interstellar plasma. This heliosphere only seems to have characteristic dimensions of the order of $10^2 AU$ and is determined in its configuration by the flow of the ambient interstellar MHD plasma relative to the solar system.

Since the existence of the heliosphere is physically tightly connected with the relative motion of the solar system with respect to the ambient interstellar medium, the question can be raised how far inwards towards the solar corona do the properties of the interplanetary plasma reflect the conditions of the Local InterStellar Medium (LISM). It seems evident that a solar wind evaporating from a nonrotating, spherically symmetric solar corona without magnetic influences prefers a radially symmetric outflow geometry with a spherical sonic surface. The asymptotic wind according to Parker's hydrodynamical description (Parker, 1958, 1963) thus will represent a supersonic, radially symmetric plasma flow.

When such a wind system is at rest with respect to the LISM there exists a homogeneous scalar LISM pressure distribution at large solar distances to which the solar wind eventually has to adapt. This situation then naturally leads to the formation of an outer spherical heliospheric shock front Γ where the wind becomes subsonic. In numerical simulations for this case of a hydrodynamical interaction of the solar wind system with the LISM at rest Matsuda et al. (1989) could, however, show that both the heliospheric

shock front and the heliopause are unstable and no steady solution for the plasma outflow can be reached. Nevertheless, it is clear that, averaged over an appropriate time-period, no net momentum is transported out of the shock surface Γ, i.e. $(-\boldsymbol{K}_s) = \int_\Gamma \rho\, \boldsymbol{v}\, (\, \boldsymbol{v} \cdot d\boldsymbol{O}\,) = \boldsymbol{0}$, and thus no net force is acting back on the wind-producing corona.
It is now interesting to ask whether or not this situation changes when the sun is in motion relative to the LISM and when the asymptotic pressure to which the solar wind eventually has to adapt its flow neither is homogeneous nor scalar. Would the solar wind system adapt to this situation such that then a net force $\boldsymbol{K}_s \neq \boldsymbol{0}$ may be acting ? If the latter would not be the case it had to be concluded that a wind-driving star could migrate through the thickest LISM walls without being decelerated. If, on the other hand, the physics involved in this context should arrange a deceleration of the wind-driving star then it has to be asked how the information on the outer LISM flow is propagated downwind towards the corona through the supersonic solar wind plasma flow. In our view there are three different mechanisms which need to be discussed in this respect showing that the vacuum solar wind solutions could be accordingly changed by the counterflow of interstellar medium such that a deceleration of the solar system would result from it:
a) The solar wind has to be treated in its process of adaptation to the outside LISM pressure conditions which requires a set of differential equations including time-derivatives. This system can only be integrated in time towards the asymptotically stationary case (if there exists any at all?) when solved as a boundary value problem with thermodynamic conditions also determined on some outer boundary surface.
b) The solar wind within which species of interstellar origin are diffusing, like pick-up ions, anomalous cosmic ray particles and galactic cosmic ray particles, does not have a clearly defined sound speed. This especially means that information from the interstellar boundary can be propagated downwind even down to regions inside the conventional critical point where the proton sound velocity is reached (Fahr et al., 1990).
c) The outflow geometry of the solar wind is likely to be affected by the LISM pressure distribution and by the dynamical action of the LISM species diffusing through the solar wind plasma (Fahr and Fichtner, 1990).

In the ongoing text we shall not discuss these points as such, rather we shall broaden up the aspect here towards the problem of how the interaction of the LISM with the solar system takes place in detail. The following paper is partly based on Fahr (1989).

2 Modulation of the neutral LISM flow

The problem of the penetration of neutral atoms from the **Local InterStellar Medium (LISM)** into the heliosphere has been treated in a series of papers in the recent past (Ripken and Fahr, 1983, Wallis, 1984; Fahr and Ripken, 1984; Błeszinski, 1987; Fahr, 1986, 1990). The main result of these papers was that the neutral LISM components when passing over the perturbed plasma interface regions ahead of the solar system will experience specific changes of their velocity distribution functions by means of charge exchange processes and electron impact ionizations. As a consequence it is expected by all above mentioned authors that in the inner heliosphere all moments of the distribution functions of these LISM neutrals, like the particle density, the bulk velocity and the temperature, will deviate specifically from their unperturbed LISM values, however different in magnitude these deviations may have been judged by the different authors. Here we only would like to present the main theoretical lines along which such results can be gained. We follow the kinetic approach by means of the Boltzmann equation used by Ripken and Fahr (1983) and Fahr (1986, 1990):

In the plasma interface region ahead of the inner "supersonic" heliosphere plasmas of solar and interstellar origin are present with organized MHD-determined plasma flows. At a place $\boldsymbol{r}$ the distribution function $f(\boldsymbol{r},\boldsymbol{v})$ of any neutral species in the stationary case is given by the following form of Boltzmann's integro-differential equation:

$$\partial f(\boldsymbol{r},\boldsymbol{v})/\partial s \;=\; (1/v)\,[P(\boldsymbol{r},\boldsymbol{v}) \;-\; L(\boldsymbol{r},\boldsymbol{v})] \qquad (1)$$

where s is the line element on the trajectory of particles reaching $\boldsymbol{r}$ with a velocity $\boldsymbol{v}$, and where the functions P and L in Eq.(1) describe the rates by which particles at $\boldsymbol{r}$ with a velocity $\boldsymbol{v}$ are produced and removed, respectively, by charge exchange processes with the interface plasma. Whereas the function P usually contains a velocity-space-integral over the distribution function $f(\boldsymbol{r},\boldsymbol{v})$ itself, the function L in Eq.(1) allows a representation with f only occurring as a factor in the form:

$$L(\boldsymbol{r},\boldsymbol{v}) \;=\; f(\boldsymbol{r},\boldsymbol{v}) \;\cdot\; \nu^{-}(\boldsymbol{r},\boldsymbol{v}) \qquad (2)$$

with ν^{-} being the total destruction frequency for particles in question. Consideration of the physical properties in the interface plasma (electron density $n_e \;\cong\; 10^{-2}cm^{-3}$, temperature $T_e \;\cong\; T_p \;\cong\; 10^4 K$) reveals that the relevant productive or destructive processes there are due to resonant charge exchanges with the LISM hydrogen ions (i.e. $\nu^{-} = \nu_{\mathrm{ex}}^{-}$). Counting the particles appearing at $\boldsymbol{r}$ according to consecutive hierarchies i, and introducing f_i as the distribution function for particles which resonantly exchanged their charge i-times before appearing at $\boldsymbol{r}$ (with the evident demand that

$f = \sum f_i$), one can convert the integro-differential equation (1) into a hierarchy of differential equations of the form:

$$\partial f_i/\partial s = (1/v)[P_{i-1} - f_i \cdot \bar{\nu_{\text{ex}}}] \qquad (3)$$

where the function P_{i-1} now is defined by:

$$P_{i-1}(\boldsymbol{r},\boldsymbol{v}) = f_p(\boldsymbol{r},\boldsymbol{v}) \int^3 \sigma_{\text{ex}}(v_{\text{rel}}) v_{\text{rel}}(\boldsymbol{r},\boldsymbol{v},\boldsymbol{v}') f_{i-1}(\boldsymbol{r},\boldsymbol{v}') d^3v' \qquad (4)$$

and thus is independent on f_i. The solution of the inhomogeneous differential equation (3) therefore can be given in the form:

$$f_i(\boldsymbol{r},\boldsymbol{v}) = \qquad (5)$$

$$f_{i,\infty}(\boldsymbol{v}_\infty) \exp(-\int_\infty^s \nu'_{\text{ex}} ds'/v') + \int_\infty^s P'_{i-1} \exp(-\int_{s'}^s \nu''_{\text{ex}} ds''/v'') ds'.$$

Here σ_{ex} and v_{rel} denote the charge exchange cross section and the relative velocity between neutrals and protons , and $f_{i,\infty}$ is the unperturbed LISM distribution only existing for i = 0, (i.e. $f_{i,\infty} = 0$ for $i > 0$).
Using now as a basis for the description of the interface plasma Parker's subsonic plasma-plasma interaction model (Parker, 1963, pp. 115-120) one is advised to represent the problem in cylindrical coordinates z, ρ, ϕ with the z-axis oriented parallel to the LISM wind *velocity* (see Fig.1 for illustrative purposes). The proton distribution function $f_p(\boldsymbol{r},\boldsymbol{v})$ can be represented in a reasonably well approximation as a Maxwellian shifted by the local bulk flow $\boldsymbol{V} = \left(V_z, V_\rho, 0\right)$ of the protons with the flow vector $\boldsymbol{V}$ given by Parker (1963) in the following form:

$$V_z = V_\infty(1 + \bar{z}/\bar{r}^3), \qquad V_\rho = V_\infty(\bar{\rho}/\bar{r}^3). \qquad (6)$$

Here the bars on top of the variables indicate a normalization with the heliopause stand-off distance L(e.g. see Fahr, 1986).
Thus with Eq.(6) the function $f_p(\boldsymbol{r},\boldsymbol{v})$ would attain the form:

$$f_p(\boldsymbol{r},\boldsymbol{v}) = (m/2\pi K T_p)^{3/2} \exp(-m/2KT_p[(v_z - V_z)^2 + (v_\rho - V_\rho)^2 + v_\phi^2]). \qquad (7)$$

Here $T_p = T_p(r)$ is the local proton temperature which in Parker's interface is determined by Bernoulli's law through (Fahr and Neutsch, 1983):

$$T_p(\boldsymbol{r}) = m/K[1/2 \; V_\infty^2 + KT_\infty/m - 1/2 \; (V_z^2 + V_\rho^2)]. \qquad (8)$$

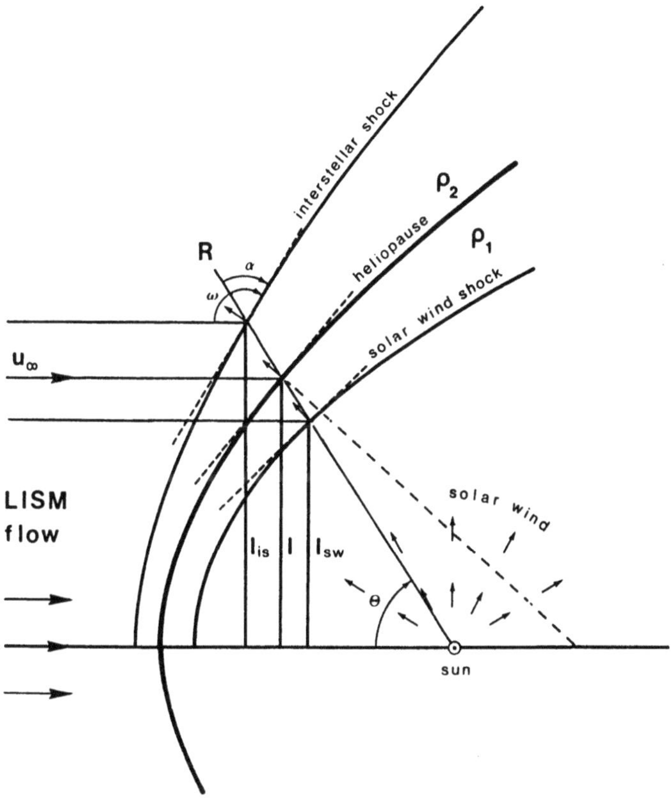

Fig. 1. Illustration of the plasma flow and denotation of variables

We now consequently approximate the production term P_{i-1} in the following form:

$$P_{i-1}(\boldsymbol{r}, \boldsymbol{v}) \;\cong\; \sigma_{\mathrm{ex}}(\bar{v}_{\mathrm{rel}})\bar{v}_{\mathrm{rel}} n_p(\boldsymbol{r}) n_{i-1}(\boldsymbol{r}) f_p(\boldsymbol{r}, \boldsymbol{v}), \qquad (9)$$

where n_p and n_{i-1} are the local densities of protons and of neutral particles of hierarchy (i-1), respectively, and where $\bar{v}_{\mathrm{rel}}(\boldsymbol{r}, \boldsymbol{v})$ is the average relative velocity of all particles with respect to protons of the velocity $\boldsymbol{v}$. In a satisfactory approximation this velocity is given by:

$$\bar{v}_{\mathrm{rel}}(\boldsymbol{r}, \boldsymbol{v}) \;=\; |(\boldsymbol{v}_p - \boldsymbol{v})| \;\cong\; [(v_z - V_z)^2 + (v_p - V_p)^2 + v_\phi^2]^{1/2} \,. \qquad (10)$$

With Eqs. (5) through (10) one is prepared to calculate the distribution $f(\boldsymbol{r}, \boldsymbol{v})$ by consecutively proceeding from lower to higher order indices i, thereby at each step i taking advantage of the predetermined distribution function f_{i-1}.

3 Neutral LISM distribution in the inner heliosphere

Due to their large resonant charge exchange cross sections with respect to protons, H- and O-atoms, amongst the neutral LISM species, are especially subject to pronounced charge change interactions with the interface protons. For hydrogen the production terms P_i given in Eq.(4) are very important. As a consequence of charge exchange reactions of O- or H-atoms with interface protons neutralized protons, i.e. new H-atoms, reappear as reaction products. To the contrast for oxygen atoms the relevance of the production term P_i is completely different. Oxygen atoms are only produced from charge exchange reactions with O^+-ions. Due to their very low cosmic abundance relative to hydrogen ($\alpha_{O/H} = 6.7 \cdot 10^{-4}$) these ions are very rare, and due to their large masses ($m_O = 16 m_H$) their thermal velocities compared to the flow velocity $\boldsymbol{V}$ in the interface plasma are fairly small. Therefore the velocity vector bringing O-atoms into the inner part of the heliosphere only exists in the plasma distribution function in a very limited region around the stagnation line.
Thus a reasonably good number for the oxygen transmissivity in the interface plasma can be obtained neglecting O-production terms at all there. On the basis of this assumption in Fig.2 as function of the LISM proton density $n_{p\infty}$ comparative results for the hydrogen and oxygen transmissivities Tr_{HP} in the interface plasma ahead of the heliopause are shown. With the most probable values for the LISM proton density of $n_{p\infty} \cong 10^{-2} cm^{-3}$(Cox and Reynolds, 1987; Reynolds, 1989) from this Fig.2 one draws the conclusion that about 50 percent of the initial LISM hydrogen and only 10 percent of

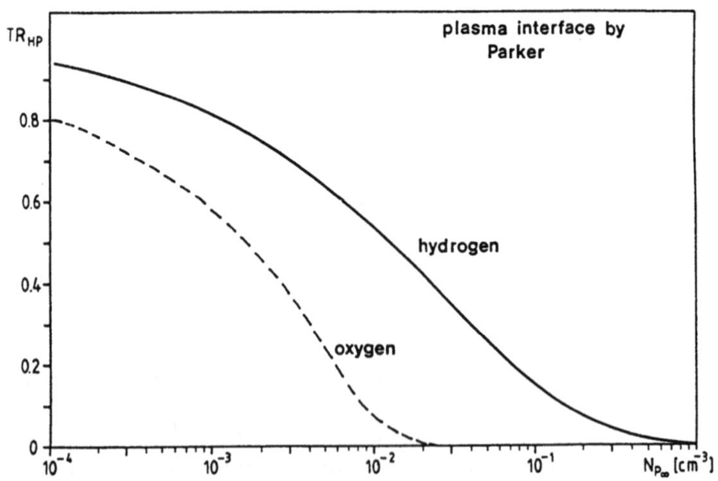

Fig. 2. Shown as function of the LISM proton density $N_{p,\infty}$ is the transmission probability of LISM H- and O- atoms on their way from the "local infinity" to the heliopause ($L = L(N_{p,\infty})$ according to Parker (1963)).

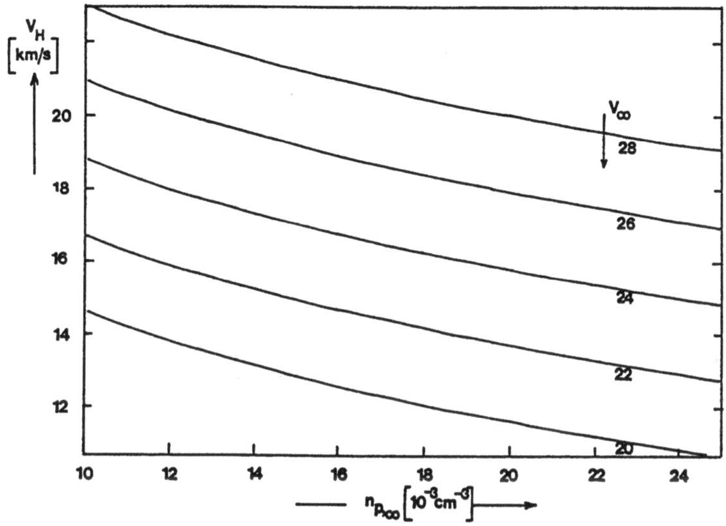

Fig. 3. Shown as function of the LISM proton density $n_{p,\infty}$ are the reduced bulk velocity of LISM hydrogen atoms with an initial bulk velocity V_∞ after penetration of the interface plasma ahead of the heliopause.

the initial LISM oxygen flow after penetration through the interface plasma shall reach the heliopause and then enter the heliosphere.

Also as a function of $n_{p\infty}$ we have given in Fig.3 the reduced velocity of the bulk flow of hydrogen atoms arriving at the heliopause and having experienced an imprint from the stagnating LISM plasma flow by means of a charge exchange coupling.
A very interesting, since analytic solution for the density distribution of oxygen in the heliosphere can be obtained due to some straightforward simplifications that are possible for this species according to what was discussed above (see Fahr, 1990). Realizing that oxygen atoms with a mean thermal velocity of only $3km/s$ compared to the bulk speed of $V_\infty = 23\ km/s$ can reasonably well be approximated by a flow of monoenergetic particles with $\boldsymbol{v}_\infty = \boldsymbol{v} = \left(V_\infty, 0, 0\right)$, and neglecting production processes, i.e. $P_i = 0$, Eq.(5) shows that the only contribution to f_O which is different from 0 is for i = 0, i.e. $f_O = f_{O,0}$. This function f_O can then also be integrated to yield the oxygen density n_O which is given by:

$$n_O(z,\rho) = n_{O,\infty} \exp[-n_{p\infty} L \int_{-\infty}^{z} \sigma_{\mathrm{ex}}(\bar{V}_\infty/\bar{r}^2)(1/\bar{r}^2)d\bar{z}] \tag{11}$$

Taking an average value $\bar{\sigma}_{\mathrm{ex}}$ which is characteristic for the interface one then even can explicitly carry out the integration in Eq.(11) and finally comes to:

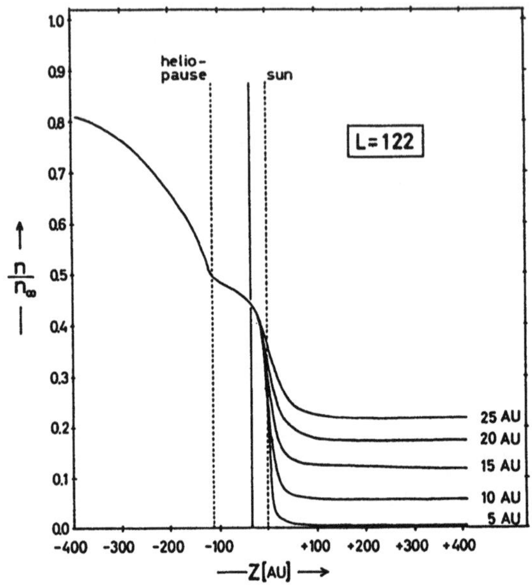

Fig. 4. Shown are cuts parallel to the stagnation line through the LISM oxygen density profile for various values of the off-axis distance r.

$$n_O(z,\rho) \;=\; n_{O,\infty}\exp[-\frac{n_{p\infty}L\bar{\sigma}_{\text{ex}}}{\rho}(\;\pi/2 + \;\text{arctg}\;(z/\rho)\;)]. \qquad (12)$$

In Fig.4 we have presented as function of the coaxial coordinate z cuts through the oxygen density for different off-axis distances ρ calculated with the expression (12).
The stand-off distance $L \;=\; 122\;AU$ used for the calculations presented in Fig.4 results from the definition given in Fahr (1986) if the following LISM properties are adopted:

$n_{p\infty} \;=\; 1.35{\cdot}10^{-2}cm^{-3}$, $T_\infty \;=\; 1{\cdot}10^4K$, $V_\infty \;=\; 23\,km/s$, $B_\infty \;=\; 5{\cdot}10^{-6}$ Gauss , $n_s \;=\; 8\cdot10^{-4}cm^{-3}$, $V_s \;=400Km/s$,
where the quantities with suffix s are proton density and velocity of the solar wind at $R_{s1} \;=\; 54\;AU$.

4 Pick-up ions in the heliosphere and their role for the anomalous cosmic ray component

It is now generally taken for granted that the so-called "anomalous" cosmic ray component which essentially represents He, O, Ne, and N particles at medium energies of around 10 MeV/nucleon is due to interstellar neutral atoms moving into the inner heliosphere from the upwind side of the solar

system. There they become ionized via photo-ionization, charge exchange, or electron/proton impact ionization, and thereafter are convected outwards with the magnetic fields frozen into the solar wind flow. As suprathermal ions they are then subject to electromagnetic acceleration processes probably due to transit-time damping of magnetosonic interplanetary field fluctuations (Fisk et al., 1974; Fisk, 1976*a*/*b*/*c*). Thereby these particles on their way to the outer fringes of the heliosphere are systematically shuffled to higher and higher energies and, following Fisk (1976*b*) and Klecker (1977), at heliocentric distances of about $50AU$ may have built up a distribution function peaking at energies of about 10 MeV/nucleon.
As these particles originate from ionizations of LISM neutrals in the heliosphere they are singly ionized (i.e. $Z = 1$) and thus have large rigidities $\rho = B\, r_g = pc/Ze$ as compared to cosmic ray particles ($Z \gg 1$) of the same momentum p (or energy E). Therefore these particles with energies $E \cong 10$ MeV are considered to be able to effectively diffuse backwards towards the inner solar system and then to show up there as "anomalous" cosmic ray particles (e.g. see Fisk, 1986, or Jokipii, 1986).

The different processes relevant within this scenario, i.e. energization by nonlinear wave-particle interaction and/or transit-time damping mechanisms, convection, adiabatic cooling, if simultaneously taken into account as diffusive motions in phase space, lead to a very complicated partial differential equation of the Fokker-Planck-type for the particle distribution function $f_j(r,p,t)$ of species j. However, the theoretical representation of these processes seem to allow for the following split-up into temporal sequences (Fisk, 1976*b*, Klecker, 1977):

a) After pick-up the original interstellar ions are found at energies of $\simeq 1$ KeV/ nucleon and at such conditions diffusion processes essentially can be neglected. Thus on their way through the inner heliosphere these interstellar ions only need to be treated as being subject to pitch-angle scattering, adiabatic cooling and energization . The corresponding change in the distribution $f_i(r,p)$ with heliocentric distance, as calculated from a simplified Fokker Planck equation (Fisk, 1976*a*), was explicitly obtained by Fisk (1976*b*) and by Klecker (1977), and also by Isenberg (1987), however, with some assumptions concerning the importance of pitch-angle scattering deviating from the first mentioned *au*thors.

b) At the outer ranges of the heliosphere (i.e. at $\simeq 50AU$) the energized particles due to their high energies and diffusion coefficients develop non-negligible diffusive flows that require a different differential equation allowing for an adequate description. This is consequently now a Fokker Planck equation without acceleration terms. In principle this latter equation treats a process similar to that of the modulation of galactic cosmic ray particles diffusing downstream the solar wind flow.

Following the method of Gleeson and Urch (1971), Klecker (1977) has numerically integrated this modulation equation starting at $R_\infty = 50\ AU$ with the distribution function for anomalous cosmic ray particles derived by Fisk (1976*a*). By doing so, he was able to achieve a satisfactory fit to anomalous cosmic ray data obtained at $1AU$ by Garcia-Munoz et al. (1973, 1975) and Klecker et al. (1977) only when he started with relative abundances of 15.0 , 0.15 , and 0.05 for He^+, N^+, and Ne^+ relative to O^+ ions at $50AU$ and in addition assumed that a fraction of only $\epsilon = 2 \cdot 10^{-3}$ of each pick-up ion species present in the outer heliosphere is energized up to 10 MeV energies.

5 Brief study of a flux balance

The question to be put here in view of the above result is whether or not this picture for the anomalous cosmic ray particles (ACR) can essentially be supported in view of the responsible source and sink terms. In the framework of a stationary situation the rate by which fresh ACR are gained in the heliosphere should just balance the corresponding loss rate, mainly caused by diffusive and convective losses of ACR particles out of the outer heliosphere.

A) The ACR production process

First the gain rate for the ACR particles may be considered here. This rate is obviously due to the penetration of LISM neutrals into the heliosphere connected with a subsequent ionization there. On the basis of typical solar ionization rates, one may calculate typical interception cross sections F_{ij} for the different elements j and determine the flow of species j through it. These cross sections are easily found with that specific solar impact parameter P_{ij} of a Keplerian trajectory an interstellar neutral species j has to have such that it marginally suffers ionization at its traversal through the heliosphere. This critical impact parameter P_i can be calculated with the interplanetary extinction function $\mathrm{Ex}(\boldsymbol{r}, \boldsymbol{v})$ as given e.g. in Fahr et al. (1987) which for a time-independent loss rate and a fall-off according to $1/r^2$ attains the following form:

$$\mathrm{Ex}(\boldsymbol{r}, \boldsymbol{v}) = \exp[-\frac{\nu_E r_E^2 \theta}{P_\infty v_\infty}] \tag{13}$$

and describes the probability of a neutral LISM atom entering the heliosphere with a velocity $\boldsymbol{v}_\infty$ and an impact parameter $P_\infty(\boldsymbol{r}, \boldsymbol{v}_\infty)$ to reach the space point $\boldsymbol{r}$ with a velocity $\boldsymbol{v}$ without having been ionized before. Here ν_E is the total loss rate at $r_E = 1\ AU$ and θ is the total angle swept over by the particle from the direction $-\boldsymbol{v}_\infty$. The critical impact parameter P_i then may be defined such that with it a particle passing over the whole heliosphere may just be ionized with a probability of exp(−1). Thus from Eq.(13) one can derive the condition:

$$1 = 2\nu_E r_E^2 \theta_{\mathrm{per}} / P_i v_\infty \tag{14}$$

where θ_{per} is the angle swept over by the particle up to the perihelion of its orbit. From elementary Keplerian relations this angle is given by an implicit relation:

$$\theta_{\mathrm{per}} = \operatorname{arctg}(\, -v_\infty^2 P_i / G M_o) = \pi - \operatorname{arctg}(\, v_\infty^2 P_i / G M_o) \tag{15}$$

where G and M_o are the gravitational constant and the effective solar mass. With equations (14) and (15) one obtains:

$$\bar{P}_i = 2\nu_E r_E [\, \pi - \operatorname{arctg}\,(v_\infty^2 r_E \bar{P}_i / G M_o)\,] / v_\infty \tag{16}$$

where $\bar{P}_i$ is the critical impact parameter in r_E-units. From Eq.(16) the quantity P_{ij} for the different relevant LISM atoms j can be determined if the corresponding ionization frequencies ν_{Ej} are known. Using available data then accordingly the following results can be obtained (Fahr, 1989):

Table 1. Critical impact parameters in *AU*

Element	Solar maximum	Solar minimum
He	3.11	1.67
O	21.27	13.10
Ne	6.88	3.55
N	14.10	8.88
C	115.52	55.00

To determine the total production per unit of time of suprathermal ions j available in the heliosphere by a stationary conversion into anomalous cosmic ray particles of this species, one simply has to calculate the number of particles collected per unit of time by the solar ionization target $F_{ij} = \pi\, P_{ij}^2$. These numbers are given by:

$$\Pi_j = n_\infty V_\infty \alpha_j (1 - \xi_j) \tau_j F_{ij} \tag{17}$$

where V_∞ is the LISM wind velocity relative to the sun and n_∞ is the total LISM density (neutral and ionized component). The quantity α_j is

the relative cosmic abundance of the element j with respect to hydrogen, ξ_j is the ionization degree of this element, and τ_j denotes the probability by which neutrals of the species j are removed at their traversal through the heliospheric plasma interface (Ripken and Fahr, 1983; Fahr and Ripken, 1984; Wallis, 1984; Błeszynski, 1987).
What concerns the values for α_j we adopt the values given by Cameron (1973) for the solar system abundances which we take to be representative also for the LISM. Then we arrive at the following picture:

Table 2.

Element (j)	Cosmic abund. (α_j)	Ionization potential (eV)	Fractional ionization $(1-\xi_j)$
H	1.0	13.6	0.54
He	0.07	24.6	1.00
N	$1.18 \cdot 10^{-4}$	21.6	0.75
O	$6.76 \cdot 10^{-4}$	13.6	0.68
Ne	$1.08 \cdot 10^{-4}$	14.5	0.71
C	$3.71 \cdot 10^{-4}$	11.3	0.35

As is obvious from recent papers by Frisch et al. (1987, 1989) or Reynolds (1989) there exists a very high degree of uncertainty about the state of our local interstellar medium, especially what concerns the electron density and the ionization state. Thus, in view of all these uncertainties, in the above Table 2 we have taken, as an idea for the values of $(1-\xi_j)$, the Saha-Eggert equilibrium ionization fractions ξ_j for an interstellar environment characterized by $n_e = 10^{-2} cm^{-3}$ and $T_e = T_o = 10^4$ K.

B) The ACR loss processes

The determination of the ACR losses from the heliosphere is not a very straightforward problem. A deep theoretical knowledge on the physics of the ACR modulation, especially in the outer heliospheric regions close to the heliospheric shock, would be necessary for this purpose. Despite the present uncertainties in the theoretical description of ACR modulation one may feel encouraged to derive a fairly reliable number for the ACR losses based on present knowledge in this field starting from the ACR particle stream $S(R,p)$ describing the flux of ACR particles at R with momentum p in a fixed reference system like that of the sun. As explained in many papers (see e.g. Jokipii, 1971) this flux is given by:

$$S(R,p) = U(R,p)VC(R,p) - \kappa(R,p)\partial U/\partial R \qquad (18)$$

where $U(R,p)$ is the pitch angle averaged differential particle density at a solar distance R with momentum p. The solar wind velocity is denoted by V. The function $C(R,p)$ denotes the Compton-Getting function explained in detail e.g. by Gleeson and Axford (1968*a*, *b*), and $\kappa(R,p)$ is the ACR space-diffusion coefficient.

In order to determine losses one needs an expression for the total flux of ACR particles out of the heliospheric shock surface. Thus one has to integrate Eq.(18) over the momentum space. For this integrated flux a conservation law may be expected to be fulfilled at the heliospheric shock region and hence one may use the following equivalence:

$$\int p^2 dp[U(R,p)VC(R,p) - \kappa(R,p)\partial U/\partial R]_1 \cong \qquad (19)$$

$$\int p^2 dp[U(R,p)VC(R,p) - \kappa(R,p)\partial U/\partial R]_2 = \Phi_{\text{out}}$$

where the suffixes 1 and 2 at the brackets mean preshock and postshock conditions. Φ_{out} is the total ACR flux per unit area leaving the heliospheric shock front.

It is now interesting to note that the quantity given in Eq.(19), at least under radially symmetric conditions, should be related to the total ACR production rate Π_{ij} given in Eq.(17) by:

$$\Phi_{\text{out},j} = \int p^2 dp\,[\,U_j V \quad C_j \quad - \quad \kappa_j \partial U_j/\partial R\,]_2 = \frac{\Pi_{ij}}{4\pi R_0^2}\,, \qquad (20)$$

where the heliospheric shock has been located at R_o .

Following Gleeson and Axford (1968*a*, *b*) the Compton-Getting function can be given in the form:

$$C(R,p) = 1 - \frac{1}{3}\frac{1}{U}\partial/\partial T(\beta T U)\,, \qquad (21)$$

where T is the kinetic energy of the ACR particles. The function β is then written as:

$$\beta(T) = (T + 2m_o c^2)/(T + m_o c^2) \qquad (22)$$

which for weakly relativistic particles (like ACR) simplifies to $\beta \cong 2$. In this subrelativistic case ($T \cong p^2/2m_o$) Eq.(21) simplifies to:

$$C(R,p) = \frac{1}{3}\,[\,1 - \frac{p}{U}\partial U/\partial p\,]_R\,. \qquad (23)$$

To calculate Φ_{out} one needs the distribution $U(R,p)$. Here use is made of the function derived by Fisk (1976*b*). At some distance R_c close to R_o this pick-up ion distribution function $U_c(R,p)$ is developed such that space diffusion starts to dominate over acceleration processes, causing both a diffusion into regions $R \geq R_c$ and the usual inward diffusion at $R_c \leq$ R. Therefore the following relation may approximately be fulfilled for the momentum-averaged quantities (denoted by brackets):

$$< | - \kappa(R,p)\partial U/\partial R| >_{R_c+\Delta R} = \tag{24}$$
$$< | + \kappa(R,p)\partial U/\partial R| >_{R_c-\Delta R} \cong \frac{1}{2} < S_c(R_c,p) >$$

With $U_c(R,p)$ given by Fisk (1976a, b):

$$U_j(R,p) = C_j n_j R^4 \exp[\, -VB/3R(p^2/D_{pp})] \tag{25}$$

and the relation (24) we can now calculate the flux $\Phi_{\rm out}$ in Eq.(21) and, keeping in mind that the coefficient C_j in Eq.(25) contains the atomic mass number A_j in the form: $C_j = C/A_j^2$ one then obtains (see Fahr, 1989):

$$\Phi_{{\rm out},j} = \int S_{cj}(R_c,p)p^2dp = \Phi_1 + \Phi_2 + \Phi_3 + \Phi_4 \tag{26}$$

with the following definitions:

$$\Phi_1 = (10^7/1.5A_j)4p_r^3 \int_0^\infty \chi^4 U_{cj}(\chi)d\chi \tag{27}$$

$$\Phi_2 = (10^7/1.5A_j)\frac{Z}{A_j}p_r^3 \int_0^\infty \chi^{3.5} U_{cj}(\chi)d\chi \tag{28}$$

$$\Phi_3 = (4 \cdot 10^7/3)p_r^3 \int_0^\infty \chi^2 U_{cj}(\chi)d\chi \tag{29}$$

$$\Phi_4 = (4 \cdot 10^7/3)\frac{1.5Z}{A_j}p_r^3 \int_0^\infty \chi^{1.5} U_{cj}(\chi)d\chi \tag{30}$$

where here distribution function U_{cj} of species j according to Eq.(25) is given by:

$$U_{cj}(\chi) = Cn_j\frac{1}{A_j^2}\exp(\, -\frac{Z}{A_j}\chi^{1.5}\,). \tag{31}$$

Introduction of $\xi = (Z/A_j)\ \chi^{1.5}$ and the Gamma function defined as $\Gamma(x) = \int_0^\infty e^{-\xi}\xi^{x-1}d\xi$ then yields the following expression for $\Phi_{{\rm out},j}$:

$$\Phi_{{\rm out},j} = Cn_j(p_r^3/A_j^2)\cdot 10^7 \cdot [(8/4.5)A_j^4/Z^5\Gamma(10/3)+ \tag{32}$$
$$(2/1.5)A_j^{4.5}/Z^{5.5}\Gamma(13/3) + (8/9)A_j^3/Z^3\Gamma(2) + (8/6)A_j^{3.5}/Z^{3.5}\Gamma(3)]$$

which after numerical evaluation of the Gamma functions gives:

$$\Phi_{{\rm out},j} = \tag{33}$$
$$Cn_j(p_r^3/A_j^2)\cdot 10^7\,[0.5\ A_j^4\ +\ 1.0\ A_j^{4.5}\ +\ 0.23\ A_j^3\ +\ 0.54\ A_j^{3.5}\,]$$

With Eqs.(33) and (20) one finds the ACR particle density of species j with:

$$n_j = \frac{\Pi_{ij}}{4\pi R_c^2} \frac{(0.5A_j^4 + 1.0A_j^{4.5} + 0.23A_j^3 + 0.54A_j^{3.5})^{-1}}{Cp_r^3 10^7} A_j^2 \qquad (34)$$

The observationally more relevant ratios of ACR particle densities n_j and n_k consequently are represented by the following expression:

$$(n_j/n_k) = (\Pi_{ij}/\Pi_{ik})A_{jk}^* \qquad (35)$$

where the A_{jk}^* is just a constant factor given in the form:

$$\underline{A_{jk}^*} = \frac{(A_j/A_k)^2(0.5A_k^4 + 1.0A_k^{4.5} + 0.23A_k^3 + 0.54A_k^{3.5})}{(0.5A_j^4 + 1.0A_j^{4.5} + 0.23A_j^3 + 0.54A_j^{3.5})} \qquad (36)$$

Using now the values given in Tables 1 and 2, and adopting no filtering at the heliospheric interface plasma (i.e. $Tr_j \equiv 1$), with Eqs.(35) and (36) one arrives at the following ACR density ratios:

Table 3. ACR density ratios at 50 *AU*

Ratio	Solar maximum	Solar minimum	Klecker (1977)
He/O	85.51(58.15)	65.0 (44.19)	15.
Ne/O	0.01(0.0068)	0.007(0.0047)	0.05
N/O	0.11(0.1)	0.12 (0.102)	0.15

Here the values given in brackets are valid for an unionized interstellar medium (i.e. $(1-\zeta_j) \equiv 1$), whereas the ones without brackets are calculated for LTE-ionization conditions leading to values $(1-\zeta_j)$ given in Table 2.

As one astonishing result in Table 3 it can be concluded that especially for the (Ne/O)- ratio the solar cycle variation is shown to be fairly small, whereas for the other species it appears more pronounced. One should keep in mind, however, that the above values were calculated both for solar maximum and minimum conditions with identical values for the ACR modulation coefficients κ and the spectral index γ. Due to a likely solar cycle variation in these quantities, possibly also a slight variation of the coefficients A_{jk}^* in Eq.(36) may be caused, however, not of great relevance for the results concerning ratios. More relevant these variations could be for the absolute values of the ACR densities which we do not aim at here.
Comparing the values in Table 3, both those for maximum and for minimum conditions, with those given by Klecker (1977) one may recognize substantial discrepancies. These are smallest for the (N/O)-ratios where 0.11 instead of 0.15 is obtained. Of definitely nonnegligible relevance these discrepancies,

however, seem to be what concerns the results for the (He/O)- and the (Ne/O)- ratio, the first being much larger, the latter being much smaller than corresponding values by Klecker (1977). On the other hand, the values chosen by Klecker (1977) for him served as the basis for a fit to ACR energy spectra measured at $1AU$, and thus should give some serious constraints to the ACR modulation theory.
The easiest way to reconcile the theoretical ratios presented here with those given by Klecker could be seen in taking more adequately into account the differential transmissions Tr_j through the heliospheric interface. Due to the large charge exchange cross-sections with respect to heliospheric protons mainly oxygen and nitrogen would be influenced by this effect, whereas helium and neon would hardly be affected.
Trying to ascribe the discrepancy factor of about 1.5 between the (N/O)-ratios given by us and by Klecker to the differential filtering effect, and extracting a factor 1.83 as typical for the difference between solar wind charge exchange rates for oxygen and nitrogen from experimental values given by Siscoe and Mukherjee (1972), one is then led to the relation:

$$Tr_N/Tr_O \; = \; \exp(-\int N_p \sigma_{HN} v_{\rm rel} \frac{ds}{v_\infty})/\exp(-\int N_p \sigma_{HO} v_{\rm rel} \frac{d}{v_\infty}) \; = \; 1.5 \qquad (37)$$

where N_p is the proton density in the heliospheric interface, $v_{\rm rel}$ their mean relative velocity with respect to LISM $O-, N$-atoms, v_∞ the LISM wind velocity, and σ_{HN}, σ_{HO} the corresponding charge exchange cross sections. Then from Eq.(37) with the above mentioned relation $\sigma_{HO} \cong 1.83\sigma_{HN}$ one then arrives at the result:

$$Tr_O \; \cong \; (1.5)^{-1/0.45} \; = \; 0.41. \qquad (38)$$

Interestingly enough the above number for Tr_O does not look unreasonable in view of an interface proton density $N_{p\infty}$ of the order of $10^{-2}cm^{-3}$(Reynolds, 1989) and an adequate consideration of the differential transmissivities of the heliospheric interface thus would appear to offer a perfect remedy of the small discrepancy seen in the (N/O)-ratio.
One may, however, run into much harder troubles when looking for a remedy of the discrepancies in the other ACR ratios of Table 3. Since helium is not likely to suffer any relevant losses in the interface plasma (i.e. $Tr_{He} \; = \; 1.0$), the ratio in Table 3, when applying $Tr_O \; = \; 0.41$ for oxygen, then is even raised to: $(n_{He}/n_O) \; = \; 85.51/0.41 \; = \; 208.56$ (!) which is even farther off Klecker's value. Contrary to that for neon the existing discrepancy by application of the transmissivity factor for oxygen would nicely be diminished, since adopting also for neon no interface effect (i.e. $Tr_{Ne} \; = \; 1.0$) and applying $Tr_O \; = \; 0.41$ would reconcile our (Ne/O)-ratio with that by Klecker within a factor of 2.

In fact, the discrepancy really hard to overcome is in the (He/O)-ratios. As it seems to us the only way out is to assume a principal difference in the acceleration of He^+-pick-up ions to ACR-typical energies compared to that of the other LISM pick-up ions. Klecker (1977) has adopted a global efficiency of $\epsilon = 2 \cdot 10^{-3}$ for all ion species to become ACR particles. There may, however, be good physical reasons why helium behaves different from the other ions in this respect.
This could most probably be connected with the fact that LISM helium atoms are ionized only fairly close to the sun. On their much longer way out to the heliospheric shock front these ionized atoms in form of He^+-pick-up ions may have a better chance to assimilate to the solar wind ion population and thus to escape the process of acceleration to ACR energies. Also these ions are convected outwards preferably into the downwind region of the heliosphere as shown by observations (Möbius et al., 1986, 1988) suggesting the possibility that in this region the acceleration efficiency is lower, perhaps because of the fact that there the distance to the heliospheric shock (i.e. R_c) is larger by a factor of 3 to 6 compared to its upwind distance as indicated by Baranov (1990), Baranov and Lebedev (1989), Shima et al. (1986), and Matsuda et al. (1989).

References

Baranov,V.B. and Lebedev,M. (1989): COSPAR-Symposium on "Physics of the outer heliosphere", Warsaw, Sept. 1989, Pergamon Press, in preparation
Baranov,V.B. (1990): Space Science Reviews, **52**, 89-120
Brandt,J.C. (1964): Icarus, **3**, 253
Błeszinski,S. (1987): Astron.Astrophys., **180**, 201
Cameron,A.G.W. (1973): Space Sci.Rev., **14**, 392
Cox,D.P. and Reynolds,R.J. (1987): Ann.Rev.Astron.Astrophys., **25**, 303
Fahr,H.J. (1980): Mitt.Astron.Gesellschaft, **47**, 233-276
Fahr,H.J. (1986): Advances Space Res., **6(2)**, 13
Fahr,H.J. (1989): "Filtration of the interstellar neutrals at the heliospheric interface and their coupling to the solar wind ", COSPAR-Symposium on "Physics of the Outer Heliosphere", Warsaw (Poland) Sept. 1989, Pergamon Press, in preparation
Fahr,H,J. (1990): "Local interstellar oxygen in the heliosphere: its analytic representation and observational consequences", Astron. Astrophys., in press
Fahr,H.J. and Fichtner,H. (1990): "Causes and consequences for a three dimensional structure of the heliosphere", Space Science Reviews, in preparation
Fahr,H.J. and Neutsch,W. (1983): Astron.Astrophys., **118**, 57
Fahr,H.J. and Ripken,H.W. (1984): Astron.Astrophys., **139**, 551
Fahr,H.J., Fichtner,H., and Neutsch,W. (1990): Planet.SpaceSci., **38**, 383
Fahr,H.J., Rucinski,D., and Nass,H.U. (1987): Ann.Geophys., **A5**, 255
Fisk,L.A. (1976*a*): J.Geophys.Res., **81**, 4633
Fisk,L.A. (1976*b*): J.Geophys.Res., **81**, 4641
Fisk,L.A. (1976*c*): Astrophys,J., **206**, 333
Fisk,L.A. (1986): In "The sun and the heliosphere in three dimensions", pp. 401, Ed. by R.G. Marsden, D.Reidel Publ.Comp., Doordrecht Holland
Fisk,L.A., Kozlovsky,B., and Ramaty,R. (1974): Astrophys.J., **190**, *L*35
Frisch,P.C., York,D.G., and Fowler,J.R. (1987): Astrophys.J., **320**, 842

Frisch,P.C., Welty,D.E., York,D.G., and Fowler,J.R.(1989): Astrophys.J. (submitted)
Garcia-Munoz,M., Mason,G.M., and Simpson,J.A. (1973): Astrophys.J., **182**, *L*81
Garcia-Munoz,M., Mason,G.M., and Simpson,J.A. (1975): Astrophys.J., **202**, 265
Gleeson,L.J. and Axford,W.I. (1968*a*): Astrophys.J., **154**, 1011
Gleeson,L.J. and Axford,W.I. (1968*b*): Astrophys.Space Sci., **2**, 431
Gleeson,L.J. and Urch,I.H. (1971): Astrophys.Space Sci., **11**, 288
Isenberg,P.A. (1987): J.Geophys.Res., **92**, 1067
Jokipii,J.R. (1971): Rev.Geophys.Space Res., **9**, 27
Jokipii,J.R. (1986): In "The sun and the heliosphere in three dimensions" pp. 375, Ed. by. R.G. Marsden, D.Reidel Publ.Comp., Doordrecht Holland
Klecker,B. (1977): J.Geophys.Res., **82**, 5287
Klecker,B., Hovestadt,D., Gloeckler,G., and Fan,C.Y. (1977): Astrophys. J., **212**, 290
Matsuda,T., Fujimoto,Y., Shima,E., Sawada,K., and Inaguchi,T. (1989): Progress of Theoret. Physics, **81**, 810 - 822
Möbius,E.D., Hovestadt,D., Klecker,B., Scholer,M., Gloeckler,G., Ipavich,F.M, and Lühr,H. (1986): Nature, **318**, 426
Möbius,E.D., Klecker,B., Hovestadt,D., and Scholer,M. (1988): Astrophys. Space. Sci., **144**, 487
Parker,E.N. (1958):Astrophys. J., **128**, 664
Parker,E.N. (1963): In "Interplanetary dynamical processes", Interscience Publishers, New York
Reynolds,R.J. (1989): "Lower limit on the ionization fraction of the very local medium", Preprint, Symposium on COSMIC GAS DYNAMICS, Moscow (USSR), Oct. 1988
Ripken,H.W. and Fahr,H.J. (1983): Astron.Astrophys., **122**, 181
Siscoe,G.L. and Mukherjee,N.R. (1972): J.Geophys.Res., **77**, 6042
Shima,E., Takuya,M., and Inaguchi,T. (1986): Monthly Not.Roy.Astron. Soc., **221**, 687
Strömgren,B. (1939): Astrophys. J., **89**, 856
Wallis,M.K. (1984): Astron.Astrophys., **130**, 200
Williams,K.G. (1965): Astrophys. J., **142**, 319

Turbulence in the Solar Wind

E. Marsch

Max-Planck-Institut für Aeronomie,
Postfach 20, W-3411 Katlenburg-Lindau, Fed. Rep. of Germany

Abstract

The solar wind is, apparently, a turbulent medium permeated by fluctuations on a broad range of scales characterizing the plasma as a fluid and kinetic entity. Here we briefly review some recent developments and observations on MHD fluctuations in the frequency range between some 10^{-6} and 10^{-2} Hz. Spectral analysis reveals that solar wind fluctuations represent ongoing turbulence, often with high Alfvénic correlations of coronal origin, which undergoes considerable spatial and spectral evolution in the radially expanding wind and is coupled to large-scale plasma structures and magnetic field inhomogeneities. Theoretical attempts to model the transport of magnetohydrodynamic fluctuations are discussed. Finally, the dissipation of turbulence and concurrent heating of the wind are addressed.

1. Introduction

Early in-situ observations indicated that the solar wind is a turbulent medium (*Coleman*, 1968) permeated by fluctuations on all observed spatial and temporal scales. These fluctuations may extend in frequency from the solar rotation frequency, $f_\odot = 4.3\ 10^{-7}$ Hz, to the ion gyrofrequency of several Hz and beyond to the plasma frequency of the electrons at several kHz for typical plasma parameters as measured at 1 AU. While fluctuations in the kinetic regime do not influence the large-scale dynamics of the flow, the lower-frequency fluctuations do affect the overall energy and momentum transport of the wind. Fig. 1, taken from *Mangeney et al.* (1990), schematically illustrates the specific power associated with each type of random fluctuation. Power in the range 10^{-7} to 10^{-4} Hz is due to variations in the solar wind stream structure (corotating fast streams, medium-sized plasma flow tubes). Between 10^{-4} and 10^{-1} Hz we have magnetohydrodynamic fluctuations (and Alfvén and magnetoacoustic waves) and the regime of MHD turbulence with which the remainder of this article will be concerned.

Adjacent to the MHD realm, the kinetic domain extends from some Hz to several kHz. The corresponding waves and fluctuations often drain their excitation energy from larger-scale fluctuations and thus influence their dissipation by means of interactions with the plasma particles. These kinetic phenomena are reviewed by *Marsch* (1991a) and by *Gurnett* (1991), who discuss in detail the power spectra, wave features, and particle velocity distribution functions. The high-frequency kinetic turbulence is certainly of important interest in its own right because of its relevance to the anomalous transport in the almost collisionless solar wind. Here, however, we shall focus on magnetohydrodynamic turbulence which is believed to have a more direct bearing on solar wind heating and acceleration.

2. Structures and Alfvénic fluctuations

Fig. 1 illustrates the very large number of scales excited in the interplanetary solar wind. Whereas the high-frequency noise is mostly of local origin, the MHD fluctuations are predominantly of solar origin and are convected by the flow to the point of in-situ observation. Of course, the large-scale structure of the solar wind itself is generated at the Sun and it reflects coronal boundaries and the large-scale morphology of the Sun's magnetic field. The resulting complex topology of the interplanetary field and plasma stream structure make the solar wind a rather inhomogeneous medium. The kinetic energy in the streams is the only interplanetary energy reservoir for generation of fresh fluctuations and stirring of turbulence by shear-induced plasma instabilities.

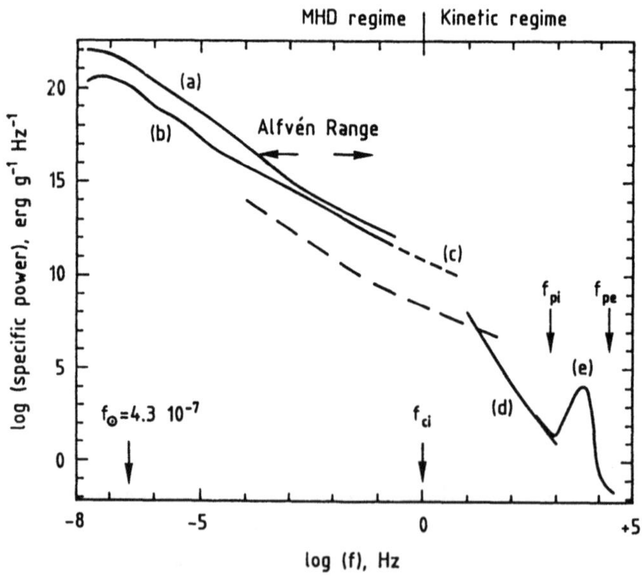

Fig. 1. Schematic outline of the specific power spectrum of fluctuations observed in the solar wind. The energy per unit mass per Hz is plotted versus frequency in the spacecraft frame. The various regimes relate to (a) structures, (b) Alfvénic fluctuations, (c) density fluctuations, (d) Whistler mode, (e) electrostatic noise (after *Mangeney et al.*, 1990).

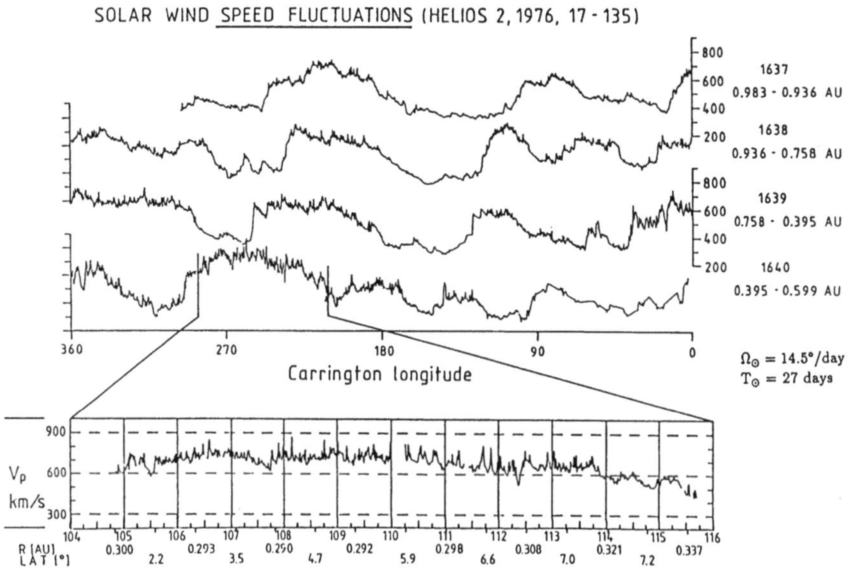

Fig. 2. Solar wind streams and speed fluctuations as measured by Helios 2 in 1976 between 0.29 and 1.0 AU. Four subsequent solar rotations are shown with speed displayed versus Carrington longitude. The enlargement shows the detailed structure of a large corotating stream, consisting of medium-sized flow tubes and Alfvénic fluctuations observed during the perihelion passage of the probe (after *Marsch*, 1991b).

The stream structure, as it appeared typically near and during the solar activity minimum of the Helios epoch in the years 1975 to 1976, is shown in Fig. 2 after *Marsch* (1991b). The flow speed is displayed for four subsequent solar rotations versus Carrington longitude. An enlargement of a high-speed stream is shown below the main frame, indicating by its fuzzy appearance a possible composition of meso-scale flow tubes and superimposed genuine fluctuations. The steady corotating stream structure indicates stationary source conditions in the solar corona from which a flow emanates with a pattern recurring about every 27 days as seen from the Earth's orbit. The overall turbulence level seems to increase with heliocentric distance decreasing from 1 to 0.3 AU at the perihelion of Helios 2. By and large one finds the slow wind to be less noisy than the fast wind. This suggests a strong correlation between the intensity of the fluctuations and the stream structure, which is itself generally closely associated with the coronal magnetic field (e.g. *Bruno and Bavassano*, 1987; *Whang and Sheeley Jr.*, 1990). It should be remembered that the stream structure changes during the solar cycle (see e.g. *Schwenn*, 1990; *Hoeksema*, 1986). However, the association of the turbulence with the field topology remains unchanged. *Roberts et al.* (1987a,b) have shown that intensive Alfvénic fluctuations are ubiquitious, independent of the phase of the solar cycle, and always occuring in magnetic sectors related to open coronal fields.

Since the early classical studies by *Belcher and Davis* (1971) much work has been done on investigating Alfvénic fluctuations in the solar wind. The older literature has been reviewed by *Barnes* (1979,1981,1983). More recent observations and related theoretical considerations can be found in the review by *Marsch* (1991b). The most striking features of observed Alfvénic fluctuations is that they often obey, to a high degree, the correlation of an ideal Alfvén wave,

$$\delta \mathbf{V} = \pm \frac{1}{\sqrt{4\pi\rho}} \delta \mathbf{B} \tag{1}$$

for the velocity and magnetic field fluctuations. Here, as usual, ρ denotes the plasma mass density and the $\pm$ sign corresponds to propagation antiparallel or parallel to the average magnetic field. A nice illustration of this correlation can be found in *Denskat* (1982) using Helios data. The theoretical relation (1) only applies to a single Alfvén wave of any amplitude propagating in a uniform or weakly inhomogeneous medium. However, in the solar wind the fluctuations appear as a complicated superposition of many "wave trains" of variable length and amplitude, with random behaviour of the time series of the vector components of the flow velocity $\mathbf{V}$ and magnetic field $\mathbf{B}$. Furthermore, *Dobrowolny et al.* (1980a,b) noted that fairly well developed power-law spectra of observed Alfvénic fluctuations imply the existence of nonlinear interactions which are in conflict with the picture of simple waves. These authors suggested that we are dealing with magnetohydrodynamic turbulence in the case of the solar wind and they adopted concepts of the phenomenology of fluid turbulence to describe the fluctuations. In particular, the theoretical rephrasing of turbulence problems in terms of *Elsässer* (1950) variables turned out to be useful for subsequent theoretical studies and data analyses. The Elsässer fields are defined as

$$\mathbf{Z}^{\pm} = \mathbf{V} \pm \mathbf{V}_A \tag{2}$$

where the Alfvén velocity is $\mathbf{V}_A = \mathbf{B}/\sqrt{4\pi\rho}$. In terms of these variables, the equations of plasma and field motion for ideal incompressible MHD are expressed as

$$\left(\frac{\partial}{\partial t} + \mathbf{Z}^{\mp} \cdot \nabla\right) \mathbf{Z}^{\pm} = -\nabla(p_T/\rho) \tag{3}$$

where $p_T = p + B^2/8\pi$ is the total plasma pressure and p the thermal pressure. Decomposition of $\mathbf{Z}^{\pm}$ into background $\mathbf{Z}_o^{\pm}$ and fluctuations $\delta\mathbf{Z}^{\pm}$ yields a simple linearized wave equation

$$\left(\frac{\partial}{\partial t} + \mathbf{Z}_o^{\mp} \cdot \nabla\right) \delta\mathbf{Z}^{\pm} = 0 \tag{4}$$

whence, by Fourier transformation, the standard dispersion relation

$$\omega - \mathbf{k} \cdot \mathbf{V}_o = \mp \mathbf{k} \cdot \mathbf{V}_{Ao} \tag{5}$$

of Alfvén waves and the correlation (1) readily follow. Here $\mathbf{k}$ denotes the wave vector. A more detailed normal mode analysis, also for magnetoacoustic waves, is given by *Marsch and Mangeney* (1987). For pure Alfvén waves either $\delta\mathbf{Z}^+$ or $\delta\mathbf{Z}^-$ vanishes and nonlinearity disappears. However, if the correlation (1) is not ideal, as is usually the case in the observations, then both fields have finite amplitudes and are coupled by the convective derivative in equation (3). This coupling is essential for the occurrence of nonlinear processes and the evolution of turbulence involving a large number of fluctuations on disparate scales.

3. MHD turbulence, concepts and observations

In the seventies solar wind fluctuation studies were somewhat mislead by the "waves versus turbulence" controversy, which produced many attempts to identify idealized wave properties in spacecraft data. It now appears that plasma and magnetic field fluctuations can be meaningfully viewed as representing MHD turbulence. The existence of power-law spectra support this idea and suggest the action of an energy cascade requiring both inward and outward oriented "Alfvénic fluctuations" $\delta\mathbf{Z}^\pm$ (possibly including structures) and their interactions. However, the turbulence is complicated by the frequent predominance of outward propagating fluctuations with high correlations as in (1), which imitate Alfvén waves and correspondingly modify the nonlinear couplings as compared to a situation with equipartition of both Elsässer fields. For a more detailed up-to-date discussion of this topic see e.g. *Mangeney et al.* (1990).

In several fundamental papers, *Matthaeus and Goldstein* (1982a,b, 1983) and *Matthaeus et al.* (1986) convincingly showed that solar wind small-scale fluctuations could be described as examples of stationary and homogeneous turbulence. In their data analysis they used the ideal invariants of incompressible MHD (*Woltjer*, 1958) as principal quantities to characterize the observed plasma state; these adequately describe the overall turbulent energy, as well as the degree of Alfvénic correlations and the helical winding of the magnetic fluctuations. In terms of Elsässer fields, these invariants can be expressed by means of the specific energies

$$e^\pm = \tfrac{1}{2} < (\delta\mathbf{Z}^\pm)^2 > \tag{6}$$

where the brackets denote an ensemble average (see *Matthaeus et al.*, 1986). By integrating $e^\pm$ over the whole volume containing the plasma one can show (e.g. *Marsch and Mangeney*, 1987) that these integrals do not change with time if there is no dissipation in the closed system under consideration. The turbulence spectra of $e^\pm$ are easily derived from the autocorrelation functions of $\delta\mathbf{Z}^\pm$ (e.g. *Tu et al.*, 1989). The wave-number spectrum $e^\pm(k)$ yields important clues to the dynamical evolution of a magnetofluid and to the distribution of turbulent energy at the various scales.

In the older literature two equivalent quantities have usually been considered which, according to the definition (2), are obtained by linear combinations of the Elsässer-field energies, the total energy

$$e = \frac{1}{2}(e^{+} + e^{-}) = \frac{1}{2} < \delta \mathbf{V}^2 + \delta \mathbf{V}_A^2 > \tag{7}$$

and the cross-helicity

$$h_c = \frac{1}{2}(e^{+} - e^{-}) =< \delta \mathbf{V} \cdot \delta \mathbf{V}_A > \tag{8}$$

Note that e must be positive while h_c need not, a notion manifesting itself in different and selective dissipative decays of these invariants (*Pouquet et al.*, 1976; *Grappin et al.*, 1982, 1983). The cross-helicity normalized by the total energy may vary between +1 and -1; thus its spectrum provides a measure of the prevailing Alfvénic correlations (see equation (1) again) as a function of wavevector or frequency. A third quantity of interest is the cross-correlation energy of $\delta \mathbf{Z}^{+}$ and $\delta \mathbf{Z}^{-}$ or the residual energy

$$e^R = \frac{1}{2} < \delta \mathbf{Z}^{+} \cdot \delta \mathbf{Z}^{-} >= \frac{1}{2} < \delta \mathbf{V}^2 - \delta \mathbf{V}_A^2 > \tag{9}$$

which measures the excess of kinetic versus magnetic energy.

During the past few years, solar wind fluctuations have been extensively analysed in terms of Elsässer variables. See, for example, *Grappin et al.* (1989, 1990); *Tu et al.* (1989, 1990); *Marsch and Tu* (1990a,b). These studies established that the intensity, spectral features and occurrence of the turbulence in the inner heliosphere is intimately related to the large-scale stream structure and consequently to the interplanetary magnetic field topology. This observation and the basic spectral features of $e^{\pm}$ are illustrated in Fig. 3, after *Tu et al.* (1990), showing the morphology of the spectra during the passage through a typical high- and low-speed stream observed by Helios near 0.3 AU.

Each spectrum in Fig. 3 covers one day, indicated by hatched bars in the top panel showing the solar wind speed. For comparison the Kolmogorov slope of -5/3 is also given in each frame. This law is obeyed by e^{+} and e^{-} in the slow wind adjacent to the heliospheric current sheet (second frame). The spectral signatures clearly change in the body of the fast stream. Here one finds, over a broad range in frequency $e^{+} >> e^{-}$ in accordance with (7) and (8), a normalized cross-helicity of order unity and consequent dominance of outward propagating Alfvénic fluctuations, always with a slight admixture of the inward mode, however. This characteristic pattern was first established by *Grappin et al.* (1990) who also found the intriguing result that the turbulent energy e^{+} was well correlated with the proton thermal speed c_p, whereas e^{-} was better correlated with the daily mean amplitude of relative density fluctuations $\delta n/n$. Furthermore, for the spectral slopes of $e^{\pm}$ a quantitative relationship could be derived with c_p and $\delta n/n$ which allowed a detailed prediction of observed spectral features by numerical fits.

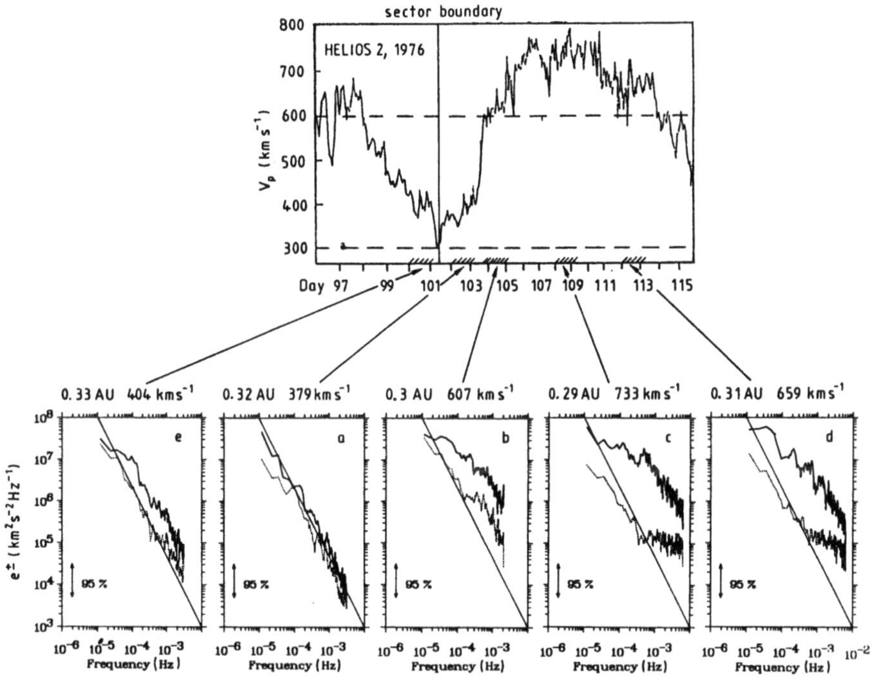

Fig. 3. Sequence of $e^{\pm}$ power spectra across a high- and low-speed stream near 0.3 AU. The upper panel shows the wind speed profile, the five lower panels the spectra (e^+ solid lines) and, for comparison, the Kolmogorov -5/3 slope which is attained by $e^{\pm}$ in slow wind adjacent to the heliospheric current sheet. Alfvénic fluctuations ($e^+ >> e^-$) prevail in the body of the fast stream (after *Tu et al.*, 1990).

It appears that there is no abrupt change but a gradual transition between the two kinds of "Alfvénic" and non-Alfvénic solar wind; this, presumably, reflects variations of the coronal sources of the turbulence rather than interplanetary evolution.

All the spectra are observed to steepen (*Marsch and Tu*, 1990a) as the plasma radially expands. A comprehensive discussion of the radial gradients of the slopes and other turbulence characteristics can be found in the above references and in *Grappin et al.* (1990), *Mangeney et al.* (1990), and *Marsch* (1991b). *Tu and Marsch* (1990a) have evaluated average spectra of $e^{\pm}$ for various flow conditions and solar distances. It has clearly been shown that, while steepening, the e^+ spectrum is also radially decreasing in intensity, although the e^- spectrum seems to remain almost constant below some 10^{-4} Hz; it also decreases at higher frequencies, however. Observationally, it appears that the "Alfvén waves" of coronal origin are mostly damped while propagating away from the Sun. In contrast, e^- seems to represent an essentially invariant background spectrum the nature of which needs further study.

4. Modelling the evolution of solar wind turbulence

In the previous section we have discussed some key observations of MHD turbulence in the solar wind. The observations may be summarized as follows. The solar wind is a strongly inhomogeneous medium which largely reflects the variability of the coronal magnetic field and plasma sources during the solar magnetic activity cycle. Within the corona, a wide range of turbulent conditions can be inferred from interplanetary observations. The type of spectra observed range from fully-developed Kolmogorov-like spectra near the heliospheric current sheet (related to coronal streamers) to comaratively flat spectra, with index close to -1, which are associated with Alvénic fluctuations emanating from magnetically open regions on the Sun (coronal holes). The small-scale turbulence strongly interacts with the large structures and $e^{\pm}$ tends ultimately to attain a spectral slope of -5/3 at large heliocentric distances (*Marsch and Tu*, 1990a). The question of spectral and spatial evolution of MHD fluctuations in the solar wind is certainly of general interest for an understanding of plasma heating and acceleration by turbulence in stellar winds and other astrophysical systems.

In recent years a number of authors have embarked on the difficult task of modelling solar wind fluctuations and the transport of MHD turbulence in weakly inhomogeneous media (*Tu et al.*, 1984; *Tu*, 1987, 1988; *Zhou and Matthaeus*, 1989; *Matthaeus and Zhou*, 1989a,b; *Marsch and Tu*, 1989; *Velli et al.*, 1989a,b, 1990; *Mangeney et al.*, 1990; *Tu and Marsch*, 1990b; *Hollweg*, 1990). A key problem is that of nonlinear interactions and triple correlations. Following *Kraichnan* (1965) and *Dobrowolny et al.* (1980b), the Kolmogorov phenomenology of fluid turbulence has been adopted for the Elsässer fields to model the nonlinearity by a cascade in wavenumber space. The turbulent energy $(\delta Z_\ell^{\pm})^2$ associated with spatial scale ℓ is assumed to be cascaded, by means of a sequence of eddies of ever decreasing scales, to the dissipation regime with scale ℓ_D. The corresponding transfer rate is

$$\epsilon_\ell^{\pm} \approx (\delta Z_\ell^{\pm})^2 / \tau_\ell^{\pm} \tag{10a}$$

where the eddy turnover time is given, according to the convective derivative in (3), by the relation $\tau_\ell^{\pm} = \ell / \delta Z_\ell^{\mp}$. Insertion of this into (10a) yields

$$\epsilon_\ell^{\pm} \approx (\delta Z_\ell^{\pm})^2 (Z_\ell^{\mp}) / \ell \tag{10b}$$

Contact with the wavenumber spectrum is established by noting that $k \approx 1/\ell$ and $e_k^{\pm} k \approx (\delta Z_\ell^{\pm})^2$. For scale invariance and a fully developed cascade we find $\epsilon_\ell^{\pm} = \epsilon_{\ell_D}^{\pm} = \epsilon^{\pm}$, where $e^{\pm}$ is the dissipation rate in units of $erg/(gs)$. Thus scale invariance implies that

$$\epsilon^{\pm} \approx e_k^{\pm} (e_k^{\mp})^{1/2} k^{5/2} \tag{11}$$

The unique solution to (11) is given by a Kolmogorov-type spectrum for each Elsässer field

$$e_k^{\pm} \propto k^{-5/3} \tag{12}$$

Matthaeus and Zhou (1989b) have proposed an extended inertial range spectral law interpolating between the Kolmogorov and Kraichnan (index -3/2) scaling. Presently it seems that observations in the solar wind favour (12) as the equilibrium spectrum. For a more detailed discussion see also *Mangeney et al.* (1990). The nonlinear changes of the spectra may be modelled as

$$\frac{\partial e_k^{\pm}}{\partial t}\Big|_{NL} = -\frac{\partial}{\partial k}\left(\alpha^{\pm} e_k^{\pm}(e_k^{\mp})^{1/2} k^{5/2}\right) \tag{13}$$

Here the spectral flux is proportional to (11) with a constant $\alpha^{\pm}$ to be determined from the measurements (e.g. *Tu et al.*, 1989b).

Spectral transfer equations combine this nonlinearity with the other changes of the spectral densities due to convection and coupling to large-scale structures. For the details we must refer the reader to the references cited at the beginning of this section. The general structure of these equations is represented by

$$\frac{d^{\mp}}{dt} e_k^{\pm} + (e_k^R - e_k^{\pm})(\mathbf{Z}_o^{\pm} \cdot \nabla) ln\sqrt{\rho_o} = \frac{\partial}{\partial t} e_k^{\pm}\Big|_{NL} - S_k^{\pm} \tag{14}$$

where the first term describes convection with the derivative $\frac{d^{\mp}}{dt} = \frac{\partial}{\partial t} + (\mathbf{Z}_o^{\mp} \cdot \nabla)$; the second relates to WKB-type effects of wave propagation in inhomogeneous media and is of the kind familiar from geometrical optics. There is also a coupling through the residual energy (9) spectrum e_k^R; the physics of this is still an issue and a matter of ongoing discussion (see *Hollweg*, 1990; *Zhou and Matthaeus*, 1990a). In addition to the nonlinear term we have sources and sinks, which are linear in the autocorrelation matrix of $\delta\mathbf{Z}^{\pm}$ but involve couplings to large-scale inhomogeneities like velocity shear, electric currents and density gradients (*Marsch and Tu*, 1989). Spectral transfer equations of the type (14) are presently used to model the spectral and spatial evolution of solar wind MHD turbulence (*Tu and Marsch*, 1990b,c; *Velli et al.*, 1990).

Finally, we address the dissipation problem briefly. *Tu* (1987,1988) has shown, by using an earlier more simplified version of (14), that the temperature profiles of interplanetary protons as observed by the Helios solar probes in the inner heliosphere, can be explained by using the cascading rate $\epsilon = \frac{1}{2}(\epsilon^+ + \epsilon^-)$ for the total turbulent energy as a heating rate in the ion internal energy equation. Results of his calculations are presented in Fig. 4, which shows the magnetic moment $\mu = T_{p\perp}/B$ of the protons ($T_{p\perp}$ is the perpendicular temperature) versus heliocentric distance in AU. The points with standard deviation bars represent the measurements (*Marsch et al.*, 1982; *Schwartz and Marsch*, 1983) while the straight line gives the numerical calculations of *Tu* (1988) according to the equation

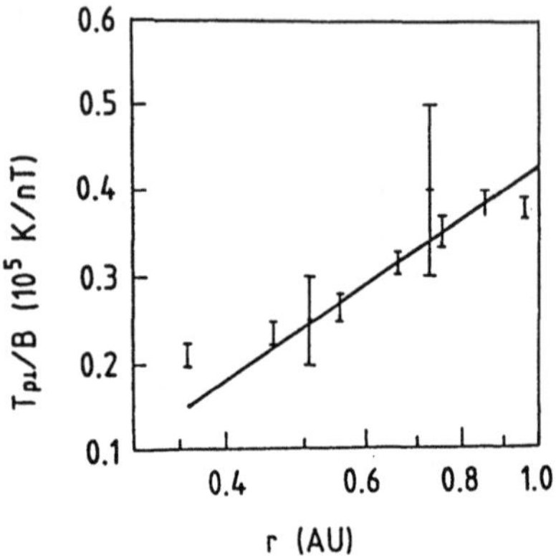

Fig. 4. Magnetic moment of the protons versus heliocentric distance. The measured points are from Helios after *Marsch et al.* (1982) and *Schwartz and Marsch* (1983). The straight line indicates numerical results by *Tu* (1988) which explain the nonadiabatic expansion of the protons by heating due to a nonlinear cascade of Alfvénic fluctuations.

$$V \frac{d}{dr} \mu = \frac{m_p}{k_B T_{p\perp}} \epsilon \tag{15}$$

One finds a typical heating rate ($\rho\epsilon$) of several 10^{-16} erg cm^{-3} s^{-1} at 1 AU, a value sufficient at this point to compensate adiabatic cooling and to even maintain nonadiabatic temperature profiles as observed in the radial evolution of a single solar wind plasma parcel. In conclusion, the turbulent energy cascade represents a promising and feasible approach to understanding the observed dissipation (*Villante and Vellante*, 1982; *Bavassano et al.*, 1982) of fluctuations in the solar wind and the concurrent heating of ions. The microscopic processes involved in the dissipation on a kinetic level are rather complicated and still not well understood. They have been reviewed by *Marsch* (1990b).

Finally, we should mention the important role that MHD fluctuations and Alfvén waves are believed to play in the generation of high-speed streams (*Leer et al.*, 1982) and stellar winds (*Holzer et al.*, 1983). The effects of turbulence on the background flow can be described in the momentum equation by the divergence of the "wave pressure" tensor $\underline{\underline{P}}_T$ (see e.g. *Marsch*, 1986, and references therein). In the case of high Alfvénic correlations we may approximate $\underline{\underline{P}}_T$ by the turbulent energy density times the unit matrix and thus obtain the acceleration force given by the negative gradient of the wave pressure, $-\nabla < \delta\mathbf{B}^2 > /8\pi$. An understanding of the radial amplitude-evolution of Alfvénic fluctuations is therefore essential for understanding the heating and acceleration of the solar wind.

5. Prospects

In this short review we have presented observations and some of the current ideas for the theoretical description and physical interpretation of MHD fluctuations in the solar wind. At present, extensive use is made of the Elsässer fields which are particularly suited to describe fluctuations with high Alfvénic correlations. Many theoretical problems remain to be solved. The nonlinear couplings in transport equations need to be better understood. The influence of spatial inhomogeneities on turbulence evolution requires further study. The dissipative processes are poorly comprehended and require sophisticated kinetic plasma physics. We have not touched upon the role of compressibility and density fluctuations which will undoutedly have an impact on Alfvénic fluctuations and their development. Finally, better models of solar wind expansion, including turbulence, need to be proposed, and numerical simulations should be carried out to study key nonlinear processes in considerable detail.

References

Barnes, A.: 1979, in *Solar System Plasma Physics*, Vol. I, ed. by E.N. Parker, C.F. Kennel, and L.J. Lanzerotti, p. 249, North-Holland, Amsterdam.

Barnes, A.: 1981, in *Solar Wind Four*, ed. by H. Rosenbauer, MPAE-Report No. W-100-81-31, p. 326.

Barnes, A.: 1983, in *Solar-Terrestrial Physics*, ed. by R. L. Carovillano and J. M. Forbes, D. Reidel, Dordrecht, p. 155.

Bavassano, B., Dobrowolny, M., Mariani, F., Ness, N.F.: 1982, J. Geophys. Res. **87**, 3617.

Belcher, J.W., Davis, L.: 1971, J. Geophys. Res. **76**, 3534.

Bruno, R., Bavassano, B.: 1987, Annales Geophys. 5A, 265.

Coleman, P.J.: 1968, Astrophys. J. **153**, 371.

Denskat, K.U.: 1982, Dissertation (PhD thesis), Technische Universität at Braunschweig, F.R. Germany.

Dobrowolny, M., Mangeney, A., Veltri, P.: 1980a, Astron. Astrophys. **83**, 26.

Dobrowolny, M., Mangeney, A., Veltri, P.: 1980b, Phys. Rev. Lett. **45**, 144.

Elsässer, W.M.: 1950, Phys. Rev. **79**, 183.

Grappin, R., Frisch, U., Léorat, J., Pouquet, A.: 1982, Astron. Astrophys. **105**, 6.

Grappin, R., Pouquet, A., Léorat, J.: 1983, Astron. Astrophys. **126**, 51.

Grappin, R., Mangeney, A., Marsch, E.: 1989, in *Turbulence and Nonlinear Dynamics in MHD Flows*, M. Meneguzzi, A. Pouquet and P.L. Sulem (Editors), Elsevier Science Publishers B.V., North-Holland, p. 81.

Grappin, R., Mangeney, A., Marsch, E.: 1990, J. Geophys. Res. **95**, 8197.

Gurnett, D.A.: 1991, in *Physics of the Inner Heliosphere*, Vol. II, ed. by R. Schwenn and E. Marsch, Springer-Verlag, Heidelberg, in press.

Hoeksema, J.T.: 1986, in *The Sun and the Heliosphere in Three Dimensions*, R.G. Marsden (Ed), D. Reidel Dordrecht, p. 241.

Hollweg, J.V.: 1990, J. Geophys. Res. **95**, 14873.

Holzer, T.E., Flå, T., Leer, E.: 1983, Astrophys. J. **275**, 808.

Kolmogorov, A.N.: 1941, Compt. Rend. Acad. Sci. U.R.S.S. **30**, 301.

Kraichnan, R.H.: 1965, Phys. Fluids **8**, 1385.

Leer, E., Holzer, T.E., Flå, T.: 1982, Space Sci. Rev. **33**, 161.

Mangeney, A., Grappin, R., Velli, M.: 1990, in *Proc. MHD Symposium*, St. Andrews, Scotland.
Marsch, E.: 1986, Astron. Astrophys. **164**, 77.
Marsch, E.: 1991a, in *Physics of the Inner Heliosphere*, Vol. II, ed. by R. Schwenn and E. Marsch, Springer-Verlag, Heidelberg, in press.
Marsch, E.: 1991b, in *Physics of the Inner Heliosphere*, Vol. II, ed. by R. Schwenn and E. Marsch, Springer-Verlag, Heidelberg, in press.
Marsch, E., Mühlhäuser, K.-H., Schwenn, R., Rosenbauer, H., Pilipp, W., Neubauer, F.M.: 1982, J. Geophys. Res. **87**, 52.
Marsch, E., Mangeney, A.: 1987, J. Geophys. Res. **92**, 7363.
Marsch, E., Tu, C.-Y.: 1989, J. Plasma Phys. **41**, 479.
Marsch, E., Tu, C.-Y.: 1990a, J. Geophys. Res. **95**, 8211.
Marsch, E., Tu, C.-Y.: 1990b, J. Geophys. Res. **95**, 11945.
Matthaeus, W.H., Goldstein, M.L.: 1982a, J. Geophys. Res. **87**, 10347.
Matthaeus, W.H., Goldstein, M.L.: 1982b, J. Geophys. Res. **87**, 6011.
Matthaeus, W.H., Goldstein, M.L., Montgomery, D.C.: 1986, J. Geophys. Res. **91**, 59.
Matthaeus, W.H., Zhou, Y.: 1989a, in *Turbulence and Nonlinear Dynamics in MHD Flows*, M. Meneguzzi, A. Pouquet, and P.L. Sulem (Editors), Elsevier Science Publishers B.V., North-Holland.
Matthaeus, W.H., Zhou, Y.: 1989b, Phys. Fluids B **1**, 1929.
Pouquet, A., Frisch, U., Léorat, J.: 1976, J. Fluid Mech. **77**, 321.
Roberts, D.A., Klein, L.W., Goldstein, M.L., Matthaeus, W.H.: 1987a, J. Geophys. Res. **92**, 11021.
Roberts, D.A., Goldstein, M.L., Klein, L.W., Matthaeus, W.H.: 1987b, J. Geophys. Res. **92**, 12023.
Schwartz, S.J., Marsch, E.: 1983, J. Geophys. Res. **88**, 9919.
Schwenn, R.: 1990, in *Physics of the inner Heliosphere*, Vol. I, ed. by R. Schwenn and E. Marsch, Springer-Verlag, Heidelberg, p. 99.
Tu, C.-Y.: 1987, Solar Phys. **109**, 149.
Tu, C.-Y.: 1988, J. Geophys. Res. **93**, 7.
Tu, C.-Y., Pu, Z.-Y., Wei, F.-S.: 1984, J. Geophys. Res. **89**, 9695.
Tu, C.-Y., Marsch, E., Thieme, K.M.: 1989, J. Geophys. Res. **95**, 11739.
Tu, C.-Y., Roberts, D.A., Goldstein, M.L.: 1989b, J. Geophys. Res. **94**, 13575.
Tu, C.-Y., Marsch, E., Rosenbauer, H.: 1990, Geophys. Res. Lett. **17**, 283.
Tu, C.-Y., Marsch, E.: 1990a, J. Geophys. Res. **95**, 4337.
Tu, C.-Y., Marsch, E.: 1990b, J. Plasma Phys., in press.
Tu, C.-Y., Marsch, E.: 1990c, J. Geophys. Res., submitted.
Velli, M., Grappin, R., Mangeney, A.: 1989a, in *Plasma Phenomena in the Solar Atmosphere*, ed. by M.A. Dubois, Editions de Physique, Orsay, in press.
Velli, M., Grappin, R., Mangeney, A.: 1989b, Phys. Rev. Lett. **63**, 1807.
Velli, M., Grappin, R., Mangeney, A.: 1990, Comput. Phys. Commun. **59**, 153.
Villante, U., Vellante, M.: 1982, Solar Phys. **81**, 367.
Wang, Y.-M., Sheeley, JR. N.R.: 1990, Astrophys. J. **355**, 726.
Woltjer, L.: 1958, Proc. Natl. Acad. Sci., USA, **44**, 489.
Zhou, Y. Matthaeus, W.H.: 1989, Geophys. Res. Lett. **16**, 755.
Zhou, Y., Matthaeus, W.H.: 1990a, J. Geophys. Res. **95**, 14863.
Zhou, Y., Matthaeus, W.H.: 1990b, J. Geophys. Res. **95**, 14881.

Dust Rings Around Planets

E. Grün

Max-Planck-Institut für Kernphysik,
W-6900 Heidelberg, Fed. Rep. of Germany

SUMMARY. Rings have been observed around all giant planets. Micron sized dust particles are constituents of most rings. Because of various dynamic effects acting on small particles dust rings are short-lived phenomena or have to be continuously replenished. Descriptions of the ring systems are given and the significance of dust is dicussed.

1. Introduction

The development of telescopes allowed astronomers the detection of material around planets. In 1659 Huygens solved the problem of Saturn's telescopic appearance by postulating that the planet is surrounded by a ring [1]. This ring system remained the only known planetray ring for centuries. The proximity to the bright planets rendered it difficult to observe any weak brightness from satellites or rings in their vicinity. The size of Saturn's ring particles could not be identified: speculations reached from micron sized grains to kilometer sized boulders in rolling contact.

Space missions made it possible to observe planetary systems from close by and at different angles of view with respect to the illuminating sun. In 1979 the Voyager 1 spacecraft flew by Jupiter and detected a narrow weak ring (Smith et al., 1979) at a radial distance of 1.8 R_J (1 R_J = 71,000 km). The brilliancy of this ring system became obvious only when Voyager's cameras looked back onto the system from the anti-solar direction. This effect immediatly pointed to mirometer sized particles making up this ring. Such small particles are most easily detected at forward scattering observation geometry.

Two years earlier Elliot et al. (1977) had detected several narrow rings around Uranus by ground based occultation observations. Attempts from the ground to find rings also around Neptune were not conclusive. Detailed views of all these rings and the detection of dust in them were only possible during the Voyager 2 fly-bys of these planets.

All rings are in or very close to the equatorial plane of the planet demonstrating the dominant effect of the central body. Most rings are located inside the Roche limit where unconsolidated bodies are disrupted by the gravitational force of the planet. Only objects which have some tensile strength will survive in that region. The radial and azimuthal structure of the rings and their close relation to near-by satellites indicate complex dynamical processes most of which are still a mystery.

Longterm observations of the evolution of these phenomena may be necessary for the development of a full understanding. Space missions like the Galileo mission to Jupiter and the Cassini mission to Saturn are under way or are being prepared and will allow us to observe these sytems for several years.

2. Physics of Dust in the Vicinity of Planets

Gravity is not the only significant force acting on small particles. Because of their increased cross section to mass ratio electromagnetic and particle radiation become important forces for smaller particles (for a review see Grün et al., 1984). Pressure excerted by the solar radiation causes the particles to spiral to the central body. This Poynting–Robertson drag decreases with heliocentric distance (Table 1).

Magnetospheric plasma is closely coupled to the planetary magnetic field which rotates fixed to the planet. Inside the distance of the synchronous orbit (i.e. a Keplerian orbit is synchronous with the planetary rotation) dust particles are slowed down and spiral to the planet. Outside this distance dust particles gain orbital energy and are driven to larger distances. Both Uranus and Neptune have extended exospheres of hydrogen which have similar effects on dust particles.

Mass is lost from dust particles by sputtering from impacts of ions and atoms of a few hundred eV energy. Especially in the Jupiter and Saturn system this is a significant loss process for small particles. Impacts by high-speed interplanetary meteoroids causes grinding or destruction of particles of all sizes. However, this latter effect also produces smaller particles form larger ones.

Impacts on larger bodies (ring particles or satellites) is an important source for small particles. Other hypothezised sources for dust are volcanoes (on Jupiter's satellite Io) or geysers (on Saturn's Enceladus or Neptune's Triton). Satellites and larger ring particles sweep up small particles form their neighborhood. If we speak about dust in this context we include ice particles since at and outside Saturn's distance sublimation of water ice becomes negligible.

Table 1: Life times (in years) of dust particles for removal from the region of the rings or for destruction. The ranges correspond to micron sized particles in the region of the dust rings.

Effect	Jupiter	Saturn	Uranus	Neptune
Poynting-Robertson drag	$\geq 10^5$	$\leq 10^6$	10^6	$>10^6$
Plasma drag	$2\cdot10^{2\pm1}$	$2\cdot10^{3\pm1}$	$10^{5\pm1}$	$10^{6\pm2}$
Exospheric drag	-	-	$10^{2\pm1}$	$10^{5\pm1}$
Sputtering	$3\cdot10^{2\pm1}$	$10^{3\pm1}$	$10^{5\pm2}$	$10^{7\pm2}$
Meteoroid collisions	$10^{5\pm1}$	$10^{6\pm1}$	$10^{6\pm1}$	$10^{6\pm2}$

3. Jupiters Ring

A faint narrow ring was detected during approach of Jupiter by the Voyager 1 spacecraft. Figure 1 shows the ring as seen from behind Jupiter. A faint toroidal halo just outside the bright ring and an even fainter extension out to 2.5 R_J have been found. Two small satellites of 10km and 15 km in radius were detected within the ring.

Impacts onto these satellites and other small parent bodies within the ring have been proposed (Morfill et al., 1980 and Burns et al., 1980) to be the production mechanisme for the ring particles. Small satellites (of a few 10 km diameter) are a prefered sources of ejecta particles since the escape velocities from their gravitational fields are small (3 - 10 m/s) and hence most ejecta can escape. Morfill et al. proposed volcanic ash from Io to diffuse through the magnetosphere until it hits the parent bodies while Burns et al. suggested interplanetary meteoroids as projectiles of the impacts which release large quantities of ejecta. A resolution may come from the Galileo mission which will study the Jovian system during its orbital tour for several years and which will investigate the ring and its dust environment.

Figure 1 Mosaic of Jupiter's ring while Voyager 2 was in Jupiter's shadow looking back towards the sun. Jupiter's limb is highlighted (thick circle) as a result of forward-scattering by hazes in the planet's atmosphere. In the left frame the foreground part of the ring enters Jupiter's shadow and is no longer illuminated. Both the bright narrow ring and the faint halo are visible (from [2]).

4. Saturn's Ring System

Various components of Saturn's ring system were identified by groundbased observations: the bright and optically thick ($\tau \sim 1$) A and B rings which are separated by the Cassini division and the faint inner C ring as well as an even fainter E ring extending out to 8 R_S (1 R_S = 60,000 km). The E ring is only visible if the ring is seen edge-on (Baum et al., 1981). Maximum brigthness of this wide dust ring is located near the orbit of the satellite Enceladus (at 4 R_S).

Figure 2 Saturn's ring system in front of the planet. Inside the B ring the optically thin C ring is visible. From the appearance of the planet through ring variations of the optical depth of the B ring can be estimated. The wide Cassini division separates the A and B rings. Near the outer edge of the A ring the narrow dark Encke division can be seen. The narrow F ring can be seen outside the A ring. The dark belts at Saturn are shadows of the thick rings. All rings except the faint C ring inside the B ring contain significant amounts of dust (from [2]).

Voyager observations showed radial structure of the inner rings down to the resolution of the camera (Figure 2). These rings are made of narrow ringlets of varying optical depths and brightnesses. Also inside the Cassini division material was found. Some of the radial structure observed was identified to be due to resonances with satellites. Centimeter to meter sized icy objects were found to be responsible for a major part of the optical cross section of the inner rings (Tyler et al., 1982).

Several new satellites and narrow ringlets were found outside the bright rings. The braided F ring which is locked up by two shepherding small satellites and the narrow G ring which has no relation to any known satellite. Most of these rings contain large quantities of small dust particles. The existence of dust in optically thick rings is indicative of effective source mechanismes which counteract the sweep-up by larger ring particles. Voyager 2 flew through the outskirts of the G ring while impacts of micron sized ring particles onto the spacecraft were directly recorded by the instruments (Gurnett et al. 1983).

Azimuthal brightness variations called spokes (Figure 3) extending over approximately 10,000 km in radius on the B ring have been observed already during Voyager's approach to Saturn. This was a major surprise since any such structure will smear out by differential

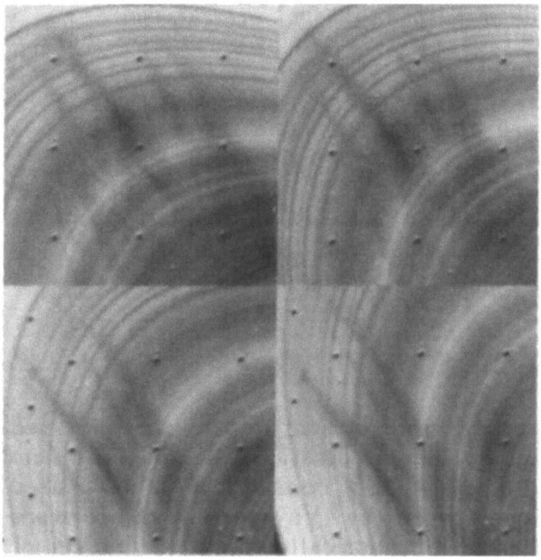

Figure 3 Sequence of images of the B ring showing spoke formation. The time difference is 5 minutes starting with upper left frame, the last frame is the lower right one. The new spoke forms radial right of the dark spoke visible in all four images. The regular dots are calibration marks of the camera (from [2]).

rotation within less than an orbital period which is approximately 10 hours.

Micron sized dust particles cause the almost radial spokes because it was noted that spokes appear dark in back scattered light but appear bright against the B ring in forward scattered illumination. This suggests that clouds of small particles elevated above the ring plane are responsible for this phenomenon. Spokes have been observed in the B ring between 1.6 and 1.9 R_S from Saturn's center (Grün et al., 1983). They commonly appear wedge shaped with a vetex at 1.87 R_S which is the distance of the synchronous orbit. The width at their base (towards Saturn) is ranges from 500 km to 12,000 km.

Several narrow spokes were observed during formation along radial lines in the sunlit portion of the ring (cf. Figure 3). The formation time is ≤ 5 minutes for a 6,000 km long spoke. The rate of formation is highest at the morning part of the B ring which just has left Saturn's shadow. Spokes are formed radially and tilt away from radial direction during the later parts of their existence. Some spoke activity seems to be correlated with Saturn's rotation rate (10.66 hours). Several recurrent spoke patterns have been observed at that rate.

The most comprehensive theory of spoke formation is that of Goertz and Morfill (1983). They show that a dense plasma cloud near the ring will charge up small dust particles which sit at the surface of larger ring particles. Once they are charged they are electrostatically lift off the ring. Such a plasma cloud contains charged dust particles which move with Keplerian orbital speed while the plasma is locked to the magnetic field. The relative speed between the two causes an azimuthal polarization of the plasma and a radial drift of the plasma cloud is the consequence. Thereby a radial dust trail is formed above the ring which after approximately half an orbital revolution settles back onto the ring.

This theory is able to account for the energetics of spoke formation, dust levitation off the ring plane and the radial alignment of young spokes. However, unsolved questions are the origin of the dense plasma clouds and the periodicity of spoke activity.

5. Uranus Rings

Nine narrow sharp edged ringlets of large optical depth ($\tau \sim 0.3$) had been identified by ground based stellar occultations between 1.6 and 2.0 R_U (1 R_U = 25,600 km). These rings consist of meter sized objects. A long exposure image taken by Voyager 2 at forward scattering geometry revealed a broad but optically thin distribution of dust within the whole ring system (Figure 4). These dust bands exhibit much more structure than was expected from the narrow rings. Only two new small satellites of 20 km radius have been identified within the ring system. High impact rates onto the spacecraft were observed while Voyager 2 crossed the ring plane at 4.5 R_U distance perhaps representing a disc like extension of the visible ring system.

Figure 4 Uranus ring system. Most of the bright rings are comprised of small particles. Several of the narrow rings observed by ground based stellar occultations are barely visible (marked by white triangles at the left margin). Due to the long expusure time background stars form short trails (from [2]).

The dominant dynamical effect acting on dust particles in the Uranian system is the drag excerted by the exosphere of hydrogen which escapes from the atmosphere. This effect causes a fast decay (~ 100 years) of dust orbits and a bias towards larger particles.

Schematic models of ring evolution (Colwell and Esposito, 1990) include creation of dust particles from meteoroid impacts onto rings and ring moons, liberation of regolith material from larger ring particles by mutual collisions, sweep up and accretion of dust by optically thick rings and satellites, spreading of narrow rings and orbital decay of dust due to exospheric and Poynting-Robertson drag and finally capture of dust by the upper atmosphere. Large amounts of unseen material (i.e. small satellites or optically thin rings of meter sized objects) are required to explain the structure and stability of the dust ring.

6. Neptune's Ring

Two narrow rings at 2.1 and 2.5 R_N (R_N = 24,800 km) distance from the center of the planet and two broad rings one at ~ 1.7 R_N and a broad plateau extending outward from the inner narrow ring out to 2.4 R_N circle the planet (Figure 5). The optical depth of the narrow rings is approximately 0.01 whereas the optical depth of the broad rings is ~ 10^{-4}. The fraction of microscopic dust in the rings and ring arcs which contributes to the optical depth ranges from 0.2 to 0.8. This value is significantly larger than that found in the main (thick) rings of Saturn and Uranus.

The outermost of these rings has three optically thicker (τ ~ 0.1) arc segments clustered together within 30 degrees of longitude. Since

Figure 5 Two images showing the Neptunian ring system in forward scattering geometry. The detected brighter ring arcs in the outermost ring are not captured in either of these frames (from [2]).

these arcs were suspected from earlier ground based stellar occultations their life times must be larger than five years. Four small satellites (of 30 to 90 km radius) have been found within the ring system. However, none of them seem to be able to shepherd the outer ring and, thereby, causing the arcs.

High rates of micron sized particle impacts have been recorded by Voyager's instruments during both ring plane crossings at 3.5 and 4.2 R_N distance. Significant impact rates were also detected more than 20,000 km above and below the ring plane, as well as over the northern polar region. These data suggest that a dense, thin disk of dust extends along the equatorial plan, surrounded by a tenuous wide halo on either side of the equatorial plane.

Planetary rings exhibit a large variety of phenomena most of which we have not yet disentangled. Small dust particles are removed from the rings within short times, therefore, their mere existence in these rings give testemony of violent processes causing their formation. Most of this dust is lost from the rings in times much shorter than the age of the planetary system, therefore, shorter ages of the rings themslves are required as well.

References

[1] Planetary Rings (R. Greenberg and A. Brahic, Eds.), Univ. of Arizona Press, Tucson (1984)

[2] Special issues on Voyager's planetary encounters:
Jupiter: Science **204** 945-1008 (1979) (VGR 1) and Science **206** 925-996 (1979) (VGR 2)
Saturn: Science **212** 159-243 (1981) (VGR 1) and Science **215** 499-594 (1982) (VGR 2)
Uranus: Science **233** 39-109 (1986)
Neptune: Science **246** 1417-1501 (1989)

Baum W A, Kreidl T, Westphal J A, Seidelmann P K and Pascu D: Icarus **47** 84 (1981)

Burns J A, Showalter M R, Cuzzi J N and Pollack: Icarus **44** 339 (1980)

Colwell J E and Esposito L W: Icarus 86 **530** (1990)

Elliot J L, Dunham E W and Millis R L: Nature **267** 328 (1977)

Goertz C K and Morfill G E: Icarus 53 **219** (1983)

Grün E, Morfill G E, Terrile R J, Johnson T V and Schwehm G: Icarus **54** 227 (1983)

Grün E, Morfill G E and Mendis D A: in [1] (1984)

Gurnett D A, Grün E, Gallagher D, Kurth W S and Scarf F L: Icarus **53** 236 (1983).

Morfill G E, Grün E and Johnson T V: Planet. Space Sci. **28** 1101 (1980)

Smith B A, Soderblohm L A, Ingersoll A P, Collins S A, Shoemaker E M, Hunt G E, Masursky H, Carr M H, Davies M E, Cook A F, Boyce J, Danielson G E, Owen T, Sagan C, Beebe R F, Veverka J, Strom R G, McCauly J F, Morrison D, Briggs G A and Suomi V E: Science **204** 951 (1979)

Tyler G L, Marouf E A, Simpson R A, Zebker H A and Eshelman V R: Icarus **54** 189 (1982)

Asteroid-Asteroid Interactions – Dynamically Irrelevant?

M. Hoffmann

Observatorium Hoher List der Universitäts-Sternwarte Bonn, W-5568 Daun, Fed. Rep. of Germany

Structural investigations of the distribution of the asteroids have so far concentrated on the interactions with the major planets, especially Jupiter, and the resulting effects in element space. Here some less known structures of their distribution in actual space are shown: the radial number density and the orthogonal orbit density. A closer look at the orbit densities at different length scales reveals structures that are difficult to explain by gravitational influences from the major planets. This is particularly conspicious if special subgroups like the original Hirayama families are considered. A further limit for inter-asteroid-interactions comes from the distribution of locations of actual close encounters in space. A comparison with the gravitational cross section densities of the terrestrial planets and their influences on the population of planet crossing asteroids indicates that some interactions with the asteroids should lead to observable effects but the involved timescales are difficult to estimate because of the inhomogeneity of the asteroid distribution.

Introduction

Cosmogonic and evolutionary processes in the planetary system and among its minor constituents can only be understood if all aspects of their actual configuration are known. Besides a considerable contribution by collisions the distribution of the asteroids is dominated by the dynamical influence of Jupiter and some degree by the terrestrial and outer planets. These lead e.g. to the Kirkwood gaps, secular resonances, and chaotic regimes in space. The resulting phenomena in actual space are less conspicious. Dynamical interactions between asteroids are seemingly negligible, as the mass even of the most massive asteroid is only 10^{-7} times that of Jupiter. Consequently observable effects of asteroid gravity can only by expected on very short distance scales.

The distribution of asteroids in space

From the 4600 presently numbered asteroids the elements of 3494 objects of 1987 July 24 were taken to calculate several properties of the actual asteroid distribution. First, the actual radial number density of these objects was calculated for intervals of 0.1AU. Contrary to the resonance-gap modulated distribution of semimajor axes the longer presence close to perihelion (if the eccentricity is below 0.5, which is mostly the case) and even more close to aphelion leads to an entirely smooth distribution with a maximum of 276 objects/ 0.1AU at 2.8AU.

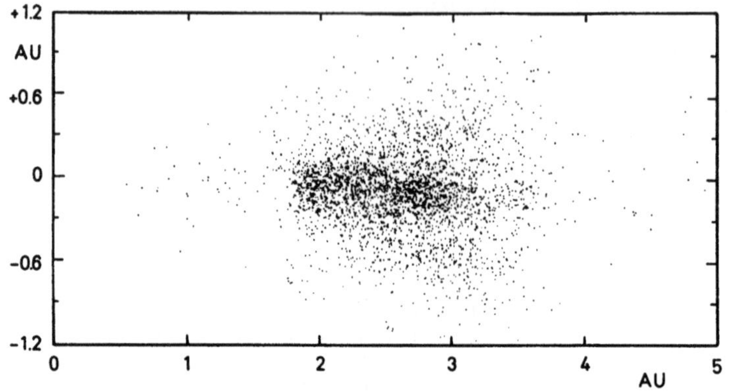

Fig. 1 Intersections of asteroid orbits with the plane of longitude 0°.

Generally the distribution of asteroids in longitude and perpendicular to the invariable plane is assumed to be structureless except a tendency for an alignment of the lines of apsides to that of the orbit of Jupiter and a mild concentration of orbits to the plane with a mean inclination of 9.°4 +/- 6.°1 (Ip, 1977). In fact, a look at the intersection points of the orbits of the first 3494 numbered objects with all planes (spaced by 20° in longitude) perpendicular to the invariable plane, and parallel to the radial direction towards the sun show some unique structures. Besides the resonance gaps in radial direction (fig. 1) also sharp edges in orthogonal direction become visible. They are not present at heliocentric distances below 2.6AU (fig. 2), where a probably gaussian-like smooth distribution is present, but quite pronounced at heliocentric distances between 2.6AU and 3.25AU(fig. 3). There a structure that can be called a ring core is superimposed to the bell-shaped general distribution. It has to be attributed to the compact Koronis and Themis families, while the Eos family causes only a gradual transition region. Also beyond 3.25AU this ring core remains present, as shown for the distances interval 3.55AU to 4.1AU (fig. 4). The longitudes of the planes selected for the figures 2 to 4 are representative for all other longitudes. While this sharp discontinuity may be maintained by some less known secular resonance or other external influences, its effects on the encounter and collision probability distribution is not negligible at all. In this circumstance the locations of the IRAS zodiacal dust bands, which coincide with the boundaries of the ring core, should be mentioned (Sykes, 1990). It can be concluded that the asteroid belt with its structures resembles much more a planetary ring than previously interpreted. Mayo (1979) investigated the effects of the asteroid belt assuming a disk-like potential. Williams (1984) pointed out that the potential of the asteroid belt at any time is lumpy, because the majority of its mass is tored in the largest objects. As the masses are small only effects accumulated over longer time scales will reach noticable amounts. In this case the potential of the asteroids can be thought as smeared over their whole orbit or, in case of commensurate orbital periods, over some finite arcs.

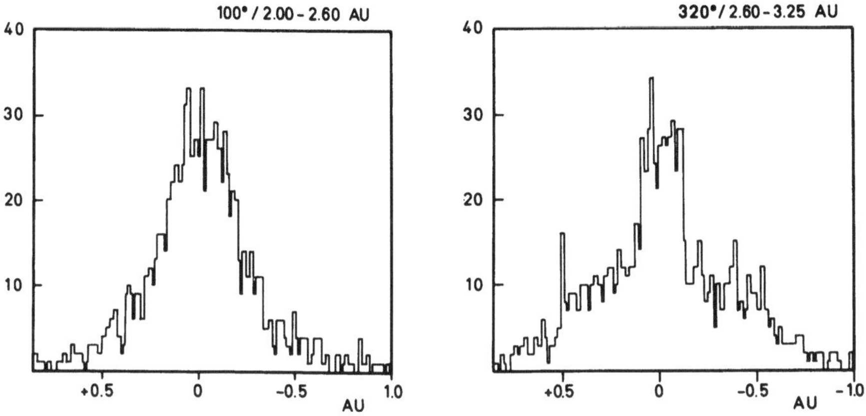

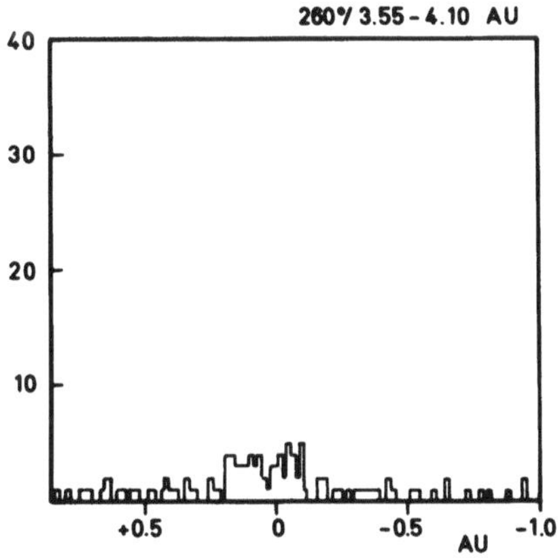

Fig. 2-4 Orthogonal profile of orbit frequencies for the plane of longitude 100° and the heliocentric distance interval 2.0AU to 2.6AU (above), longitude 320°, heliocentric distances 2.6AU to 3.25AU (upper right), and longitude 260°, heliocentric distances 3.55AU to 4.1AU (lower right).

Longitude dependence of mutual asteroid orbit distances

Each 20° of longitude all distances between the intersection points of the orbits of the 3494 first numbered asteroids were counted in a cumulative way for specific upper distance limits. The limits 0.25AU and 0.01AU proved to be quite instructive. The former is small enough to exclude severe boundary problems of the finite belt dimensions, but large enough to be counted as a "global" distance in terms of the belt. The smaller interval is small enough to show the summary of local, orbit dependent properties but large enough to avoid problems with the statistics of small numbers. Figure 5 shows the results of these counts depending on the longitude. A wave-line with a maximum at 0° longitude is present in both distance intervals. This agrees fairly well with the longitude of the perihelion of Jupiter. The belt seems to be compressed, therefore, by the giant planet. The minimum, however, occurs at the longitude 260°. This is not far from the aphelion of the orbit of Saturn, but if this distant and less massive planet is able to cause such a strong asymmetry in the asteroid belt, may be doubtful. In any case the asymmetry belongs to the global properties of the belt. This distribution is unlike that of the locations of the least distances between all orbits, which is

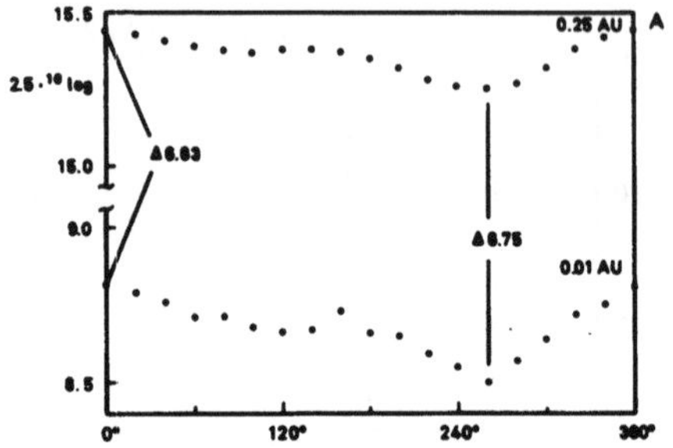

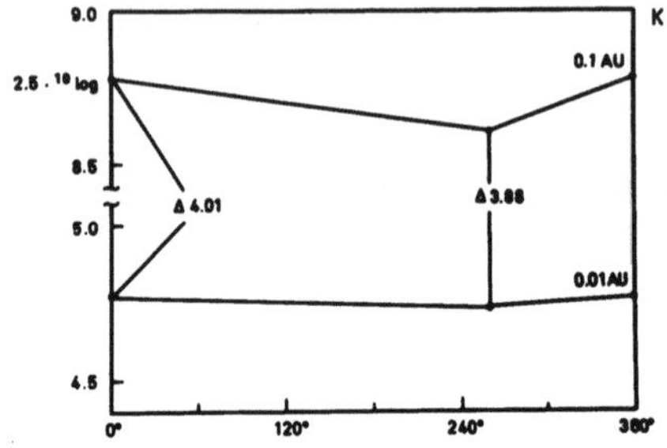

Fig. 5-7 Cumulative numbers of orbit distances at given longitudes for upper distance limits 0.25AU and 0.01AU of the first 3494 numbered asteroids (upper left); for numbers of the Koronis family only: Upper distance limit: 0.1AU (upper right); for encounters of members of the Koronis family with those of the Themis family only: Upper distance limit: 0.1AU (lower right).

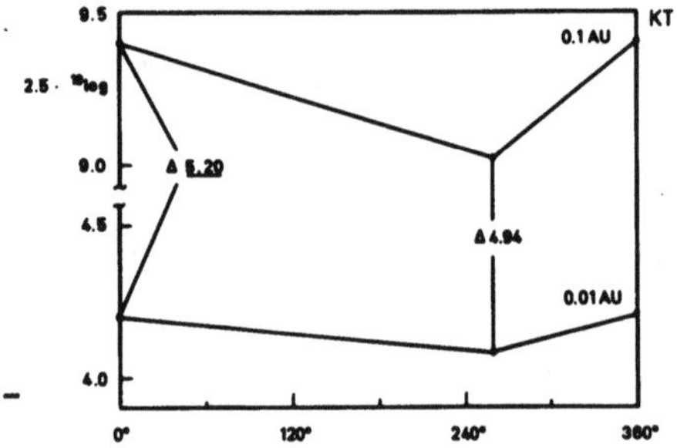

much more symmetric (Kuzmanoski, 1990). Figure 5 also shows that the asymmetry is significantly stronger for small distances of all regions of the belt are involved, an explanation by influences of the large planets is not obvious. The ratios of the counts of the intervals at the two extreme longitudes indicate an underabundance of the small distances at 260°, not the opposite at the other longitude.

If internal dynamical processes are detectable in the asteroid belt, they should be most pronounced for interactions with the most massive objects and populations that have extraordinary high probabilities for efficient encounters. The three original Hirayama families represent quite favourable test groups of that kind, because the Koronis and Themis families are populous and confined to a small range of low inclinations. Therefore close, efficient encounters with small relative velocities should occur frequently. The asteroids of the Eos family are also quite homogeneous with respect to their orbital elements, but their much greater inclinations preclude such frequent, slow encounters. This time the wide and narrow distance intervals chosen are 0.1AU and 0.01AU. Comparing the numbers of encounters up to the given distance limits it becomes obvious that the ratio of the numbers of encounters at 0° longitude is significantly greater than at 260° for the Koronis family and the combination of the Koronis with the Themis family (figs. 6 and 7). The ratios at 0° are too great to be expected for an equal distribution of the orbits, and can only by reached, if the orbits avoid small mutual distances just at the longitude where the orbit density not only of the asteroids in general, but also of the family objects in particular, is the greatest. This tendency is absent for the less compact Themis and the higher inclination Eos family.
The largest and most massive asteroid, 1 Ceres, is located in the central region of the asteroid main belt and has no

obvious orbital properties that could be termed atypical for average main belt asteroids. Again cumulative counts of orbit distances were made for a wider and a narrower interval (0.1AU and 0.01AU). To reduce small number effects that are difficult to avoid for encounters with a single object, the data of each 9 neighbouring planes were combined that leave two different hemispheres for a comparison. If the longitude intervals 120°-280° and 300°-100° are considered, which approximately represent the hemispheres with the greatest differences if all asteroids are taken into account, the numbers are 1221 and 1213 orbits within 0.1AU, and 5 versus 18 orbits within 0.01AU. While the numbers of the wide interval are essentially equal, those of the smaller one differ almost by a factor of 4 in the same sense as the higher amplitude of the distribution wave of figure 5. Still this difference is statistically only marginally significant, and it may depend on the reader if this result is accepted as an indication for a non-random process.

The distribution of the locations of very close encounters

Kuzmanoski and Hoffmann (1990) have collected a list of known close asteroid encounters in space. These involve mostly at least one object with a diameter of more than 100km. Their dependence on longitudes and heliocentric distances was investigated. The distribution of the longitudes of encounters with mutual distances <0.01AU is approximately the number density of asteroids but shows an unusually great scatter. This distribution will be discussed below in more detail.

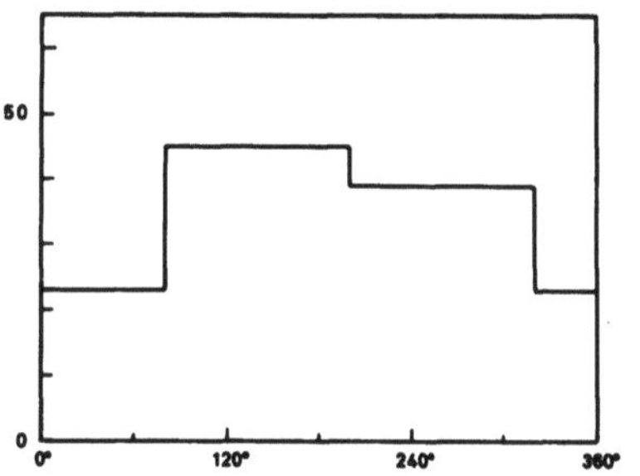

Fig. 8 Numbers of encounters of asteroids in space with mutual distances <0.01AU depending on the longitude arranged in 3 intervals of 120°

To estimate the quantitative efficiency of asteroid-asteroid interactions those between near-Earth asteroids and the terrestrial planets can give helpful informations. A direct criterion on the dynamical relation of an object to a massive perturber is the Tisserand-invariant (Kresak, 1972).
Usually it is given for the restricted three-body system Sun - Jupiter - test object. It is defined as

$$T = a_J / a_A + 2 * SQR (a_A * (1 - e_A^2) / a_J)$$

with a_J, a_A denoting the semimajor axes of Jupiter and the test asteroid, e_A its eccentricity (SQR representing the square root sign).
A value close to 3 indicates the closest dynamical relationship to the orbit of the secondary object. In the case of Jupiter this is typical for the Trojan and Hilda asteroids. If we

substitute Jupiter by the Earth and compare the so modified Tisserand invariants of the near-Earth asteroids with their semimajor axes, there is a group of objects with T very close to 3 which is separated from those with smaller T by a marked gap. Furthermore the number of objects with perihelia or aphelia close to the orbit of the Earth and the other terrestrial planets is much too low with respect to a uniform distribution thus showing their dynamical influences. Therefore, a comparison of the gravitational cross sections of the asteroids will yield a quantitative basis for the dynamical influences inside the asteroid belt, if the involved scales of the life-times in a given environment (dynamical and collisional) are taken into account.
The term "gravitational cross section" is a matter of definition, as there is no distance limit for changes of orbital elements by perturbations of other planetary objects. Therefore several definitions for a so called sphere of action exist. Here the Tisserand definition (Öpik, 1976) shall be adopted, which can be written as

$$W / V \sim ((m_A / m_S) ** 0.4 * R) ** 3 =$$
$$(d ** 0.4 * r^{1.2} * R) ** 3$$

with the volume V of a shell given by the heliocentric distance interval that is used for the norming of the results, m_A and m_S the masses of the perturbing asteroid and the sun, R the heliocentric distance, r the radius of the perturbing asteroid, and d its density.
One important feature of this equation is the exponent of the dependence of the cross section volume on the radius of the perturbing planet. An exponent smaller than 2 means that a given mass distributed over several objects produces a larger gravitational cross section than it would result from a concentration of the same mass in only one object. This is in favour of the efficiency of the asteroids where we have a small total mass scattered in many objects. On the other hand, an exponent larger than one means that the fraction of the gravitational cross section filled by the collisional cross section will increase for smaller masses until the radius of the gravitational cross section coincides with the radius of the objects themselves. This limiting radius depends on the choice of the other parameters. Assuming a density equal to that of Mars and a cross section radius of Jupiter r=0.323AU it becomes 47km at a heliocentric distance of 1.5AU and 26km for 3.0AU. Some modulation of the distribution of the cross section densities with respect to the heliocentric distance will originate in the local variation of the power law that describes the frequency of the radii of the asteroids. Figure 9 shows a comparison of the number and gravitational cross section densities depending on the heliocentric distance.
In this distribution steps and peaks can be noticed that originate in the resonant structure of the distribution of the most massive asteroids and which act as (small) potential wells.
Figure 10 compares the significance of the gravitational cross section densities of the asteroids with that of the terrestrial planets. The maximum cross section density by the asteroids is about 1/200 that by Mars. Therefore the gravitational effects by the asteroids are indeed small, but in spectral cases they are obviously not negligible, e.g. in cases of resonant interactions.

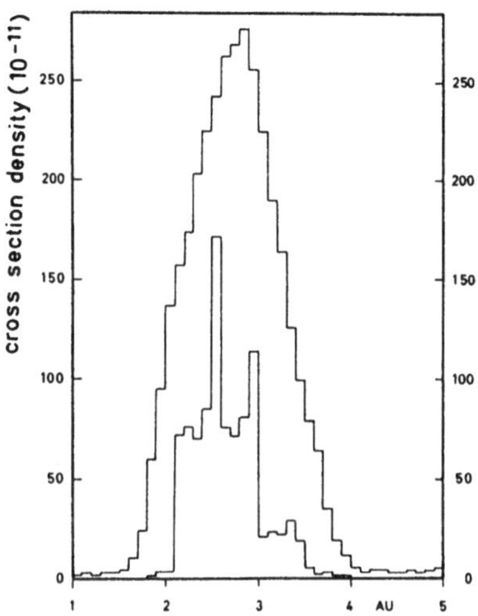

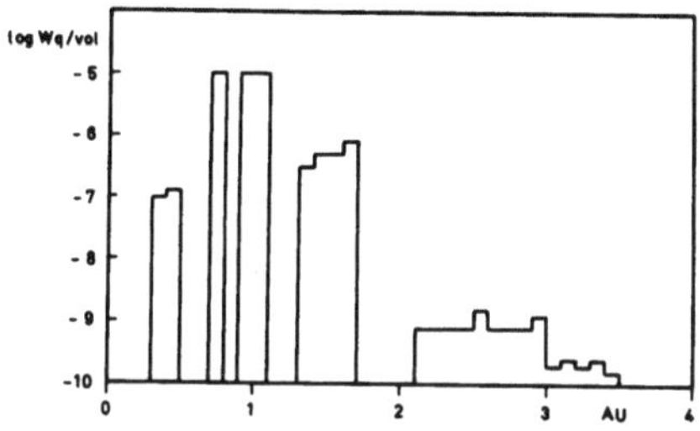

Fig. 10 Distribution of the gravitational cross section densities (logarithmic scales) of the terrestrial planets and asteroids.

Fig. 9 Heliocentric distribution of the gravitational cross section density of the asteroids (lower curve) and their number density (upper curve).

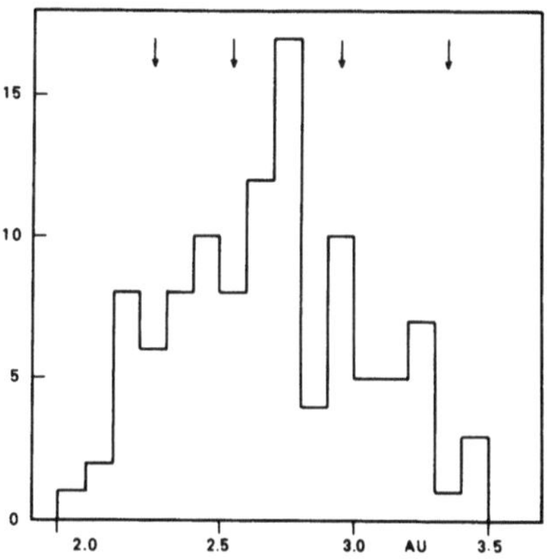

Fig. 11 Heliocentric distribution of encounters of asteroids in space with mutual distances <0.01AU. Arrows indicate the local maxima of the gravitational cross section density.

Figure 11 shows the heliocentric distribution of the very close encounters mentioned above. At the locations marked by arrows the gravitational cross section densities show maxima in fig. 9. Strikingly these two distributions are highly anticorrelated. Similarly anticorrelated are the cumulative asteroid counts and the longitudes of the actual close encounters (figs. 5 and 8).

Generally it can be concluded that under conditions that should lead to an enhanced probability of dynamically efficient asteroid-asteroid interactions the observed number of such events is systematically too low. It is obvious that such volumes in element space, where the probability of those elements is especially high, will be depleted much faster. The individual orbits are then subject to a "mild" chaos, which precludes studies of long term evolutions by numerical integrations. On the other hand, statistical treatments of the problem, in the sense of a scattering process promise informations if the individual scattering functions are calculated explicitly. The dynamical efficiency of such processes is also enhanced by mean motion commensurabilities, e.g. 1:1 for members of the same familiy. Because of the diversity of the kinematical properties of the asteroids a homogeneous relaxation time scale cannot be defined for them. The different ratios of collisional versus gravitational cross section should also have consequences on the fragmentation, cratering and regolith formation on asteroids of very different size. A further open problem to be attacked by the assumption of gravitational scattering between asteroids is the exchange between dynamically isolated regions of element space, in perticular the transfer of larger objects into the Phocaea and Pallas regions. Still the significance of such investigations is influenced by the possibility of rare encounters of the asteroid region with very large comets.

References

Ip, W.-H. 1977, Icarus 32, 378
Kuzmanoski, M. 1990, personal communication
Kuzmanoski, M., Hoffmann, M. 1990, in preparation
Mayo, A. P. 1979, Celest. Mech. 19, 317
Sykes, M. V. 1990, Icarus 84, 267
Williams, J. G. 1984, Icarus 57, 1

First Results from the X-Ray Astronomy Mission ROSAT

B. Aschenbach

Max-Planck-Institut für Extraterrestrische Physik,
W-8046 Garching, Fed. Rep. of Germany

1. Introduction

X-Ray astronomy is now moving towards the end of the third decade of observational research since the discovery of the first non-solar cosmic X-ray source in 1962. Collimated counters have been used exclusively in the first 16 years, and they are still the only way to perform X-ray observations at energies above about 20 keV. At lower energies imaging telescopes offer the opportunity to take true pictures of celestial sources at a significantly reduced level of background noise and angular resolution of a few arcseconds up to now. In this respect X-ray astronomy has come close to the performance delivered by ground-based optical telescopes and radio telescopes.

In 1978 the Einstein Observatory was launched and the first imaging X-ray observations from a satellite were performed. At the end of the mission it had been demonstrated that X-ray emission from cosmic sources is not the exception but is the rule. Almost every class of astronomical objects emits X-rays, a fact which promises a rich scientific return from an all-sky survey with an imaging telescope.

On June 1 1990, 17:48 local time, after about 15 years of research, development and construction the German X-ray astronomy satellite ROSAT was launched from Cape Canaveral by a Delta-II rocket into a circular near-Earth orbit of 580 km altitude and 53° inclination. The satellite carries the largest and most precise X-ray telescope built so far with 3 focal plane instruments. A second telescope, the Wide Field Camera, extends the energy range of the X-ray telescope from the soft X-ray band longward into the XUV and EUV. On June 16 the X-ray telescope saw first light during a set of pointings at regions in the Large Magellanic Cloud. A detailed programme of observations on a variety of different sources followed to establish the calibration and to verify the performance of the instruments and the various operational modes. On July 30, 1990 the commissioning phase ended and the all-sky survey started with both the X-ray telescope and the

Wide Field Camera co-aligned. The survey will be completed about half a year later and from then on the pointing observations of the guest observer programme will be carried out.

The general summary of the past few months of ROSAT operations is that both the spacecraft and the scientific instruments perform as anticipated, sometimes even better. The scientific return, originally in the form of about 700 Mbit/day science data, is overwhelming, and it is clear that at present no full account of the science can be given. Instead, I will restrict the presentation to a few highlights and I will try to give something like a preview on what we expect ROSAT will contribute to astronomy and astrophysics.

2. ROSAT: the Spacecraft and Scientific Payload

ROSAT is the biggest scientific satellite built so far in Western Europe. The total mass of the satellite amounts to 2.4 tons with a maximum length and diameter of 4.5 m and 2.4 m, respectively. An artist view of ROSAT is shown in Fig. 1. The satellite is electrically powered by a solar array which generates about 1 kW (see Fig. 2). The attitude of the spacecraft is measured and controlled in three orthogonal axes by means of sun sensors, 2 CCD-based star cameras and gyro packages. The torques produced by a set of 4 reaction wheels are used to stabilize the spacecraft counteracting residual disturbing torques which are imposed on the spacecraft by the drag of the upper atmosphere, the solar radiation pressure and the gravity gradient. It is also used, of course, to manouvre the spacecraft during the scanning and slewing.

ROSAT carries two independent but co-aligned imaging telescopes. The X-ray telescope (XRT) (see Fig. 3), working between 0.1 to 2.4 keV, consists of a 4-fold nested Wolter type I mirror assembly of 2.4 m focal length and a maximum aperture of 835 mm, the design of which was made at the MPE (*Aschenbach* [1]). The mirror assembly has been built by Carl Zeiss, and extraordinary performance values have been achieved. The angular resolution is estimated to 3.6 arcsec, and the residual microroughness of the mirror surfaces is 2.8 Å. These data still await in orbit confirmation.

The focal plane instrumentation, developed and built at the MPE, includes a carousel with three X-ray detectors, each of which can be rotated into the focus of the telescope. Two detectors are redundant Position Sensitive Proportional Counters (PSPC's), which are operated in gas flow mode (*Pfeffermann et al.* [2]). The PSPC provides spatial and spectral resolution over the full field of view of 2° which vary slightly with photon energy E. The angular resolution is limited by the PSPC to about 25 arcsec on-axis and degrades to a few arcminutes for field angles larger than about 20 arcmin

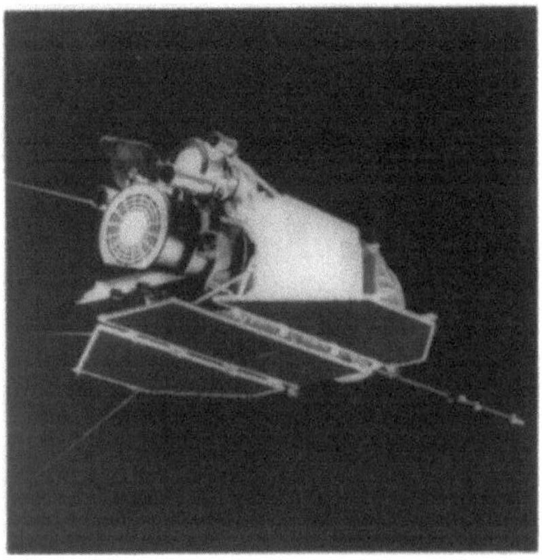

Fig. 1. The X-ray astronomy satellite ROSAT (photograph Dornier)

Fig. 2. View on the unfolded solar array of ROSAT (photograph Dornier)

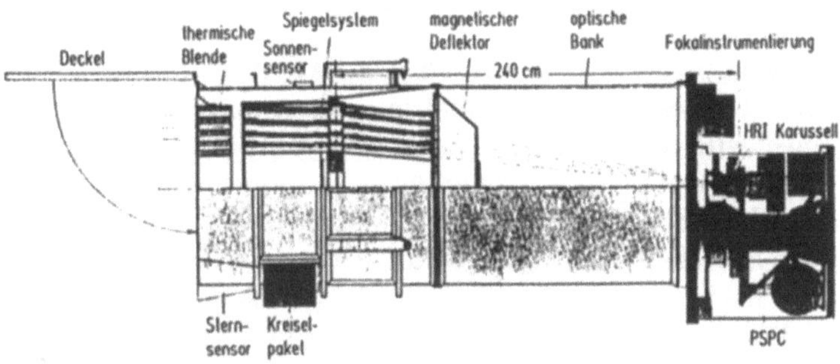

Fig. 3. Cross section of the ROSAT X-ray telescope (photograph Dornier)

due to the geometric aberrations of the telescope. The energy resolution is $\Delta E/E = 0.41/\sqrt{E_{keV}}$. The effective collecting area including the PSPC efficiency is about 220 cm^2 at 1 keV for on-axis rays and decreases slightly with increasing field angle.

The third detector, the High Resolution Imager (HRI) built at the SAO Cambridge, USA, is a microchannel plate detector with an intrinsic spatial resolution of about 20 μm. With the HRI in the focus of the telescope the angular resolution is dominated by the point spread function of the mirror assembly. A value close to 4 arcsec is expected in orbit. However, up to now no measurements have been made with the HRI to assess the ultimate angular resolution of the telescope.

A photograph of the complete XRT, which is 4 m long and has a mass of about 1.5 tons, is shown in Fig. 4. The XRT has undergone extensive ground test and calibration measurements, including full aperture X-ray tests, in MPE's 130 m long X-ray facility. During the ground calibration an upper limit of 5.6 arcsec for the on-axis angular resolution has been measured, which is however dominated by the finite distance of the X-ray

Fig. 4. The flight unit of the ROSAT X-ray telescope

source. More detailed information about the ground calibration is given in the paper by *Briel et al.* [3].

The second telescope is the Wide Field Camera (WFC) furnished by the United Kingdom. The telescope is a 3-fold mirror assembly of the Wolter type I arrangement with a focal plane turrent from which one of two identical microchannel plate detectors can be selected. A set of 8 filters defines the spectral wavelength passbands which cover the range from about 17 to 210 eV. The telescope has a focal length of 525 mm, a maximum mirror diameter of 576 mm and a field of view of 5° diameter. A more detailed description of the WFC and an account of the first successful observations have been presented by *Pounds and Wells* [4].

A comprehensive and more detailed description of ROSAT and the instruments can be found for instance in the papers by *Trümper* [5].

3. In-Orbit PSPC Performance

One major factor, which determines the sensitivity of astronomical instruments to detect point like sources and to measure brightness distributions of extended objects, is the amount of unwanted background. For X-ray astronomy measurements two components of background radiation are to be considered, which are due to the photons of the diffuse celestial X-ray background emission and charged particles, the latter of which mimic X-ray photons in the detector. Whereas the study of the diffuse X-ray background is an astrophysically interesting objective in itself, the detector background induced by charged particles is usually not, and the rate of events accepted by the detector is minimized by anti-coincidence techniques and shielding.

Fig. 5 shows the pulse height spectrum of the charged particle background and the diffuse X-ray background measured with the ROSAT PSPC in orbit during a pointing at β Leonis. The spectrum of the particle background is essentially flat, and the total amount of accepted background events amounts to 3×10^{-5} counts s^{-1} $arcmin^{-2}$. This very low background means that pointed PSPC observations with a typical resolution element of 25 arcsec diameter are not background limited for exposures less than 10^5 sec. For the survey the resolution element is about 2 arcmin in diameter, and PSPC observations become particle background limited only for times exceeding 10^4 sec, which is available only in regions close to the ecliptic poles. Therefore the majority of the ROSAT PSPC observations both in the all-sky survey and the pointing programme are essentially free of particle background contamination, and the majority of the recorded events are true X-ray photons.

The spectrum of the diffuse X-ray background differs from the spectrum of the charged particle background and has a significantly higher brightness for energies lower than about 2 keV (see Fig. 5), so that the mapping of the X-ray background by ROSAT is essentially limited by the brightness of the X-ray backround emission itself. It is estimated that the morphology of the diffuse emission can be mapped at an angular resolution of about 6 arcmin over the whole sky. The energy spectrum (see Fig. 5) clearly shows two components, at least. The prominent bump at energies below 0.5 keV is attributed to galactic emission, whereas the flux above 1 keV is very likely to be of extragalactic origin with some mix for energies in between.

It is mentioned that under bad viewing conditions of the telescope close to the Earth horizon scattered solar X-rays provide a third source of background.

The spectral resolution of the ROSAT PSPC is illustrated by the following three sample spectra. Fig. 6a, 6b and 6c show the pulse height spectrum of Cyg X-2 [6], HZ43 [6], and AR Lac [7], respectively. These spectra have been chosen to demonstrate the capability of the PSPC to discriminate between different source temperatures. Cyg X-2 is a low mass X-ray binary, containing a mass accreting neutron star, so that a hard X-ray spectrum is emitted. The spectrum is photolectrically absorbed at low energies by interstellar matter, so that the resulting spectrum peaks at about 1.1 keV.

HZ43 is a hot DA white dwarf with a source temperature of 50,000 K and very low interstellar absorption equivalent to 6-16$\times10^{17}$ atoms cm^{-2} [8]. Therefore the PSPC spectrum peaks at about 100 eV.

AR Lac is a nearby active RSCV system. The ROSAT PSPC spectrum shows two pronounced peaks at about 180 eV and 900 eV, and the spectrum can be interpreted to originate from a two temperature plasma with log T_{low} = 5.8 and log T_{high} = 7.1 [7]. Clearly a multi-temperature plasma with an emission measure continuously varying with temperature would fit the data even better (see Fig. 6c).

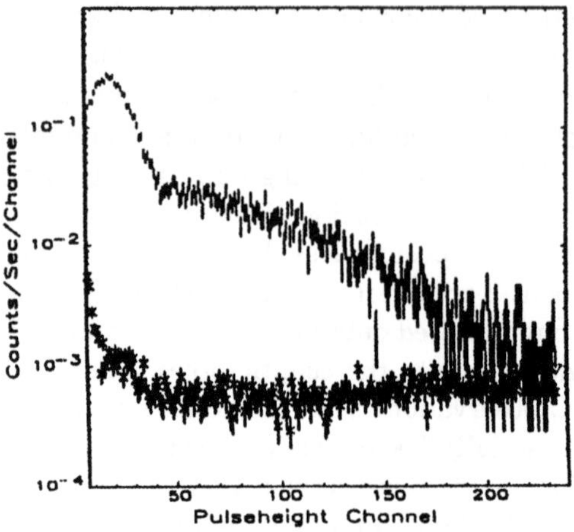

Fig. 5. PSPC pulse height spectra of the diffuse X-ray background (top data) and the charged particle background (bottom data); 1 channel~10 eV

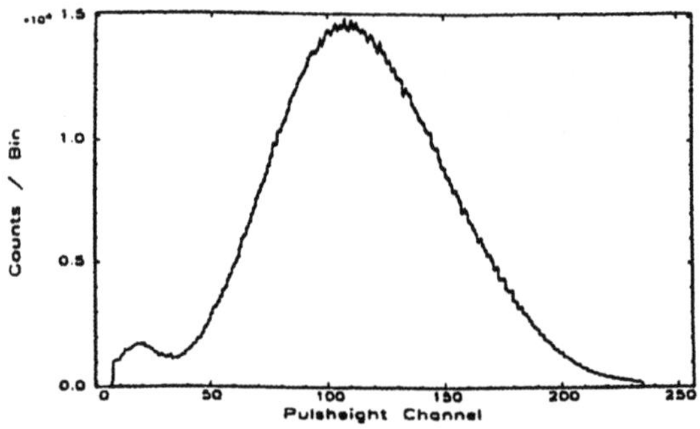

Fig. 6a. PSPC pulse height spectrum of Cyg-X2; 1 channel~10 eV

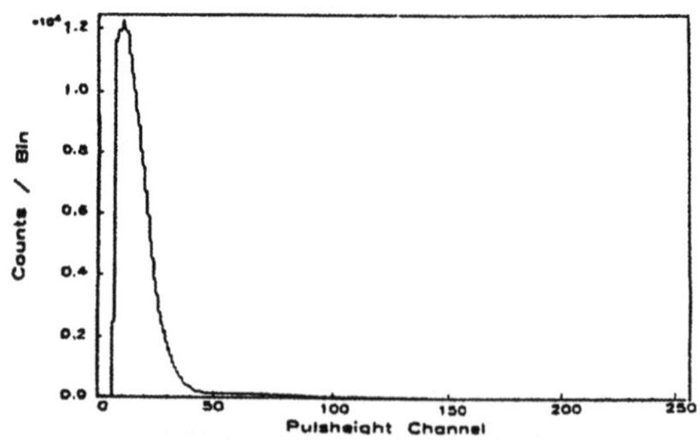

Fig. 6b. PSPC pulse height spectrum of HZ43; 1 channel~10 eV

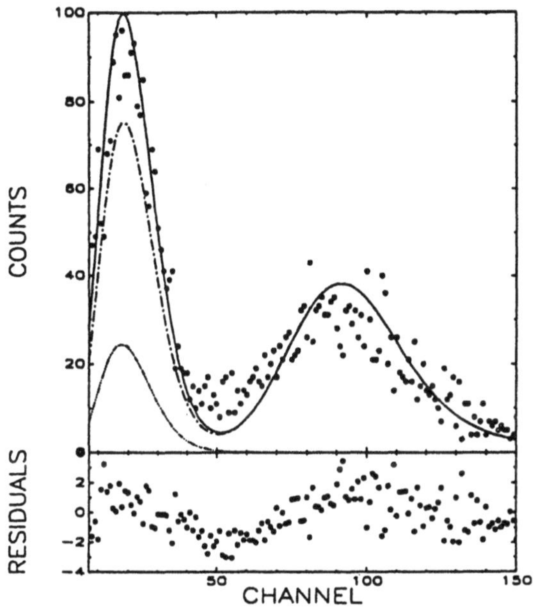

Fig. 6c. PSPC pulse height spectrum of AR Lac; 1 channel~10 eV

4. Astrophysical Results

The mission profile of ROSAT consists of three phases. The first phase was devoted to calibration and performance verification of the various instruments; during the second phase the all-sky survey with the PSPC is performed, and the third phase is devoted to pointing observations carried out exclusively in the frame of a guest observer programme. Observational results from the first two phases are now available and a minute fraction of the data is presented below. Of course, the data processing is not yet final and therefore the results are preliminary only.

4.1 The All-Sky Survey

The survey is performed by scanning the sky along great circles in ecliptic longitude at a rate of about 360° in 96 minutes, which is the orbit period of the satellite, so that the sky is imaged in form of stripes 360° long and 2° wide. The width of the stripe corresponds to the field of view of the XRT/PSPC. By the motion of the Earth around the Sun the great circle is advanced by about 1° per day completing the all-sky coverage in half a year.

Both at the north ecliptic pole and the south ecliptic pole all scan circles from every orbit overlap with each other. Fig. 7 shows an early X-ray image

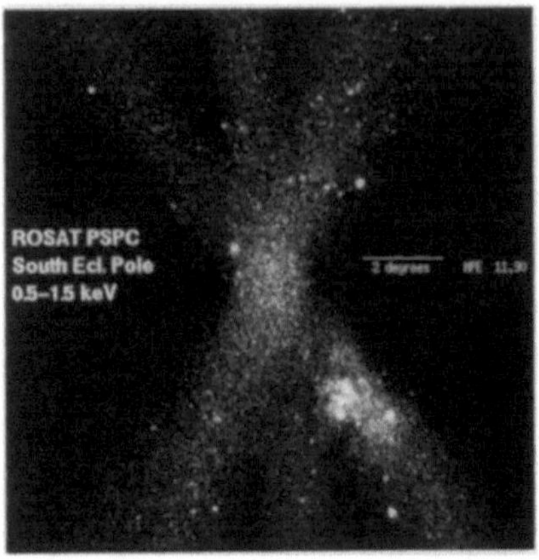

Fig. 7. PSPC all-sky survey stripe through the south ecliptic pole

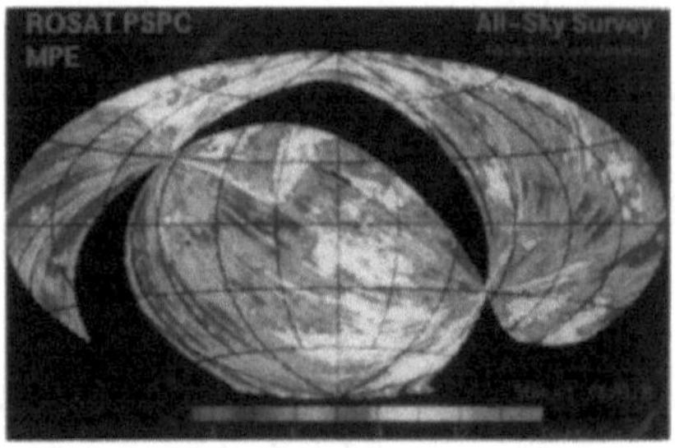

Fig. 8. X-ray brightness of the sky in galactic coordinates from the ROSAT all-sky survey; status 1.1.1991

of a section around the south ecliptic pole. The various scans approaching the pole are clearly visible [6]. The picture contains quite a number of point like sources but also some diffuse emission. Off-set by about 3 degrees to the south-west the LMC is visible.

ROSAT images are constructed in the following way. The X-ray telescope operates in photon counting mode, which means that each recorded count is labeled with detector x-/y-position, pulse height and arrival time. Using the instantaneous viewing direction of the spacecraft obtained from its Attitude Measurement and Control System the position of each recorded X-ray event is projected back on a frame of sky coordinates. Although the intrinsic field of view is only 2° in diameter the size of the effective field is virtually unlimited due to the scan motion of the telescope.

Fig. 8 illustrates the status of the sky coverage as of the 1. of January 1991 [9]. About 83% of the celestial sphere have been scanned at that time, and the figure shows the distribution of the X-ray flux measured per field of view in galactic coordinates. Large scale structures of diffuse emission are visible. They originate from the hot ($\sim 10^6$K) component of the interstellar medium, which is energized primarily by supernova explosions. On top of the diffuse emission a few prominent individual sources are visible, like the Pleiades, the Crab nebula, sections of the North Polar Spur, Her X-1, the LMC, etc. The image is neither background (scattered solar X-rays) nor exposure corrected, which explains the appearance of stripe-like structures.

4.2 The Large Magellanic Cloud

During the verification and calibration phase a series of pointings, including the "1. Light" of the XRT, have been made at the LMC, which have been merged into a PSPC image covering a field of about 8 square degrees

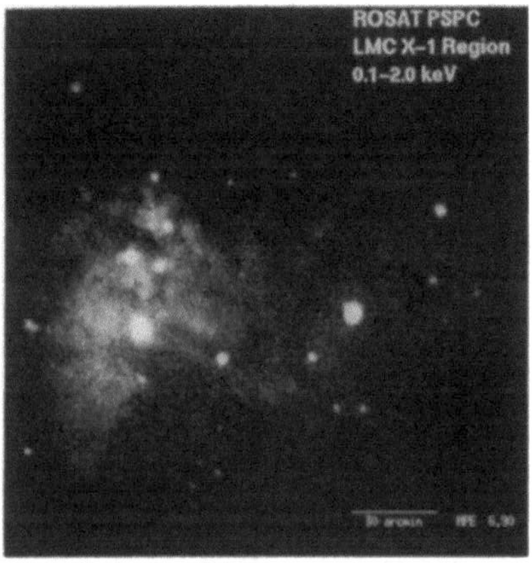

Fig. 9. Collection of PSPC images of the Large Magellanic Cloud

[10] (see Fig. 9). 45 individual sources have been found, 30 of which have been known from the Einstein observations, but 15 new objects have been discovered. The picture is dominated by LMC X-1, a bright X-ray binary, which is a candidate object for a black hole. Somewhat higher and to the left in the image the 50 msec pulsar source can be seen, which is very similiar to the Crab Nebula. The brightest source in the right portion of the image is the supernova remnant N132D. Apart from this remnant another 8 new candidates for supernova remnants have been found. Optical follow-up observations in H_α and [OIII], initiated at the Australian Mount Stromlo Siding Springs Observatory, will help to settle the classification.

From the position of the supernova SN1987A no X-ray emission has been detected and the upper limit is 2.5×10^{34} erg s^{-1}. Thermal X-ray emission is expected from the interaction of the supernova shock wave with the circumstellar material, which has basically been delivered from the stellar wind of the progenitor star. The ROSAT upper limit is consistent with a mass loss rate over wind velocity of $8.8\times10^{-6} M_\odot$/year per 550 km s^{-1}, which is comparable with the highest values observed for galactic blue supergiants. Typical values will produce an X-ray luminosity about 2 orders of magnitude lower than the ROSAT limit. An increase in X-ray emission will not happen before the shock front has reached the denser material associated with the wind of the red giant phase of the progenitor star.

Non-thermal soft X-rays from a potential neutron star / pulsar are photoelectrically absorbed in the expanding envelope of the progenitor star due to its high metallicity.

Apart from individual sources the X-ray image shows extended diffuse emission which is particularly strong in the eastern half of the LMC and in the 30 Doradus region, which shows up in the upper part of Fig. 9. As in our galaxy this emission is due to a hot (a few million degrees) interstellar medium. Because of the very low particle background, the PSPC image reveals for the first time fine details of the spatial structure. Faint and bright regions change from one to the other, whereby this transition seems to be narrower than about 1 arcmin. There is a strong correlation with the structures observed in visible light. The X-ray emission is strong in HII-regions, and it is low in dark cloud regions.

Obviously these images bear much more information than has been analysed so far.

4.3 Galactic Supernova Remnants

Supernova remnants have been known for a long time as strong X-ray sources. In fact, X-ray emission dominates the total radiation loss.

4.3.1 Tycho Fig. 10 shows the ROSAT HRI image of the Tycho supernova remnant [11], which is generally considered as the archetype of a remnant resulting from a type I supernova. The object clearly has a shell type appearance, which is one of the characteristic features of a thermally radiating remnant. The bright shell is preceeded by a faint limb, the boundary of which against the outer background - but also and even more impressively - against the inner bright shell structures is extremely sharp. This kind of image contrast has become possible because of the very low scattering of ROSAT mirror surfaces. The image shown in Fig. 10 still includes some residual attitude jitter, so that the structures in the image will become even more pronounced after their elimination. Especially, the study of the size distribution of the sub-structures will help to determine the physical mechanism, by which the initially homogeneous shell has broken up into clumps [11].

4.3.2 Cas-A Unlike Tycho, Cas-A is generally considered to have resulted from a type II supernova explosion. The ROSAT PSPC image, taken during the calibration phase, is shown in Fig. 11. Although the image is roughly

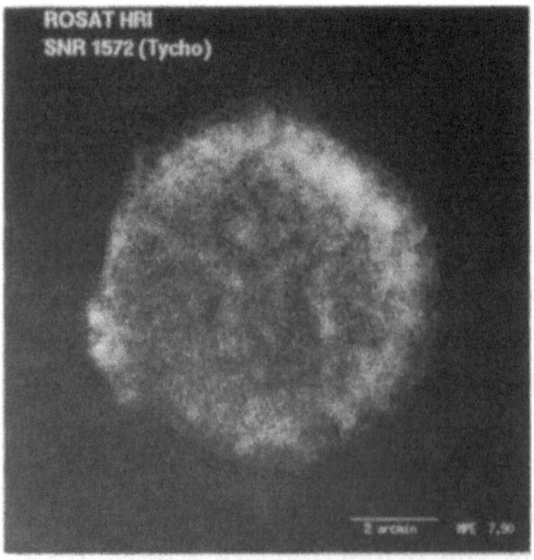

Fig. 10. HRI image of the Tycho supernova remnant

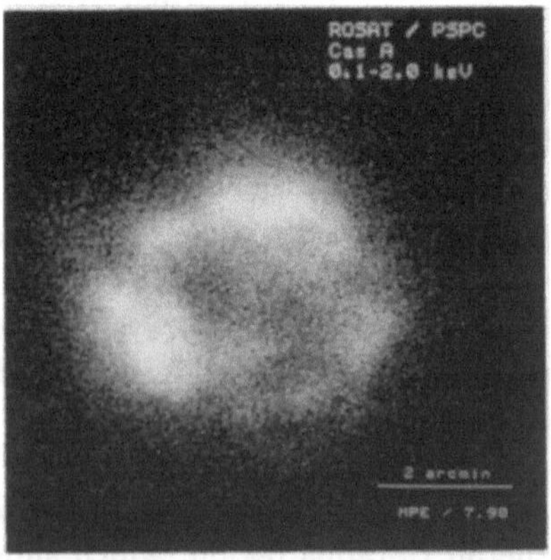

Fig. 11. PSPC image of the supernova remnant Cas-A

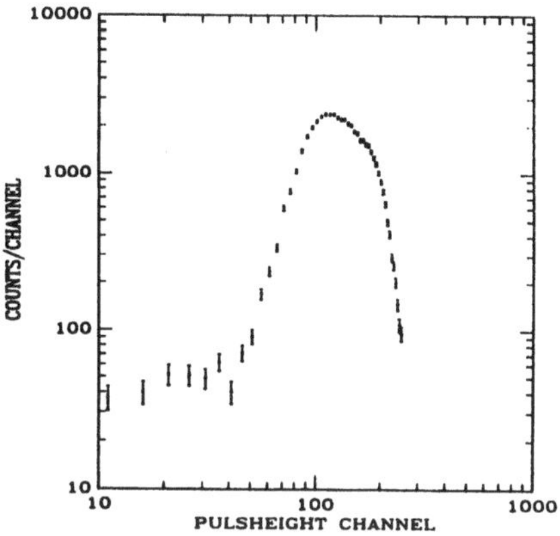

Fig. 12a. PSPC pulse height spectrum of the inner (R<3 arcmin) region of Cas-A; 1 channel~10 eV

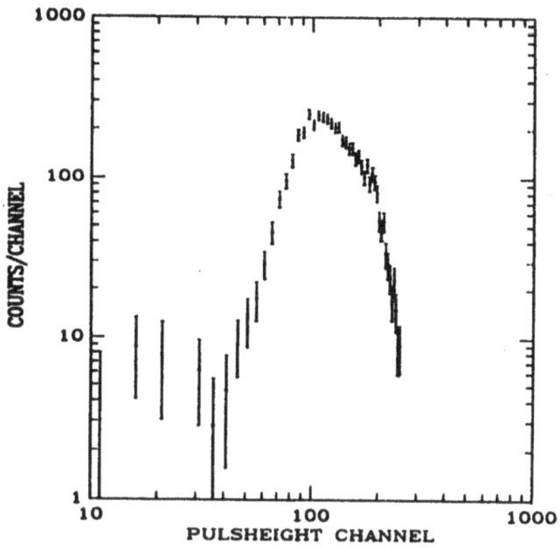

Fig. 12b. PSPC pulse image of the outer rim region (3<R< 6arcmin)of Cas-A; 1 channel~10 eV

circular, it is dominated by the emission from larger clumps, which are particularly bright at the north and the south-east. The emission from these clumps has been suspected for quite some time to be emission from the stellar ejecta heated by a reverse shock. In contrast, the emission from the outermost rim originates from the adjacent interstellar medium heated by the supernova blast wave. Since the ejecta should be enriched with heavy elements, spatially resolved spectra should allow to discriminate the locations of the ejecta against the interstellar medium.

Fig. 12a and 12b show two PSPC pulse height spectra, one of which has been extracted from a region less than 3 arcmin in radius, including the bright clumps, and the second one from a region between 3 and 6 arcmin in radius, excluding the bright portions of the remnant. The R $<$ 3 arcmin spectrum shows a small bump at about 1.8 keV, which is the convolution of K-emission lines from hydrogen and helium like silicon ions. This feature is absent in spectrum of the outer region. Secondly, the spectrum of the inner region shows a marked excess above the spectrum of the outer region in the energy range between 0.9 and 2 keV. For a thermal plasma this spectral range is dominated by emission lines of oxygen, neon and iron, which, since not spectrally resolved, form a broad bump. These two PSPC spectra prove for the first time that the bright interior of Cas-A is due to radiation from enriched elements. The next step will be to build an image in the silicon lines and determine the elemental abundance of silicon across Cas-A [12].

4.3.3 A ROSAT Discovered Galactic Supernova Remnant A previously unknown supernova remnant has been discovered at $\alpha(2000) = 4^h59^m7^s$, $\delta(2000) = +51^o46'35"$ during the ROSAT X-ray all-sky survey [13]. Fig. 13 shows the superposition of a few all-sky survey stripes across the Auriga constellation. Within that stripe the well-known supernova remnant HB9 can be seen as well as Capella, the cluster of galaxies 3C129, and, very close to the north-west part of the rim of the supernova remnant, extended emission which is associated with the radio galaxy 3C130 possibly embedded in a galaxy cluster.

The new remnant is so bright that it was easily detected in the quicklook data. It is of almost perfect spherical shape with a diameter of 108 arcmin and has an X-ray flux of 2.7×10^{-10} erg cm^{-2} s^{-1} between 0.1 - 2.4 keV, ranking it among the 10 brightest galactic supernova remnants. Based on the Sedov analysis the progenitor star went off 2.4×10^{4} yrs ago expanding into a very low density medium of 0.012 atoms cm^{-3}. Both the low interstellar matter density and the intervening photoelectric absorption of 8.8×10^{20} cm^{-2} result in a distance estimate of about 2.8 kpc, which would place the remnant in the inter-arm region between the local Orion arm and Perseus arm suggesting a type I event for the supernova.

The classification as a galactic supernova remnant has been meanwhile confirmed by follow-up radio observations, which revealed a radio surface brightness about a factor of 4 lower than that of any remnant known so far. If low radio brightness is common to old remnants radio surveys may have missed a substantial number, impacting on previous estimates about the galactic supernova rate and the energy input to the interstellar medium.

4.4 The Abell Cluster of Galaxies A2256

Clusters of galaxies are known to be strong X-ray sources. The X-ray emission is from interstellar gas, which has been expelled from the galaxies by galaxy-galaxy collisions and thereby heated to temperatures of about 10 keV. Fig. 14 shows a pointed PSPC image of A2256 taken with an exposure time of 17,000 sec during the calibration phase.

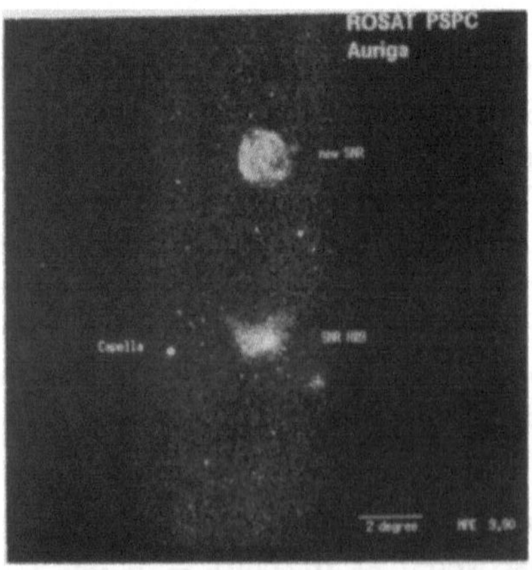

Fig. 13. New supernova remnant discovered by ROSAT during the all-sky survey

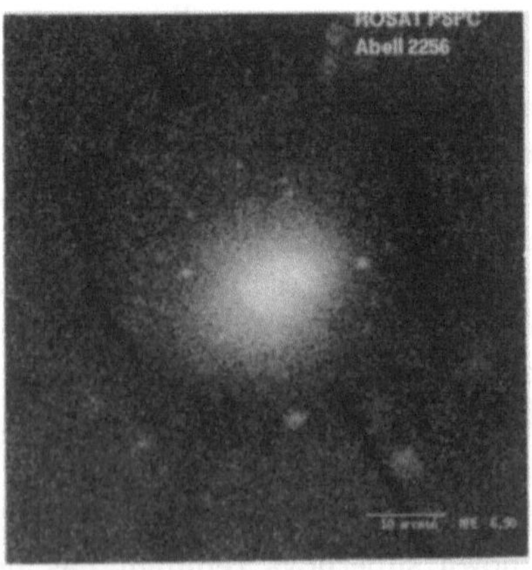

Fig. 14. PSPC image of the Abell cluster of galaxies A2256 resolved by ROSAT in two objects

This cluster of redshift z = 0.055 was observed by the Einstein observatory but the ROSAT PSPC observations resolve for the first time the central part of the cluster and reveal a double structure, the peaks of which are 3.5 arcmin apart [14]. The eastern peak coincides with the optically brightest galaxy, which is a cD-galaxy in the cluster centre. The X-ray emission is very likely to be due to a second separate cluster or a galaxy group, which are interacting with each other and are possibly merging. The hypothesis that the X-ray emission is indeed from two different objects is strengthened by the fact that the eastern and western sections show different temperatures of 9 and 6 keV, respectively. In addition to the diffuse emission X-rays from individual cluster member galaxies have been detected. They show luminosities about a factor of 10 higher than ordinary galaxies.

4.5 The Extragalactic X-Ray Background

As old as observational X-ray astronomy is the discovery of the diffuse X-ray background emission, which for energies above about 1 keV is of extragalactic origin. From the very beginning two possible scenarios have been discussed. Either the background is of truly diffuse character and arises from diffuse intergalactic hot gas, or, alternatively, the background results from the superposition of numerous indiviual point sources. If the latter hypothesis is correct, the diffuse background should increasingly break up into individual sources with improving angular resolution.

Because of the very low non-Xray background, the ROSAT PSPC is the almost perfect instrument to image the X-ray background to search for point sources in deep exposures. A few fields have been observed with exposure times between 10,000 and 50,000 sec during the calibration programme. Fig. 15 shows a series of images obtained during a 13,000 sec pointing centered on HZ43 [6]. A few other pointings have been made at the immediate vicinity but of less exposure. The central field contains 52 sources per square degrees. The deepest X-ray image ever taken so far is from a ROSAT PSPC pointing towards the north ecliptic pole. 101 sources per square degree have been found with an X-ray flux greater than 10^{-14} erg cm^{-2} s^{-1}.

Two main conclusions have been drawn from these observations [6]. Firstly, the cumulative number-count relation shows a significant flattening for X-ray fluxes less than about 3×10^{-14} erg cm^{-2} s^{-1}, which could mean that the number density of sources starts to decrease with increasing distance or that the X-ray luminosity shows a decrease because of evolutionary effects. Cosmological space-time effects have been discussed as well. Secondly, the accumulated flux of the isolated point-like sources makes up a fraction of about 40% of the total flux contained in the deepest exposure image. This means that a large fraction of the X-ray background originates from individual sources and deeper exposures, which have been scheduled for the up-coming pointing programme, will certainly increase this fraction.

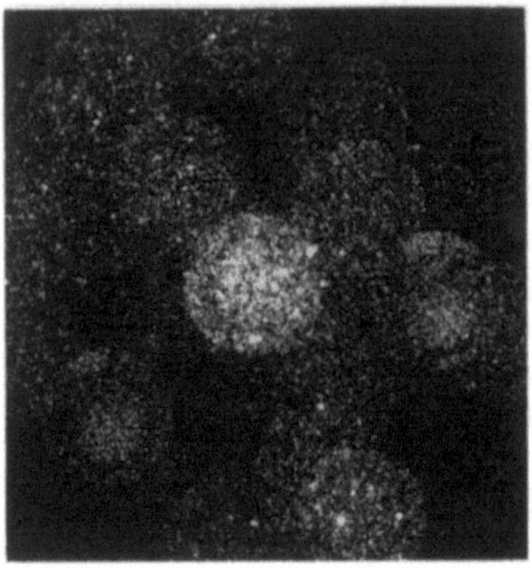

Fig. 15. A series of PSPC pointings at the region centered on HZ43; E>0.5 keV data only

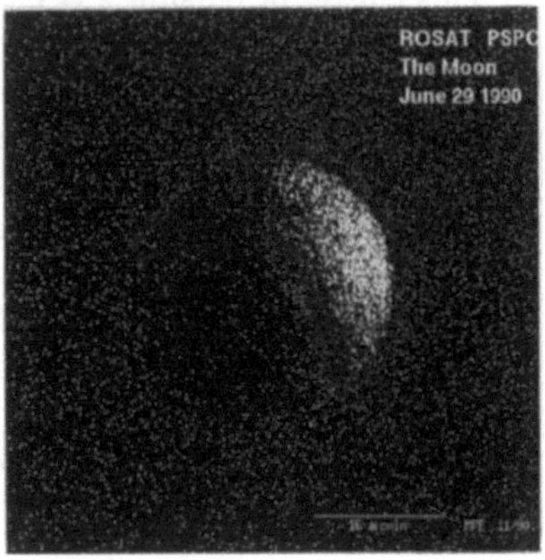

Fig. 16. PSPC image of the Moon and the occultation of the X-ray background

4.6 Soft X-Rays from the Moon

X-ray astronomy came to life in 1962 by the attempt to detect X-rays from the Moon [15]. The experiment failed at that time and it took 28 years to make the first soft X-ray image of the Moon. Fig. 16 shows the PSPC image of the Sun-lit crescent of the Moon taken during the calibration phase [16]. Like in visible light the Moon is bright because the surface reflects solar X-rays. The X-ray surface brightness of the dark Moon is lower than the surrounding diffuse emission, which means that the Moon occults the X-rays, which therefore cannot be local but come from deep space. And finally, the dark Moon is not really dark but shows a residual brightness, which is still higher than the PSPC particle background. The X-rays from the dark Moon are suspected to be bremsstrahlung emission from solar wind electrons impacting on the Moon's surface. These electrons can reach the back-side of the Moon only since they are guided by the interplanetary spiral magnetic field of the Sun.

5. Concluding Remarks

At the time of typing the final manuscript of the present paper the ROSAT all-sky survey is finished with a coverage of 97% . The remaining 3% will be filled in at some time during the next months. More than 160 proposals for scientific research projects using survey data have been put forward. More than 50 projects include internatonal collaboration.

The guest observer pointing programme for the first half year is under way. More than 400 observations are contained in the mission timeline, and will hopefully be carried out successfully.

Acknowledgement:
The ROSAT project is supported by the Bundesministerium für Forschung und Technologie (BMFT). I like to thank many people in BMFT, NASA, DLR/DARA, industry, scientific institutes and my colleagues at the Max-Planck-Institut für Extraterrestrische Physik. In particular, I am grateful to U. Briel, W. Brinkmann, H. Fink, G. Hasinger, E. Pfeffermann, P. Predehl, J. Schmitt and S. Snowden, who provided parts of the presentation material and were involved in many fruitful discussions.

References

1. B. Aschenbach: Appl. Optics **27** 1404 (1988)
2. E. Pfeffermann, U.G. Briel, H. Hippmann, G. Kettenring, G. Metzner, P. Predehl, G. Reger, K.H. Stephan, M.V. Zombeck, J. Chappell, S.S. Murray: Proc SPIE **733** 519 (1986)
3. U.G. Briel, E. Pfeffermann, G. Hartner, G. Hasinger: Proc. SPIE **982** 401 (1990)
4. K.A. Pounds, A.A. Wells: COSPAR Proc. *Adv. Space Research*, The Hague (1991)
5. J. Trümper: Adv. Space Res. **2**, No. 4, 241 (1983);
   J. Trümper: Phys. Scripta **T7** 209 (1984);
   J. Trümper: Phys. BL. **46**, Nr. 5, 137 (1990)
6. G. Hasinger: private communication (1990)
7. J.H.M.M. Schmitt: COSPAR Proc. *Adv. Space Research*, The Hague (1991)
8. M.A. Barstow: COSPAR Proc. *Adv. Space Research*, The Hague (1991)
9. S. Snowden: private communication (1990)
10. J. Trümper, G. Hasinger, B. Aschenbach, H. Bräuninger, U.G. Briel, W. Burkert, H. Fink, E. Pfeffermann, W. Pietsch, P. Predehl, J.H.M.M. Schmitt, W. Voges, U. Zimmermann, K. Beuermann: Nature **349** 579 (1991)
11. H.H. Fink: private communication (1990)
12. B. Aschenbach: to be published (1991)
13. E. Pfeffermann, B. Aschenbach, P. Predehl: to be published in Astron. Astrophys. Lett. (1991)
14. U.G. Briel, P. Henry, R. Schwarz, H. Böhringer, H. Ebeling, A. Edge, G. Hartner, J. Trümper, W. Voges: to be published in Astron. Astrophys. Lett. (1991)
15. R. Giacconi, H. Gursky, F.R. Paolini, B.B. Rossi: Phys. Rev. Lett. **9** 439 (1962)
16. J.H.M.M Schmitt, S.L. Snowden, B. Aschenbach, G. Hasinger, E. Pfeffermann, P. Predehl, J. Trümper: Nature **349** 583 (1991)

TYCHO/HIPPARCOS
A Successful Mission!
First Results and Further Improvements

A. Wicenec

Astronomisches Institut Tübingen,
Waldhäuserstr. 64, W-7400 Tübingen, Fed. Rep. of Germany

1. Introduction

The astrometric satellite HIPPARCOS has been launched by an Ariane IV rocket early in the morning of 9-August-1990. The intention of this mission to measure very accurate positions (2 milli-arcsec), trigonometric parallaxes and annual proper motions for about 118,000 selected stars seems to become impracticable after the failure of the apogee boost motor. However it turns out that the difficulties related with the highly elliptic geostationary transfer orbit can be mostly overcome with the socalled 'revised mission'. The degradation of the solar cells and the electronics due to the particles of the van Allen belts have been shown to be much less than even optimistic specialists stated during the first weeks after launch (see for example *Jahreiß* [1]). With a now expected lifetime of at least three years it is possible to get almost the same quantity and quality of input and output data as expected for the nominal mission. The additional TYCHO project will result in photometric and astrometric data for at least 500,000 possibly up to 1.000.000 stars in the range of $2 \lesssim mag_B \lesssim 11$. This is even more than expected before launch and the values given in [1] for the revised mission are fortunately completely outdated.
The satellite hardware as well as the operational principles have already been described elsewhere, for example in *M.A.C. Perryman* [2, 3, 4], here only a table is shown to give a remainder. The main part of this review is concerned with the TYCHO data, it's overall and special features and with data reduction improvements developed in the TYCHO Data Analysis Consortium (TDAC) during the implementation and verification of the revised mission.

2. The Satellite

ESA's satellite HIPPARCOS is built around an all reflective Schmidt telescope with two widely ($\approx 58\,degrees$) seperated fields of view. These two fields are superposed with the mean of a beam combiner and focused on the focal plane instrumentation. The main parts of this instrumentation are the grid system (main grid and star mapper grids on both sides), the relay optics, the image dissector tube (IDT) for the main experiment and the photomultipliers for the TYCHO experiment (for more details see for example [2]). Hereafter a table is shown to give the most prominent values on hand.

3. The TYCHO-Data

The readout frequency of the TYCHO photomultipliers of 600 Hz implies a data rate of 1200 bytes/second of scientific data (only TYCHO experiment), i.e. the countrates of the two wavelength bands are coded in one byte each onboard the satellite in order to reduce the transferred data rate. The starmapper grid geometry[1] together with the rotation rate of the satellite[2] produces a typical shape of the countrate versus sample[3] plots as can be seen in *Fig. 1.* This figure shows a star transit of the vertical slit group. This transit of a star with $mag_B \approx 9$ has a SNR of ≈ 10 whereas a star at the detection limit of $mag_B \approx 12.1$ has a SNR of 1.5. Despite the fact that most of the stars are at or near the detection limit I will refer to brighter stars because a transit of such a faint star can hardly be seen on the plots.
Obviously the detection limit is a function of the background. On the other hand the background is a sum of several independant effects

$$B30 = \sum_{n=1}^{5} B_n \tag{1}$$

where

B_1 = Sky background
B_2 = Cherenkov radiation
B_3 = Straylight
B_4 = Dark current
B_5 = other sources (see below)

[1] 4 vertical and 4 inclined slits with nonperiodic spacings
[2] $168.75\,arcsec/sec \pm 2\%$
[3] $1/600 sec$

Table 1. Hardware features of the HIPPARCOS satellite

Details of the Instrument Optics

Telescope:

• *Configuration:*	*Two regions of the celestial sphere are viewed by the telescope using a beam combiner*
• *Field of view:*	*0.9 * 0.9 degrees*
• *Basic angle:*	*58 degrees*
• *Diameter of primary mirror:*	*290 mm*
• *Focal length:*	*1400 mm*
• *Surface accuracy:*	$\frac{\lambda}{60}$ *rms*

Primary Detector:

• *Modulating grid:*	*2688 slits separated by 1.2 arcsec (8.2 μ)*
• *Detection system:*	*Image Disector Tube (IDT), sensitive aperture 37 arcsec, commandable in position to view and follow the star movements*
• *Sampling frequency:*	*1200 Hz*

Starmapper (TYCHO) Detector:

• *Detection system:*	*Photomutiplier tubes*	
• *Sampling frequency:*	*600 Hz*	
• *Frid system:*	*4 vertical and 4 inclined slits*	
• *Effective wavelengths:*	*B 426 nm*	*V 532 nm*
• *Transmission FWHM:*	*B 92 nm*	*V 96 nm*

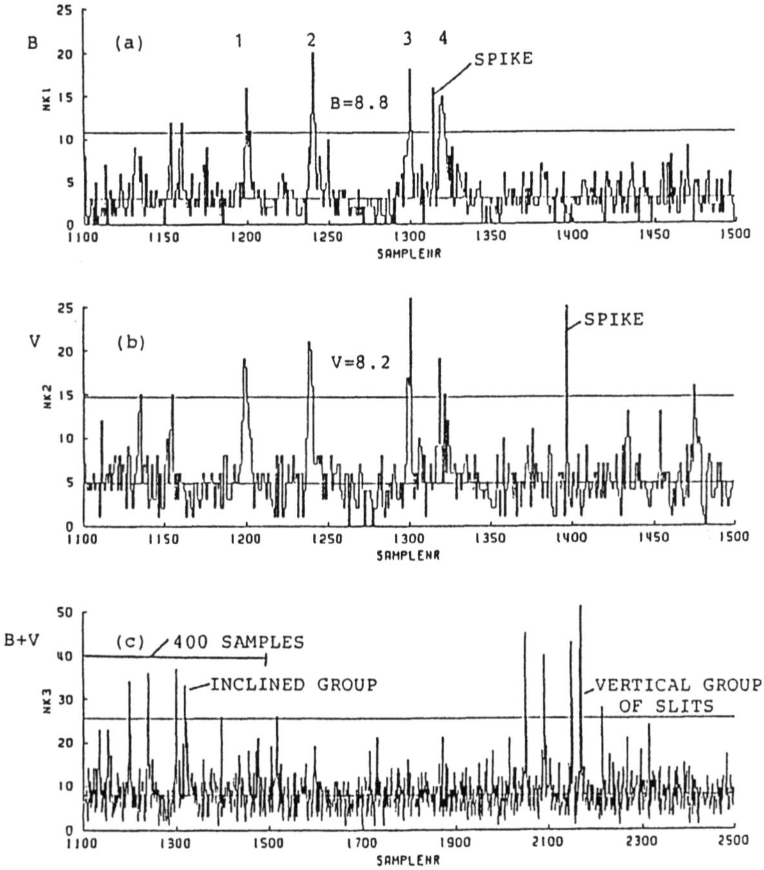

Fig. 1. TYCHO counts in B_T (a) and V_T (b) with the four peaks of a star (marked)

The expected $B30$ was about 12 counts/sample in $B_T + V_T$[4], whereas the observed $B30$ turns out to be highly variable due to the crossings of the van Allen belts during each orbit. In the apogee region where the effect of the highly energetic particles of the radiation belts is neglectable the observed $B30$ is in the range of 2! to 12 counts/sample with a most probable value of $\approx$ 5 counts/sample. The variability of $B30$ during one orbit is shown in *Fig. 2*.

The background is determined in each colour seperately from 6400 samples = 1 format and therefore called *format background*. This is done by determination of the actual distribution function and the median value therein. Because the counts are compressed[5] and digitally coded onboard the satellite it is necessary to apply a suitable interpolation in order to get more continuous background values.

[4] the variable $B30$ indicates $B_T + V_T$ hereafter

[5] semilog. compression

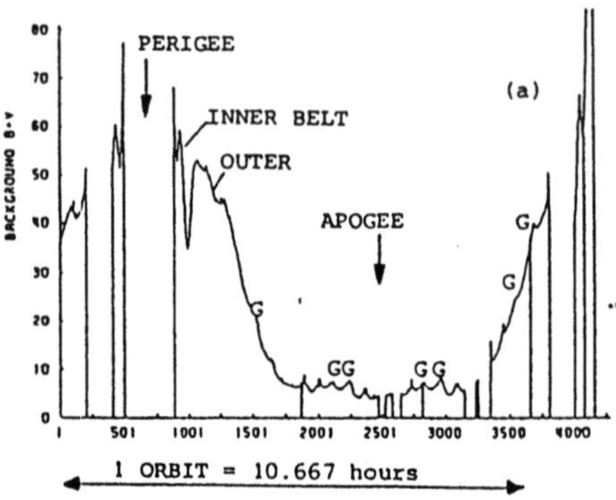

Fig. 2a. Typical B30 during one orbit = 10.667 hours. Missing signal when the telescope is pointing at Earth or Moon, or when the background is too high. Unit: 1 format=10.667 seconds

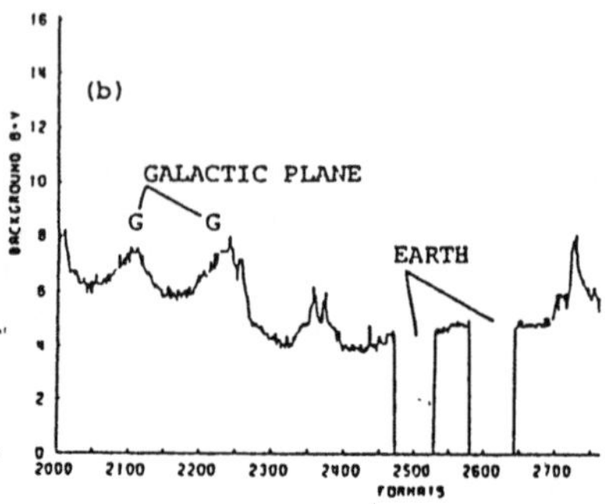

Fig. 2b. Magnification of *Fig. 2a* by a factor of 5 showing the low background region around apogee in more detail. Unit: 1 format=10.667 seconds

Another disturbing effect for the detection of stars are socalled 'spikes', i.e. narrow countrate peaks (normally 1 sample) not belonging to a star transit. These spikes are probably due to fluorescense in the glass of the photomultiplier tubes caused by high energy protons. An extreme example of disturbing spikes can be seen in *Fig. 3a* in comparison with *Fig. 3b* where only stars, (but lots of them) change the shape of the actual distribution function. The spikes seem to be only partly an effect of the elliptic orbit, because they are found throughout the data and the dependance of the spike distribution on the orbit is normally small.

4. TYCHO Data Reduction

The TYCHO data together with housekeeping and other relevant data will sum up to about 8 high density tapes per week, which are sent from ESOC[6] to TDAC, responsible for the TYCHO processing. The raw photon counts from the starmapper will undergo a sequence of processing steps. It should be noted here, that despite the data reduction uses the TYCHO input catalogue to determine the 'epoch' when a slit group crossing takes place (Predicted Group Crossing, *PGC*), the whole data stream is searched for additional transits (companion and serendipity transits, *COSY*). The only difference is the applied minimal SNR for the selection of possible transits. The evaluation of the PGC's is done at the Astronomisches Rechen Institut, *ARI* in Heidelberg. The prediction of the group crossings is done for the 3 million stars of the revised TIC by using the attitude data delivered

6 European Space Operation Centre, Darmstadt/FRG

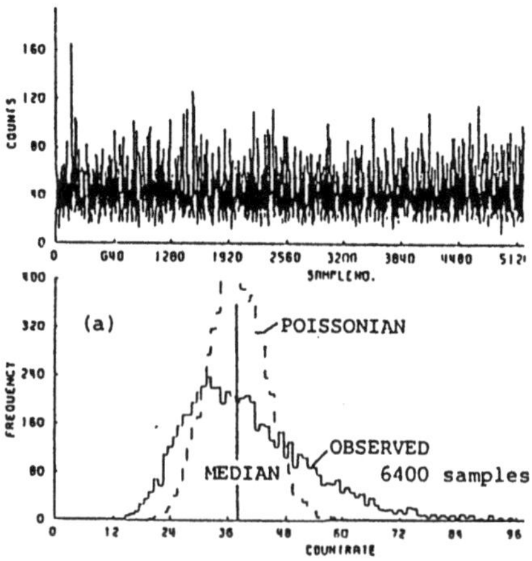

Fig. 3a. These TYCHO counts are obtained between the inner and outer van Allen belts. The poissonian distribution is strongly disturbed due to spikes.

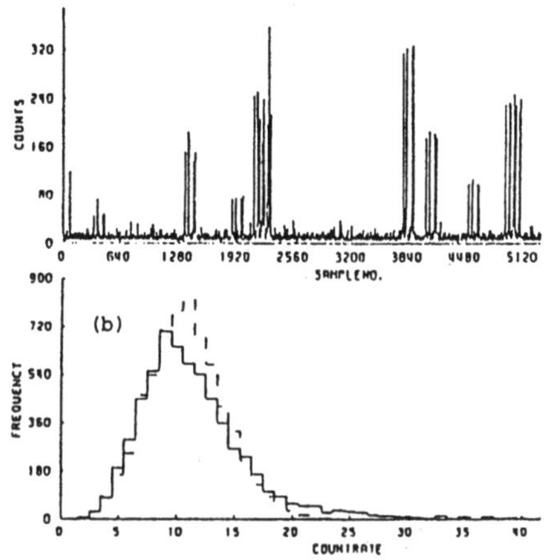

Fig. 3b. These counts are obtained between the inner and outer van Allen belts, too, but they do not show the disturbance of the poissonian distribution. Only lots of stars tend to flatten the upper wing of the distribution

by ESOC. This data together with the raw photon counts are the input for the *Detection* process. This reduction step includes the detection of statistically significant transits and the estimation of their epochs and signal amplitudes. As a 'side effect' the background is determined for the whole data, containing valuable scientific data for the monitoring of the van Allen belts and the coupling of solar wind particles to the geomagnetic field.
The two processes mentioned above are directly connected with raw data, whereas the following use the output data produced by the socalled *upstream* data reduction steps. *Detection* produces three different data streams which are delivered to three institutes working in the *downstream* data reduction:

Copenhagen University Observatory, CUO: Astrometric calibration and final *Astrometry*

Astronomisches Institut Tübingen, AIT: Photometric calibration and final *Photometry*

Centre de Donnèes Astronomiques de Strasbourg, CDS: *Recognition*[7] and production of revised TIC, *TICR*

The data streams produced by *Recognition* are sent to

[7] mapping of the transits found in *Detection* for a selection of true transits

Astronomisches Rechen-Institut Heidelberg, ARI: Evaluation of updated PGC's, *PGC Updating* with using TICR and the on ground attitude, *OGAR*.

5. Improvements

The highly variable background due to the elliptic orbit of the revised mission causes some troubles for TDAC, because the data reduction chain was not ready to select between 'good' and 'bad' data in respect to the background. Moreover the spikes found in the data have to be removed in order to reduce false transits and disturbancies of true transits. Because *Detection* is the only process in the TDAC reduction chain which is confronted with the raw data, both the background monitoring and the spike elimination have to be done there.

5.1 Background

Fig. 2 shows the results of the background determination now in use. Naturally the background has been determined in the prelaunch software, but this has been done only near transits and under use of 36 data points (samples). The advantages of the format background (see chapter 2) are:

- lower standard deviation (± 0.2 counts for $B30 = 12$)
- one background value every 10.667 seconds ($= 1$ format)
- background is independent of transits
- no disturbancies due to spikes or bright stars

It turns out, that the use of the format background is not completely satisfying in two special (rare) cases. One is the occurance of a bright star transit and the other is the periodically high gradient of the background when the satellite crosses the van Allen belts. In both cases another background is used instead which is determined in the same way as the format background. However, it uses only 200 samples to deal with local background variations even if a very bright star crosses a slit group.

5.2 Filtering method

The development of a new digital filter for *Detection* arose from the need of a method to eliminatate the spikes. The filter used up to that point in time is a strictly linear one as defined by

$$P_k = \sum_j w_j N_{k+j} \tag{2}$$

where

k: running number of the sample
N_k: photon counts (600 Hz = 1 sample) in B or V or B+V
w_j: 4-peak filter with 4 values = 1, and 0 elsewhere
P_k: result of folding N_k with a 4-peak filter

The spacings between the nonzero points of w_j are defined by the grid geometry and the measured scan velocity. The explanation of the effect of this and the new filter is shown in *Fig. 4*. This *linear* filter has some side effects leading to difficulties in the interpretation of the folded signal (P_k), which can not be explained in detail here. However some of them can clearly be seen in *Fig. 2*, e.g. the disturbancies of the transit of one star caused by another nearby star transit[8]. *Fig.4d-f* shows a well known star: ϵ Lyrae, which is in fact two double stars in a very interesting constellation: The two doubles are widely separated from each other ($\approx$ 4 *arcmin*), while the stars in the doubles are close together ($\approx$ 4 *arcsec*). Due to a projection effect of the slits only one of the pairs is resolved here, resulting in three transits. *Fig.4f* shows the effect of the linear filter, the three transits are only badly resolved.
The new filter may be described as 'nonlinear where needed' (a description of the priciples may be found in [5]). Here only the effect is described and shown in *Fig.4e*. The linear part of this filter works exactly the same way as the old linear filter, i.e. the behaviour of the data reduction at the detection limit is not changed. The nonlinear part in turn is responsible for the elimination of spikes and sidelobes of brighter stars ($mag_B \lesssim 10$). The result is a very clear folded signal (P_k) without the difficulties produced by the linear filter.

6 Conclusions

The revised HIPPARCOS mission will be able to produce the same quality of output data as the nominal mission, in some respects it will be possible to get even better results. The TYCHO input catalogue has been extended to 3 million stars, because the instrumentation has a better response and the background is lower than expected. In addition with the better background determination and the non-linear filter in *Detection* the magnitude limit of the TYCHO experiment will be about 0.5 magnitudes lower than expected.

---

[8] *Note:* 'nearby' means close in the combined field of view of the telescope, i.e. this is not necessarily a double star!

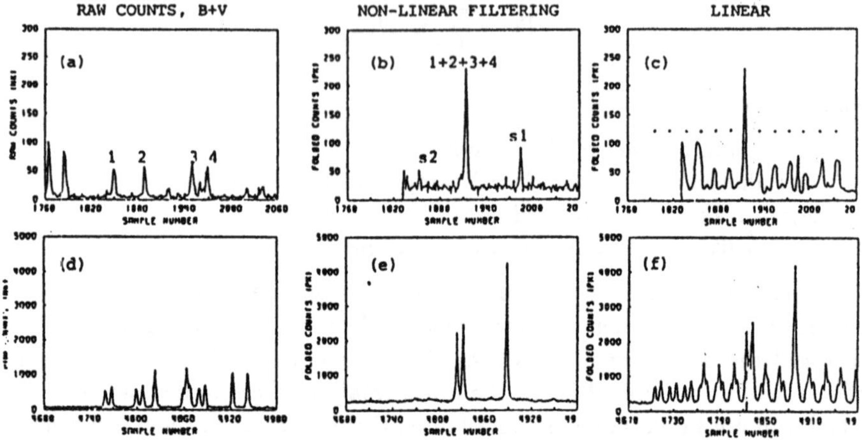

Fig. 4. Filtering of raw photon counts. Spikes and side lobes are eliminated with the new non-linear filter, but not with the old 4-peak filter.- (a) Raw counts with 4 peaks of a bright star (m=7.5). (b) Output from non-linear filter , showing the bright star and two fainter ones at s1 and s2. (c) Output from the old linear filter giving a main lobe as in (b) and 12 side lobes at positions at positions marked by dots which may disturb main lobes of other stars. (d, e, f) Similarly for a complicated triple star (ϵ*Lyrae*).

References

1 H.Jahreiß: *HIPPARCOS After Launch!? The Preparation of the Input Catalogue* in Reviews in modern Astronomy 2, Springer 1989 p.72

2 M.A.C. Perryman, H.Hassan (eds.): *The HIPPARCOS mission: Pre-launch status, Volume I. The HIPPARCOS Satellite* ESA SP-1111, 1989

3 M.A.C. Perryman, H.Hassan (eds.): *The HIPPARCOS mission: Pre-launch status, Volume II. The Input Catalogue* ESA SP-1111, 1989

2 M.A.C. Perryman, H.Hassan (eds.): *The HIPPARCOS mission: Pre-launch status, Volume III. The Data Reductions* ESA SP-1111, 1989

5 E.Høg, A.Wicenec *TYCHO Assessment*, Proceedings of the 28th COSPAR Plenary Meeting, COSPAR Symposium S7, 'Astrometry form Space' tbp.

Shock Waves in Accretion Disks

H.C. Spruit

Max-Planck Institut für Astrophysik,
Karl-Schwarzschildstr. 1, W-8046 Garching, Fed. Rep. of Germany

1. Angular momentum transport processes

Internal angular momentum transport processes are necessary to allow a gaseous disk to accrete onto the central object. For some astrophysical objects (cataclysmic variables, protostars) estimates of the efficiency of these processes are available from observed time scales. These estimates give different results, and it may well be that different processes are responsible for the angular momentum transport in different circumstances. One such process is transport by systems of global spiral shaped shock waves, proposed by Donner (1979), Michel (1984), Sawada et al. (1986), and Spruit (1987). Before discussing this in more detail, and in order to put it into perspective, I first review two other promising transport processes.

1.1. Shear turbulence

Historically, one of the oldest processes that have been proposed to allow mass in a disk to lose angular momentum and thereby accrete onto the central object is *shear turbulence* (e.g. Zeldovich, 1979; see also Von Weizsäcker, 1948). The shear in the fluid flow between neighboring orbits is similar to that in a rotating Couette flow experiment (apart from the compressibility in the astrophysical case, the equations are nearly identical). For experimental parameters (the rotation rates of the inner and outer cylinders) that correspond closest to the astrophysical case, turbulence has not been observed in the laboratory, as was pointed out by Kippenhahn and Thomas (1981). Since shear flows in the laboratory always seem to lead to turbulence at sufficiently high Reynolds numbers, it is often considered obvious (e.g. Di Prima and Swinney 1981) that this will also happen in the rotating Couette experiment, for all parameter values. In the astrophysical community the question is usually considered unresolved; the effect of rotation (not present in most laboratory shear flows) is often believed to be stabilizing

(Paczyński, 1978b; cf. Pringle, 1981). Numerical simulations of (parts of) thin accretion disks (Siregar and Léorat, 1989, Kaisig, 1989a,b, 1991) have not found the expected turbulence, though instabilities operating on a dynamical time scale have been found in thick disks (Drury, 1979; Papaloizou and Pringle, 1985; Zurek and Benz, 1986), in thin tori (Goodman et al. 1987; Hawley, 1989), and at the edges of disks (Hawley and Blaes, 1988). If thin disks are in fact unstable to shear turbulence, it will probably be found only in three dimensional calculations since it is known (e.g. Orszag and Kells, 1980; Bayly et al. 1988) that the third dimension is often essential for the instability mechanism.

Since the instability would be a dynamical one, the resulting viscosity would be high (α of the order unity), and it would operate in the same way in *all* disks, at least those where a fluid approximation is justified. It seems that observations may well be able to shed light on the question. For example, very low values of α in some disks, or strong variations in α between objects would be hard to understand in the shear turbulence picture. Do protostellar disks give us any clues? If taken at face value, the observed maximum sizes (100 AU), temperatures ($T = 300\, r_{AU}$ K, Hayashi et al. 1985), and time scales (10^6 yr or longer) imply α's less than 10^{-2}. Even a modest improvement in the reliability of these numbers would be enough to justify doubt about viscosities due to shear turbulence.

Since the observations of protostellar disks do not require very high α's (compared with the case of cataclysmic variables, Pringle et al. 1986), several proposed mechanisms are viable, especially for the inner parts of the disks. We discuss here two of these, namely angular momentum loss due to a wind from the disk, and spiral waves generated by selfgravitating processes or planets embedded in the disk.

1.2. Angular momentum loss due to a magnetic wind

The idea that a magnetized disk could lose lose angular momentum and thereby allow accretion to take place via a wind was proposed by Bisnovatyi Kogan and Ruzmaikin (1976) and Blandford (1976). The idea relies on the existence of a fairly organized vertical (perpendicular to the disk) magnetic field component of uniform sign. Such a field would evolve naturally in the inner parts of an extended disk from a random field captured by the accretion the outer parts. Any nonaxisymmetric component of this field would quickly be sheared to small length scales and dissipate away (Zeldovich et al. 1983) leaving an axisymmetric part that is concentrated inward by the accretion flow. [In addition to this field there could be an internal field, possibly stronger, generated by dynamo action. To the extent that this would fluctuate in sign on a small scale, its contribution to the acceleration of a wind would be minor, cf. Blandford and Payne 1982.] In the protostellar case, it is not even necessary to start from a random field at the disk edge, since the observations indicate that the disks are usually perpendicular to the (strong and fairly orderly) ambient cloud field (presumably because of the very strong influence of the magnetic field on the collapse of the cloud). The resulting poloidal field, kept in place by the high gas density in the disk, can accelerate a cool wind in the same way as in magnetic stellar winds (see Sakurai, 1985 and references therein). Full solutions of such winds have been obtained (Blandford and Payne, 1982; Sakurai, 1987). In these calculations, the mass loss from the disk surface is specified as a function of radius. If no dissipative processes (such as Ohmic diffusion, ambipolar diffusion or viscosity) are present in the disk, the disk can accrete with *all* the angular momentum and *all* the accretion energy being carried away by the wind. In practice, angular momentum transport by the wind may be competing with, say, a viscous process in the disk. In this case the magnetic wind will dominate only if the magnetic field accelerating the wind, and the mass loss rate are sufficiently high. The terminal speed of the wind can (if the mass loss rate is sufficiently low) be much higher than the escape speed from the point on the disk at which the wind started. This makes the process attractive for explaining jets, especially in cases (active galactic nuclei, SS433) in which high speeds are seen.

In protostellar disks, ambipolar diffusion is likely to be effective in the outer parts of the disk. Königl (1989), and Königl and Wardle (1990) have obtained detailed solutions for the field structure and flow field in the disk, taking into account ambipolar or Ohmic diffusion.

In the present state of the theory, the main uncertainty lies in the mass flux in the wind. Like in the case of stellar magnetic winds, the actual amount of mass lost from the disk is sensitive to details of the heating and cooling processes near the disk surface. As the mass loss is increased, the resulting wind becomes slower and less collimated, but it carries more angular momentum away from the disk.

2 Angular momentum transport by spiral waves

Nonaxisymmetric waves in accretion disks have a spiral shape, due to the differential rotation (axisymmetric waves are also possible but less interesting because they can not transport angular momentum). Angular momentum exchange by such waves is possible in a number of ways. A companion if present exerts gravitational torques on the nonaxisymmetric density distribution in the spirals, adding or extracting angular momentum from the disk. If the disk is massive enough for selfgravity to be important, the perturbations in the gravitational field due to the spirals produces torques on the fluid. Trailing spirals lead to an outward transport of angular momentum (Toomre 1964; Larson, 1984). But even in the absence of selfgravity or companions, angular momentum transport is possible because spiral waves effectively carry angular momentum. That is to say, the total angular momentum of a disk with waves is different from that of the same system without waves. As long as the waves are linear, no interaction takes place with the fluid, and the angular momenta of fluid and waves are conserved separately (for a discussion of wave angular momentum in disks see Narayan et al. 1987). With dissipation, for example in the form of shock waves, angular momentum is transferred from the waves to the fluid. *Trailing* waves carry a *negative* angular momentum (in the sense described). Where they dissipate, they *reduce* the angular momentum of the fluid, allowing accretion to take place. Waves generated at the outer edge of a disk by a stationary or a corotating perturber are such trailing waves. As they propagate inward, their negative angular momentum provides a sink of angular momentum for the fluid that is available as far the wave can travel inward.

It may sound counterintuitive at first that a wave source at the edge of a disk can cause matter to accrete in the inner parts of the disk, since the specific angular momentum of the fluid is large at the edge. It can be made more understandable by describing the waves in terms of wave packets or 'phonons'. Assume that at the outer edge of the disk we have a corotating source of phonons, emitted isotropically. The average momentum of these particles in the corotating frame is therefore zero. A phonon traveling inward meets an increasingly higher rotational velocity. Assuming for a moment that the phonon energy in an *inertial* frame is conserved, its energy measured in the corotating frame increases due to the Doppler-effect. At the same time, in a comoving frame the particles have an average momentum *opposite* to the direction of the fluid. Interaction between the particles and the fluid therefore will *brake* the fluid. The dissipation of the waves thus communicates to the fluid the angular velocity of the outer edge, rather than the angular momentum there. The energy released in the process comes not from the initial wave energy of the emitted phonons but from the 'Doppler- boosted' energy of the waves in the corotating frame, and therefore from the orbital motion of the fluid. The orbiting fluid feels a 'friction' that slows it down with respect to the field of phonons emitted at the edge. The continous inward increase of the orbital speed allows the phonons to continue losing energy by interaction with the fluid while gaining energy by Doppler effect, so that a steady state (in an inertial frame) with a net accretion is possible.

The same process takes place if waves are generated at the inner edge of a disk, for example by nonaxisymmetrries on a a slowly rotating central star. This hase been proposed by Michel (1984). Such waves could propagate outward, and where they dissipate they would communicate the slow rotation of the central object to the fluid. In this case however, the energy of the waves measured in the comoving frame *decreases* outward. Dissipation of the waves therefore limits their range, and the torque exerted on the disk is limited to the neighborhood of the central object. Numerical simulations of this process were done by Kaisig (1989b).

The wave-particle duality that we appeal to here can be given an exact basis by noting (Whitham, 1970; see Lighthill, 1978) that wave packets moving in a shearing flow conserve their wave action $s = e/\omega$, where e and ω are the wave energy and wave frequency measured in the fluid frame. One verifies that this conservation property corresponds to the assumed conservation of phonon energy in an inertial frame, if the phonons are treated as particles of zero rest mass. We note that the analogy holds independent of the precise nature of the waves (acoustic or internal gravity waves for example) as long as one can meaningfully build localized wave packets from them.

2.1 Selfgravitating spiral waves

In disks of sufficiently high surface density or sufficiently low temperature clumps can form by Jeans instability if the growth rate is large enough to offset the disruption by the shear in the Kepler flow. In terms of the Q parameter, $Q = c_s\Omega/(G\Sigma)$ this means $Q < 1$, where c_s is the sound speed and Σ the surface mass density. In the absence of (external or internal) heat sources, a disk would cool off by radiation and eventually reach instability. The instability will lead to internal dissipation (extracted form the shear), heating the disk up again. Paczyński (1978a) has proposed that at least certain disks might reach a quasi-equilibrium in this way, in which the heating by the instability balances cooling by radiation. Such disks would have a Q just below unity. Kosłowski et al (1979), Lin and Pringle (1987, 1990; see also Adams etal, this volume) have calculated models of disks in which the instability is taken to act as a source of effective viscosity, which can be modeled in a way similar to the α prescription. Numerical simulations have been done by Anthony and Carlberg (1988). The angular momentum transport by such instabilities can also be understood in terms of the wave packet description given above. The instability generates waves wich propagate some distance away from their origin. Since $Q < 1$, the gravitational torques between the density fluctuations in the waves and the fluid are important and produce an effective coupling between the wave packets and the fluid. At radii smaller than the origin of the waves, they extract angular momentum from the fluid in the same way as discussed above; outside the origin they add angular momentum by this interaction. The net effect is an outward angular momentum flux like that produced by a viscosity.

2.2. Angular momentum transport by spiral shocks

Suppose at the outer edge of a disk low amplitude sound waves are generated. As these propagate inward, their energy density increases, as described above, and eventually form shock waves. The dissipation in these shocks limits this increase and at the same time transfers the negative angular momentum of the waves to the fluid, so that accretion can take place.

Numerical simulations of mass transfer in a Roche geometry by Sawada et al. (1986), Matsuda et al. (1987, 1990) showed that this process can lead to a steady (in a corotating frame) two- armed shock pattern, and a corresponding steady accretion. In the outer parts of the disk formed in this way, rather strong shocks are maintained by to tidal forces (at least if a sufficiently massive companion is present). As these waves propagate inward, their strength decreases. In the inner parts of the disk the tidal effects are small, and the strength of the shocks is determined entirely by the balance between shock dissipation and the inward increase in energy density of the wave. The angular momentum transport in these inner regions is therefore *independent* of the strength of the wave excited at the outer edge (Spruit et al. 1987, Spruit, 1989a,b). Low amplitude waves amplify, high amplitude wave dissipate propagating inward until they reach the strength at which amplification and dissipation balance. Spruit (1987) described nonlinear solutions of the hydrodynamic equations that apply to this situation. The efficiency of the accretion process can be measured by comparing the radial drift speed with the standard α description for viscous disks. The effective α of the process, in regions where there is a balance between amplification and dissipation of the shocks, is of the order $0.01(c_s/\Omega r)^{3/2}$, where c_s is the isothermal sound speed. This dependence was also derived by Larson (1989). It must be stressed that this 'effective' viscosity does not imply a real viscous behavior but only measures the accretion rate. In particular, time dependent processes in spiral- shocked disks may well be very different form those in viscous ones.

The main importance of shock wave accretion lies in the fact that it provides a reliable means of angular momentum transport in non- selfgravitating disks, in places that are far from the inner and outer edges of the disk. In a typical protoplanetary case (r = 5 AU, T = $800K$) one gets $\alpha = 5\,10^{-4}$, corresponding to an accretion time scale of 10^6 yr. Though this is not a very high value, it still makes spiral shock accretion relevant for angular momentum transport in the giant planet region of a protostellar disk.

2.3. Planet- generated shock waves

A plausible source of shock waves in a disk are embedded companions, for example protoplanets. The interaction of protoplanets with a disk has been considered by Lin and Papaloizou (1979) and Goldreich and Tremaine (1980), both assuming a linear response of the disk to the tidal force. These authors derived expressions for the tidal torque exerted on the disk, and showed that the protoplanet will in general (excepting the case of a very high disk viscosity) clear a gap between its orbit and the disk's outer edge. The position of this edge is determined by a balance between the tidal forces which extract angular momentum from the edge, and the outward diffusion of angular momentum by viscosity. When the viscosity is small, shocks form at the disk edge, invalidating the assumption of linearity made. These shocks have been studied in some detail by Lin and Papaloizou (1986), though the viscosity used was still so high that the shocks did not propagate very far into the disk. If we are interested in the angular momentum transport by shocks, and therefore neglect viscosity altogether, the problem of finding the edge of the disk becomes somewhat ill posed. Shocks at the edge of the disk extract angular momentum from it and return it to the orbit of the companion by gravitational interaction. In addition, Lindblad resonances (Goldreich and Tremaine, 1980) between the disk and the companion extract angular momentum in a manner that is independent of viscosity (at low viscosity). In the absence of an additional 'spreading agent' like viscosity the disk would continue to shrink until the tidal forces become negligible. One may wonder if this behavior is consistent with the known behavior of disks in binaries. Observational information on the size of a disk compared with the orbit of the companion is available for cataclysmic variables. The disks in this case have radii of about 0.3 times the orbital separation, for systems with mass ratio of the order 0.1, in quiescence (Wood et al. 1986,1989a,b, Zoła, 1988; during outburst the disks are slightly larger). At this size, the tidal forces produce velocity variations at the edge that are of the order of the sound speed. If this is accepted as significant, it suggests that the main factor limiting the disk size is angular momentum extraction through shock waves at the outer edge. Only weak shocks would then propagate inward, and the scenario described for accretion by spiral shocks would give, for the protoplanetary case, the value $\approx 5\,10^{-4}$ quoted above. On the other hand, during their outbursts the disks in CV's definitely increase in size (e.g. Zoła, 1988) on a fairly short time scale. It is not clear at present whether this is possible by tidally generated shocks alone; perhaps these observations indicate a truly 'viscous' behavior.

Another estimate of the maximum size of a disk in the presence of a companion is that of Paczyński (1978a). He calculates the largest orbit that can fit inside a Roche lobe without intersecting other orbits. Orbits larger than this would be impossible because of the formation of strong shocks. The maximum sizes obtained are larger than the quiescent CV disks, but perhaps compatible with the disk sizes (Warner and O'Donoghue, 1988; Horne and Cook, 1985) at outburst. On the other hand it is clear that this estimate is an upper limit, since shocks may form well before orbits intersect. This could happen as soon as the tidal perturbations are of the order of the sound speed.

Numerical simulations of disks in binaries should be useful in answering these questions. The present generation of simulations (Schwarzenberg-Czerny and Różycka, 1988; Sawada et al. 1986; Matsuda et al. 1987,1990; Whitehurst, 1988; Osaki and Hirose, 1990; Różycka and Spruit, 1989) do address them, but the interpretation of the results is not entirely unambiguous. Hydrodynamical simulations without viscosity have been done for a gas with constant ratio of specific heats, ignoring radiation losses (Matsuda et al. 1990). Though these calculations start with fairly cool disks, the dissipation in the shocks quickly heats the gas up to a value near the virial temperature. The resulting disks are stationary only in a rather average sense, and don't have very well defined edges. The accretion time scale in these simulations is roughly inversely proportional to the companion mass. It remains to be seen how inviscid disks with more realistic low temperatures behave. Hydrodynamic simulations of a viscous disk have been done by Schwarzenberg-Czerny and Różycka (1988). Particle simulations of disks in binaries have been done by Whitehurst (1988) and Osaki and Hirose (1990). The way in which particle interactions model the hydrodynamic forces in these simulations prevent one from assigning a sound speed to the calculations but it is believed that they represent a highly supersonic, viscous flow. The disks in these simulations are somewhat larger than the observed quiescent disks in CV's, probably as a result of the high effective viscosity.

2.4. One armed spirals, excentric instability

Kato (1983, 1989) and Okazaki and Kato (1985) have pointed out the existence of a class of low- frequency one- armed spiral modes whose frequency (in an inertial frame) is proportional to the disk temperature. The fundamental mode of oscillation is equivalent to a slow precession of an excentric disk. In the limit of vanishing temperature (pressureless disk) a disk in this mode is a collection of nested elliptic Kepler orbits. These modes can become unstable in the presence of a companion. This instability so far has been studied in the zero- pressure limit by orbital mechanics methods (Piotrowski and Ziołkowski, 1970; Paczyński, 1977; Osaki, 1989; Osaki and Hirose, 1990). The instability is associated with the 3:1 resonance between a disk orbit and the orbit of the companion. For small companion mass, the range of unstable orbits near the resonance shrinks and the growth rate vanishes asymptotically (Hirose and Osaki, 1990). Thus, though the instability is very important for cataclysmic variables (Whitehurst, 1988), it is not certain that it is very relevant for the protoplanetary case.

The global $m = 1$ modes studied by Kato can also be destabilized by selfgravity of the disk. Such instabilities have been described by Adams et al. 1989). An important ingredient in the instability mechanism is the motion of both the central object and the disk with respect to the common center of mass. The disk mass required for instability is roughly proportional to the stability parameter Q. ($M_D/M_{tot} \approx 0.24Q$, Shu et al. 1990). For disk masses less than 0.24, $Q < 1$ is needed for instability. For $Q < 1$ the disk is also unstable to local selfgravitating instabilities discussed by Paczyński, Lin and Pringle, and Anthony and Carlberg (see above). For hot ($Q > 1$) massive ($M_D/M > 0.24$) disks the global instability would be important, but for cool ($Q < 1$) less massive disks the local selfgravitating instabilities would be more important.

References

Adams, F.C., Ruden, S.P., and Shu, F.H.: 1989, *Astrophys., J.*, **347**, 959

Anthony, D.M. and Carlberg, R.G.: 1988, *Astrophys., J.*, **332**, 637

Bayly, B.J., Orszag, S.A. and Herbert, T.: 1988, *Ann Rev. Fluid Mech.* **20**, 359

Bisnovatyi Kogan, G. and Ruzmaikin, A., 1976 *Ap. Space Sci.*, **42**, 401

Blandford, R.D.: 1976, *Mon. Not. R. astron. Soc.*, **176**, 465

Blandford, R.D. and Payne, D.G.: 1982, *Mon. Not. R. astron. Soc.*, **199**, 883

Di Prima, R.C. and Swinney, H.L.: 1981 in Swinney and Gollub (eds.) *Hydrodynamic instabilities and the transition to turbulence*, (Topics in Applied Physics **45**), Springer, New York, p.144 (second sentence from top).

Drury, L.O'C.: 1979, PhD thesis, Cambridge , UK

Goldreich, P., and Tremaine, S.:1980, *Astrophys., J.*, **241**, 425

Goodman, J., Narayan, R. and Goldreich, P.: 1987, *Mon. Not. R. astron. Soc.*, **225**, 695

Hawley, J.F. 1989, in F. Meyer et al. (eds.) *Theory of accretion disks*, Kluwer, Dordrecht, p259

Hawley, J.F. and Blaes, O.M.: 1988, *Astrophys., J.*, **326**, 277

Hayashi, C., Nakazawa, K. and Nakagawa, Y.: 1985, in D. Black and M. Matthews (eds.) *Protostars and Planets II*, University of Arizona, Tucson, p1100

Horne, K., and Cook, M.C.:1985 *Mon. Not. R. astron. Soc.*, **214**, 307

Kaisig, M.: 1989a, *Astron. Astrophys.*, **218**, 89

Kaisig, M. 1989b, *Astron. Astrophys.*, **218**, 102

Kaisig, M. 1991, in prepapration

Kato, S. 1983: *Publ. Astr. Soc. Japan*, **35**, 249

Kato, S. 1989: in F. Meyer et al. (eds.) *Theory of accretion disks*, Kluwer, Dordrecht, p173

Kippenhahn, R. and Thomas, H.-C.: 1981, in *Fundamental Problems in the Theory of stellar Evolution* (IAU Symp. 93), Reidel, Dordrecht, p. 237
Königl, A. 1989, *Astrophys., J.,* **342**, 208
Königl, A. and Wardle, 1990, *Astrophys., J.,* in press
Kosłowski, M, Wiita, P.J. and Paczyński, B., 1979: *Acta Astron.* **29**, 157
Larson, R.B.: 1984, *Mon. Not. R. astron. Soc.,* **206**, 197
Larson, R.B.: 1989, in H.A. Weaver and L. Danly (eds.) *The Formation and Evolution of Planetary Systems*, Cambridge University Press p.31
Lighthill, M.J., 1978 *Waves in Fluids*, Cambridge University Press, p.331.
Lin, D,N.C. and Papaloizou, J.C.B.: 1979, *Mon. Not. R. astron. Soc.,* **188**, 171
Lin, D,N.C. and Papaloizou, J.C.B.: 1986, *Astrophys., J.,* **307**, 395
Lin, D.N.C., Pringle, J.E.: 1987, *Mon. Not. R. astron. Soc.,* **225**, 607
Lin, D.N.C., Pringle, J.E.: 1990, *Astrophys., J.,* **358**, 515
Matsuda, T., Inoue, M., Sawada, K., Shima, E. and Wakamatsu, K.: 1987, *Mon. Not. R. astron. Soc.,* **229**, 295
Matsuda, T., Sekino, N., Shima, E. and Spruit, H.C.: 1990, *Astron. Astrophys.,* in press.
Michel, F.C.: 1984, *Astrophys., J.,* **279**, 807
Narayan, R., Goldreich, P, and Goodman, J.: 1987, *Mon. Not. R. astron. Soc.,* **228**, 1
Orszag, S.A. and Kells, L.C.: 1980, *J. Fluid Mech.,* **96**, 159
Osaki, Y. 1989: *Publ. astron. Soc. Japan* **41**, 1005
Osaki, Y., and Hirose, xx ,1990: *Publ. astron. Soc. Japan* xxx
Paczyński, B., 1977: *Astrophys., J.,* **216**, 822
Paczyński, B., 1978a: *Acta Astron.* **28**, 91
Paczyński, B., 1978b: *Acta Astron.* **28**, 253
Papaloizou, J.C.B. and Pringle, J.E.: 1985, *Mon. Not. R. astron. Soc.,* **225**, 267
Piotrowski, S.L. and Ziołkowski, K. 1970: *Astrophys. Space Sci.* **8**, 66
Pringle, J.E., Verbunt, F. and Wade, R.A.: 1986, *Mon. Not. R. astron. Soc.,* **221**, 169
Różycka, M. and Spruit, H.C.: 1989, in F. Meyer et al. (eds.) *Theory of accretion disks*, Kluwer, Dordrecht, p. 341
Sakurai, T.: 1985, *Astron. Astrophys.,* **152**,121
Sakurai, T.: 1987, *Publ. Astr. Soc. Japan,* **39**, 835
Sawada, K., Matsuda, T. and Hachisu, I.: 1986, *Mon. Not. R. astron. Soc.,* **219**, 75
Schwarzenberg- Czerny, A. and Różycka, M.: 1988, *Acta Astron.* **38**, 189
Shu, F.H., Tremaine, S., Adams, F.C., Ruden, S.P.: 1990 *Astrophys., J.,* **358**, 495
Siregar, E. and Léorat, J.: 1989, *Astron. Astrophys.,* **193**, 131
Spruit, H.C.: 1987, *Astron. Astrophys.,* **184**, 173

Spruit, H.C.: 1989a, in G. Belvedere (ed.) Magnetic Fields and Acrretion Disks in Astrophysics, Kluwer, Dordrecht, p.59
Spruit, H.C.: 1989b, in F. Meyer et al. (eds.) *Theory of accretion disks*, Kluwer, Dordrecht, p. 325
Spruit, H.C., Matsuda, T., Inoue, M. and Sawada, K.: 1987, *Mon. Not. R. astron. Soc.*, **229**, 517
Toomre, A.: 1964, *Astrophys., J.*, **139**, 1217
Warner, B. and O'Donoghue, D.:1988, *Mon. Not. R. astron. Soc.*, **223**, 705
von Weizsäcker, C.F.: 1948 *Z. Naturforschung* **3a**, 524
Whitehurst, R., 1988: *Mon.Not.R. astron. Soc* **232**, 42
Whitham, G.B., 1970: *J. Fluid Mech.* **44**, 373
Wood, J.H., Horne, K., Berriman, G., and Wade, R.A., O'Donoghue, D. and Warner, B.: 1986, *Mon. Not. R. astron. Soc.*, **219**, 629
Wood, J.H., Horne, K., Berriman, G., and Wade, R.A.: 1989a, *Astrophys., J.*, **341**, 974
Wood, J.H., et al.: 1989b, *Mon. Not. R. astron. Soc.*, **239**, 809
Zeldovich, Ya.B.: 1979, *Proc. R. Soc. Lond.* **A 374**, 299
Zeldovich, Ya.B., Ruzmaikin, A.A., and Sokoloff, D.D.: 1983, *Magnetic Fields in Astrophysics*, Gordon and Breach, New York.
Zoła, S., 1988, *Acta Astron.* **39**, 45
Zurek, W.H., and Benz, W.: 1986, *Astrophys., J.*, **308**, 123

Magnetic Field Measurements on Cool Stars

S.K. Solanki

Institut für Astronomie, ETH-Zentrum,
CH-8092 Zürich, Switzerland

Abstract: This review attempts to provide a simple introduction to the measurement of cool-star magnetic fields for the non-specialist. After a short historical overview the basics underlying the measurement of magnetic fields on cool stars are briefly introduced. The special role played by the sun in the study of stellar magnetism is pointed out and illustrated with the help of two examples. Stellar measurements in circularly polarized light are discussed. The Robinson technique for the measurement of stellar fields in unpolarized light is described and the problems associated with it, as well as possible solutions to them, are sketched out. Some trends resulting from the sum of the magnetic field detections on cool stars are discussed. A well observed star, ϵ Eri, is then considered in greater detail. The main conclusions of the talk are summarized.

1 Introduction

The first detection of a magnetic field on a cool star other than the sun is only ten years old. However, in the intervening decade a large amount of effort has been put into such work, making it impossible to cover every aspect of stellar magnetic field measurements in a single review. However, the subject has been extensively and competently reviewed in the past and more details may be obtained from the following papers: Marcy (1983), Linsky (1985), Saar (1987a, 1990), Pallavicini (1987), Robinson (1986), Gray (1988) and Mathys (1989). For reasons of brevity the present review does not consider degenerate stars or stars earlier than F0. The structure and origin of the magnetic fields on hot or degenerate stars are quite distinct from those of cool-star fields. Finally, only magnetic fields in stellar photospheres are considered, since these are the only layers in which magnetic fields have been directly measured (excluding the sun where transition region and coronal fields have also been measured). However, see also Pallavicini (1987) and Güdel and Benz (1989) for a dicussion of limits on coronal fields set by radio observations. In the following I often abbreviate the terms 'cool star' or 'late-type star' simply as 'star'.

2. Historical Overview

Let me begin by listing in chronological order the first detections of magnetic fields on different types of stars.

1908: The first detection of a non-terrestrial magnetic field was made on the sun by Hale (1908). Using the Zeeman effect — discovered only a decade earlier by Zeeman (1897) — and polarimetry he measured field strengths of approximately 3000 G in sunspot umbrae.

1947: 78 Vir (A2p) was the second star on which a magnetic field was discovered. Babcock (1947) used essentially the same approach as Hale had applied to the sun to measure a field strength of approximately 1500 G at the pole of what he correctly assumed to be an almost dipolar field. This and further measurements (e.g. Babcock 1958, cf. Mathys 1989) have established the oblique rotator model of Ap star magnetic fields due to Stibbs (1950) and Deutsch (1970).

1952: The first reliable detections of 'weak' magnetic fields of a few G (i.e. magnetic fields outside of sunspots) were made possible by the invention of the magnetograph (Thiessen 1952, Babcock and Babcock 1952, Kiepenheuer 1953), although there had been earlier hints of their presence (Hale 1922). These 'weak' fields were later shown to be spatially unresolved. Their true field strengths were found to be 1000–1500 G in the line-forming layers (Stenflo 1973).

1970: Next, magnetic fields were detected on white dwarfs. Kemp et al. (1970) derived a field strength of 10^7 G for the peculiar DB star Grw +70° 8247 from the measured continuum polarization, which they explained with the mechanism of 'grey body' magneto emission proposed by Kemp (1970).

1977: A field strength of $3 - 5 \times 10^{12}$ G was derived by Trümper et al. (1977, 1978) for the neutron star Her X-1 from quantized cyclotron emission features in its pulsed hard X-ray spectrum. Such spectral features between individual Landau levels had been predicted by Gnedin and Sunyaev (1974) and Basco and Sunyaev (1975).

1980: Finally, 72 years after the discovery of solar magnetism, a field was also detected on another lower main sequence star, ξ Boo A (G8 V), by Robinson et al. (1980). They employed a technique based on the Zeeman effect and developed by Robinson (1980). A field of strength

2400–2900 G covering 20–45 % of the stellar surface was deduced from the spectra.

1983: There followed the detection of a field on the first cool giant/subgiant λ And (an RS CVn binary with a G8 III–IV primary), by Giampapa et al. (1983) after adapting the Robinson technique to the infrared.

1990: Mathys and Lanz (1990) detected a magnetic field on an Am star, o Pegasi (A1 IV). In contrast to Ap stars, o Peg was found to have a complex magnetic field distribution reminescent of late-type stars, with a spatially averaged field strength of approximately 2000 G.

1990: Finally, a magnetic field detection on the T Tauri star TAP 35 has recently been reported by Basri and Marcy (1990). Their preliminary value for the spatially averaged magnetic field strength is 2000 ± 500 G.

3. The Case for Stellar Magnetic Field Measurements

Magnetic fields with a complex spatial structure, as produced by dynamo action, are responsible for a whole variety of effects. Some of these, together with their observational consequences are listed below.

1. Stellar coronae (i.e. gas at $10^6 - 10^8$ K), transition zones (10^5 K) and, at least partly, chromospheres (10^4 K). These features are responsible for the stellar X-ray and microwave flux, many of the UV lines, line core emission from Ca II H and K and Mg II h and k, etc.
2. Stellar suface inhomogeneities, e.g. starspots, whose presence may be deduced from modulations of the stellar luminosity, colour and spectral line profile shapes.
3. Stellar flares, i.e. sudden brightenings in the stellar X-ray, UV and microwave emission.
4. Stellar activity cycles, observed as long-term quasi-periodic changes (with periods of years) in, e.g., the Ca II H and K line core flux.
5. A considerable part of the rotational spin-down of a large fraction of all stars. This may be deduced from age-rotation relationships and from their comparison with theoretical predictions.

Phenomena of the types listed above — collectively labeled active phenomena — are exhibited by stars lying in the shaded portions of the HR diagram plotted in Fig. 1. These stars have been called 'solar like' by Linsky (1985), in that the dynamics and energetics of their outer atmospheric layers are controlled by their magnetic fields. They include main sequence stars later than approximately F0, as well as subgiants and giants between approximately F0 and approximately K3. Particularly active are members of relatively short period binaries (mainly RS CVn type stars). In order to understand the atmospheres of solar-like stars it is imperative to know the strength and structure of their magnetic fields. Unfortunately, although the magnetic field has a strong influence on stellar emission through its influence on the thermodynamics of the stellar atmosphere, its direct influence on the spectrum is small and it is difficult to quantitatively measure the field.

4. Basics of Stellar Magnetic Field Measurements: The Zeeman Effect

Although a number of physical effects allow direct measurements of stellar magnetic fields, the only one so far successfully applied to the photospheres of non-degenerate late-type stars other than the sun has been the Zeeman effect. The magnetic field manifests itself through the Zeeman effect primarily by splitting a spectral line into a number of components. In the simplest case of a magnetic field aligned along the line of sight and a 'normally' Zeeman split line there are only two components. The wavelength difference $2\Delta\lambda_H$ between these components is in Å:

$$2\Delta\lambda_H = 9.34 \times 10^{-13} g \; \lambda^2 \; B,$$

where λ is the wavelength of the line in Å, g is the Landé factor and B is the field strength in G. Another important property of the Zeeman effect is that it polarizes the individual Zeeman components. In the above example the two components are oppositely circulary polarized.

Unfortunately, for the following reasons it is generally not possible to determine the stellar field strength directly from the Zeeman splitting, $\Delta\lambda_H$.

1. In the visible part of the spectrum $\Delta\lambda_H$ is smaller than the half width $\Delta\lambda_D$ of the unsplit line for $B \lesssim 1500$ G. $\Delta\lambda_D$ includes line saturation effects and 'turbulence' velocity broadening, but in this estimate neglects

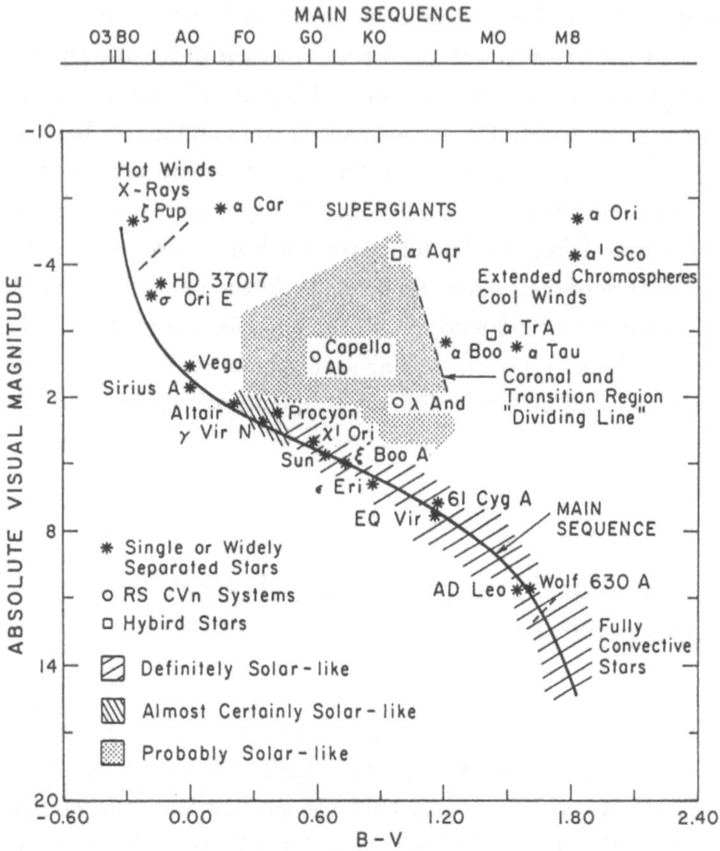

Fig. 1. An H-R diagram. The stars in the shaded areas are definitely or probably solar-like on the basis of direct or indirect indicators of strong, turbulent magnetic fields, which dominate the energetics and dynamics of the outer stellar atmospheres. Adapted from Linsky (1985).

broadening due to stellar rotation. Accordingly, for a field strength typical of a cool star the lines are not split, but only broadened. Unfortunately, line broadening is a less than ideal diagnostic of the magnetic field, since a number of other physical agents also broaden the lines. If the influence of stellar rotation on $\Delta\lambda_D$ is also taken into account, then the ratio $\Delta\lambda_H/\Delta\lambda_D$ is further decreased, making magnetic field detections even more difficult.

One solution to this problem is to make use of the is approximate proportionality of $\Delta\lambda_H/\Delta\lambda_D$ to the wavelength and to observe in the infrared. However, until recently the detectors in the infrared beyond 1μ were grossly inferiour to those in the visible.

2. In general, magnetic fields cover only a small fraction of the stellar surface. The simplest description of such an inhomogeneous field is by a model with the following two components: i) A magnetic component with a field of strength B covering a fraction f of the visible surface, where f is called the magnetic filling factor. ii) A field-free component covering the remaining fraction, $1-f$, of the visible surface. Neglecting for the moment the spherical shape of the stellar surface and related projection and field line inclination effects, we can write the measured intensity I as:

$$I = f\, I_{\mathrm{m}} + (1-f) I_{\mathrm{nm}},$$

where I_{m} is the intensity of light from the magnetic component and I_{nm} is the intensity from the non-magnetic component. If f is small then clearly the observed line profile will only be slightly affected by the magnetic field, making its detection even more difficult.
The simplest solution to this problem is to measure the spectrum in polarized light as well. For a complete description of the magnetic field vector four different states of polarization must be measured. However, in practice mainly the spectrum of the net circular polarization (Stokes V) has been measured and is the only polarization state discussed further in the present review. The Stokes V profile of a spectral line gives a measure of the line-of-sight component of the magnetic field. In particular, it can easily be shown that light coming from a field-free portion of the stellar atmosphere is free of net circular polarization, i.e. Stokes $V = 0$. Therefore, by measuring Stokes V it is possible to obtain information exclusively on the magnetic features, even if they are spatially unresolved. This property is employed extensively for the investigation of the solar magnetic field. The other relevant property of Stokes V is its dependence on magnetic polarity. The sign of Stokes V reflects the sign of the longitudinal component of the field, B_{long}:

$$V(-B_{\mathrm{long}}) = -V(B_{\mathrm{long}}).$$

This relationship implies that the presence of opposite polarity fields within a spatial resolution element leads to a partial and sometimes even complete cancellation of the Stokes V signal. For most late-type stars a cancellation of polarities over the visible stellar hemisphere hinders the use of the Stokes V signal for magnetic field detections (more details are to be found in Sect. 6).

5. The Best Studied Star: The Sun

It is no coincidence that I start the survey of stellar magnetic field measurements with a discussion of the sun. We know much more about the structure and the properties of the solar magnetic field than we do about the fields of all other cool stars put together. There are two main reasons for this, both of which have to do with the sun's relative proximity:

1. The solar surface is spatially resolved, so that the spatial structure of the magnetic field can be determined relatively easily.
2. The large photon flux allows spectra with a very high signal-to-noise ratio to be obtained (values of 10^4 in the continuum are not uncommon), making detailed interpretations of subtle spectral features possible.

Consequently the sun allows us to develop and rigorously test both observational and theoretical techniques for studying magnetic fields. Once the techniques have achieved sufficient maturity and reliability, they may often also be applied to other stars.

Another point, whose importance should not be underestimated, is that our knowledge of the solar magnetic field gives us an idea of what to expect on other stars and helps us to interpret stellar observations. Of course there are dangers inherent in this approach, since the field of other cool stars may in some respects differ qualitatively from that of the sun. However, the interpretation of some observations requires additional assumptions. In such cases taking the sun as a reference is often unavoidable.

From the large body of solar magnetic field research I have chosen two examples that appear particularly relevant to the measurement of magnetic fields on other stars.

5.1. Full-Disk Magnetograms

Figure 2 shows a magnetogram of the full solar disk obtained with the vacuum telescope at Kitt Peak. A magnetogram is simply a map of the distribution on the solar surface of the Stokes V strength in the wing of a spectral line and corresponds roughly to a map of the longitudinal magnetic field. In Fig. 2 gray regions are field-free, while white and black patches correspond to magnetic features with opposite polarity. Three points are evident at a glance:

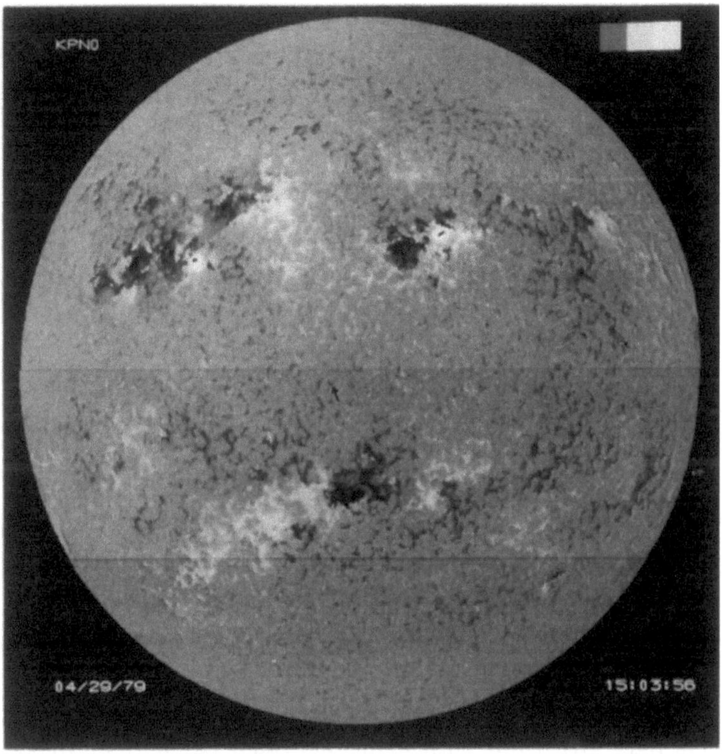

Fig. 2. Solar full-disk magnetogram obtained at solar activity maximum. Grey areas are field-free. Black and white patches represent magnetic features with opposite polarities. Courtesy National Solar Observatory, Kitt Peak.

1. The solar magnetic field has a complex spatial structure, composed of bipolar active regions and a quiet network that may either be unipolar over large stretches, or may exhibit locally mixed polarities (salt and pepper).
2. Much of the structure in the magnetogram is at small scales. The sizes of magnetic features can be followed down to the smallest spatial scales currently resolvable (approximately 200–300 km on the sun). Indirect techniques indicate the existence of even smaller magnetic features. Of importance for the interpretation of stellar spectra is that the properties of the smallest magnetic features, often called magnetic elements, determine to a considerable extent the observations averaged over a large area. Therefore, in order to correctly interpret spatially unresolved observations, we must know the physical properties of the small-scale magnetic features.

3. Almost equal amounts of opposite polarity flux are present on the solar surface. This implies that due to the cancellation effect mentioned in Sect. 4 the Stokes V signal averaged over the solar surface is very small ($10^{-5} - 10^{-6}$ in units of the continuum intensity for the broad-band signal, Kemp et al. 1987a) and little evidence for a magnetic field is obtained in this manner. Since the best currently achievable accuracy of stellar polarimetry is approximately 10^{-4} it appears unlikely that many solar-like stars can be detected if their fields show a distribution similar to that of the sun.

5.2. Small-Scale Magnetic Features

The magnetic field in the solar photosphere is mainly concentrated into so-called flux tubes, i.e. bundles of magnetic field lines passing from the solar interior into the outer atmosphere. The field is confined by the deficit in gas pressure within the flux tubes. The decrease in gas pressure with height causes the field strength to decrease and the tubes to expand until they merge in the lower chromosphere.

Photospheric flux tubes come in various sizes. The largest are visible as sunspots, i.e. dark structures with diameters often larger than 10'000 km. The field strength varies from approximately 1000 G at their outer penumbral boundary to approximately 3000 G in their core (umbra). Dark structures of intermediate size, having diameters of around 1000–2000 km are called pores and have field strengths between 1500 and 2000 G. Finally, the smallest and by far the most common of the solar magnetic features are the magnetic elements. These are bright structures with field strengths of 1000–1500 G in the line-forming layers. Their sizes are not accurately known, but for many of them are probably still below the best current spatial resolution of 200–300 km. More about the general properties of small-scale magnetic features may be obtained from Sect. 8.2 of Stix (1990) and from the reviews by Spruit and Roberts (1983), Solanki (1987b), Stenflo (1989) and Spruit et al. (1990).

To be able to predict the influence of small-scale magnetic features on the spatially averaged observed spectrum we must determine their detailed physical structure. Spectra and images in polarized light serve as the ideal empirical tools for such an undertaking. From such data it is possible to obtain, for example, the stratifications of field strength, temperature and stationary flow velocity in the magnetic features and partly also in their surroundings. In addition the data give an idea of the direction of the magnetic vector, the dynamics of the flux tubes and their interaction with the surrounding convection, etc. More details may be obtained from the following reviews devoted to the discussion of empirical results: Stenflo (1984, 1985, 1986), Solanki (1987a, 1990), Martin (1990).

Theoretical flux tube models have developed in parallel with the empirical advances. State of the art flux tube models now not only satisfy the MHD equations, but also include multi-dimensional non-grey radiative transfer as part of the energy balance. The models resulting from such ab initio calculations reproduce many of the observed features, for example, a decrease of the field strength with height, the absence of sizeable stationary flows within the flux tubes, the concentration of the tubes into downflowing intergranular lanes, the temperature enhancement within the magnetic features, etc. The correspondence between the best theoretical and empirical models has moved beyond the qualitative stage and become increasingly quantitative. The quantitative accuracy of theoretical flux tube models is expected to constitute an important aid to new techniques of stellar magnetic field measurement (cf. Sect. 7). Reviews containing additional details on the theoretical aspects of solar flux tubes have been given by Spruit (1981, 1983), Nordlund (1986), Schüssler (1986, 1987, 1990). Also of interest are various articles in Russell et al. (1990) as well as recent papers by Steiner and Stenflo (1990) and Knölker et al. (1990).

6. Circular Polarization Measurements on Cool Stars

Circular polarization measurements may appear to be attractive magnetic diagnostics, since the detection of net circular polarization can generally be equated with the detection of a magnetic field (Sect. 4). However, until the mid 1980s no positive detection of circular polarization was reported despite numerous surveys with ever-increasing sensitivity (e.g. Boesgaard 1974, Boesgaard et al. 1975, Vogt 1980, Brown and Landstreet 1981, Bonsack and Simon 1983, Borra et al. 1984). On the other hand, a number of stars showing no net circular polarization exhibit distinct signatures of a magnetic field in unpolarized light (Sect. 7). In combination these observations imply that the magnetic field on cool stars has a complex geometry, qualitatively similar to that of the solar field.

In recent years net circular polarization has finally been detected on a handful of objects. Borra et al. (1984) reported a detection at the 3.9 σ level on ξ BooA (G8 V) on a single night.

Kemp et al. (1987b) detected broad-band circular polarization on λ And on numerous nights spanning five years. Clear temporal variations on the approximate time scale of the rotation period were seen (cf. Fig. 3). There are two main conclusions to be drawn from their work. a) The temporal variation of the circular polarization and its correlation with variations of the stellar photometric brightness suggests that the field giving rise to the signal is concentrated into a few large regions that evolve over the years (Fig. 3).[1] If measurements of the solar broad-band circular polarization are taken as a guide, then an idea of the distribution and the photometric properties of large-scale magnetic features on λ And may be obtained (Mürset et al. 1988). b) The very fact that a broad-band circular polarization signal is measured is proof that on λ And the Stokes V profiles must be asymmetric, just as they are on the sun. This suggests that the mechanism giving rise to the asymmetry is not particular to solar conditions. The Stokes V asymmetry has recently been explained in terms of the interactions of magnetic fields and convection (Grossmann-Doerth et al. 1988, 1989b, Solanki 1989, Bünte et al. 1990, Knölker et al. 1990) in faculae and due to the Evershed effect in sunspots (Skumanich and Lites 1987). The observations of Kemp et al. (1987b) indicate that these or similar mechanisms may also be working on other, quite different stars (red giants).

Donati et al. (1990) have observed Stokes V line profiles on the primary of HR 1099 (RS CVn, with K1 IV and G5 V components), a very fast rotator for which standard techniques of the type described in Sect. 7 do not work. By comparing the wavelengths of the relatively narrow V profiles with the very broad Stokes I line profiles and making use of Doppler imaging results for this star (Vogt and Penrod 1983, Vogt 1988), they were able to conclude that the largest uncancelled V signal is produced outside the prominent starspots.

Very recently Elias and Dorren (1990) have measured a broad-band circular polarization signal of 1% on HD 129 333, a young solar analogue. The size of the signal is surprising and worthy of further investigation.

---

[1] Note that there may be a more uniformly distributed, more solar-like field component as well. However, due to flux cancellation it would not show up in Stokes V measurements.

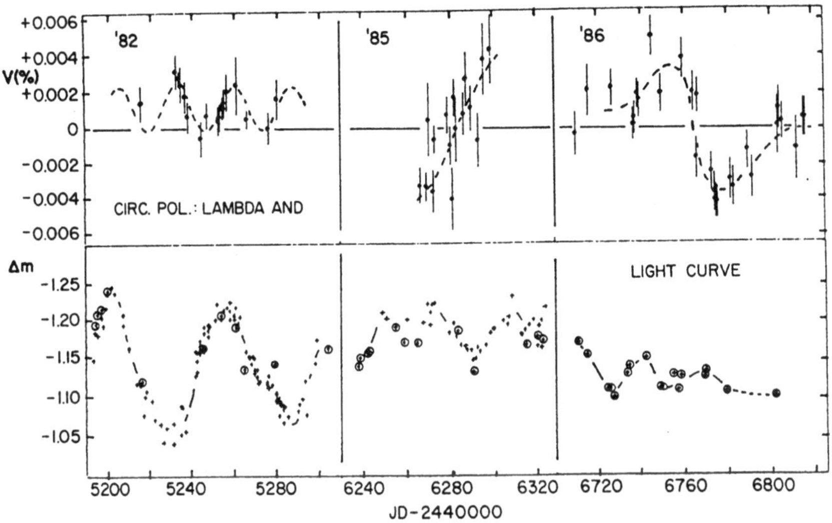

Fig. 3. Broad-band circular polarization (upper panels) and light curves (lower panels) of λ And in 3 seasons (1982, 1985, 1986). The dashed curve in the upper left panel is a least squares fit to the data, the rest are hand drawn. The rotation period of the star is approximately 54 days. After Kemp et al. (1987b).

Finally, although linear polarization is also expected to be produced by certain geometries of the magnetic field, I have not reviewed such measurements due to a current controversy about the reality of magnetic linear polarization detections on cool stars.

7. Stellar Magnetic Fields Derived from the Intensity Spectrum

7.1. Basic Idea

The combination of relatively small Zeeman splitting, small magnetic filling factor and the mixture of polarities on the stellar surface necessitates indirect techniques of detecting magnetic fields using unpolarized light. The basic idea underlying such techniques is due to Robinson (1980). The recipe is to take two spectral lines which, ideally, are identical (i.e. belong to the same ion, have the same oscillator strength and excitation potential and similar wavelengths), except for their Landé factors, i.e. their Zeeman sensitivity. Such a choice of lines ensures that in the non-magnetic part of the stellar atmosphere the profiles of the two lines are almost identical, irrespective of the thermodynamic structure of the atmosphere, surface gravity, etc. The magnetic field introduces a difference of a particular spectral shape between the two line profiles. The presence of such a spectral signature implies the probable detection of a magnetic field. The observations may then be interpreted with, for example, a two-component model, allowing the field strength B and the filling factor f to be determined.

7.2. The Real World

1. For most stars the Zeeman effect influences the line profiles only minutely, at least for spectral lines in the visible. Examples of observed (crosses) and calculated (curves) line profiles for an active star are given in Fig. 4. Note the small difference in the upper panel between the observed and synthetic profiles (calculated with $B = 0$) of the Zeeman sensitive line 8468Å. The implications are: i) The observational data must have a high S/N ratio of at least 100. Therefore, only relatively bright stars offer any hope of measuring magnetic fields. The condition of high S/N may be relaxed somewhat for very active stars observed in the infrared (e.g. dMe stars, Saar and Linsky 1985). ii) Even small blends can falsify the derived fields (Kurucz and Hartmann 1984, Hartmann 1987). Blends can work both ways, leading to spurious detections or hiding the presence of a magnetic field, depending on the details of the blending process (Gondoin et al. 1985). Blending due to atomic and molecular lines increases dramatically towards later spectral types and becomes a major problem for stars cooler than K5. Since the density of lines decreases somewhat with wavelength, the infrared is the spectral range of choice for observing the coolest stars (Saar and Linsky 1985). Another possibility is to use many spectral lines, since statistically the effects of blends are smoothed out (Mathys and Solanki 1989).
2. No 'ideal' line pairs (i.e. clean lines identical in all respects except their Landé factors) are available.[2] Since the two lines of a chosen pair often do not have exactly the same equivalent width it is dangerous to compare the two profiles directly. An indirect comparison using synthetic profiles as intermediaries is to be preferred, since it automatically

[2] The Fe I line pair 5250.2 Å and 5247.1 Å come close to being ideal, but 5250.2 Å suffers from blending in cooler stars and the Landé factors may be somewhat too similar. However, see Sánchez Almeida and García López 1990.

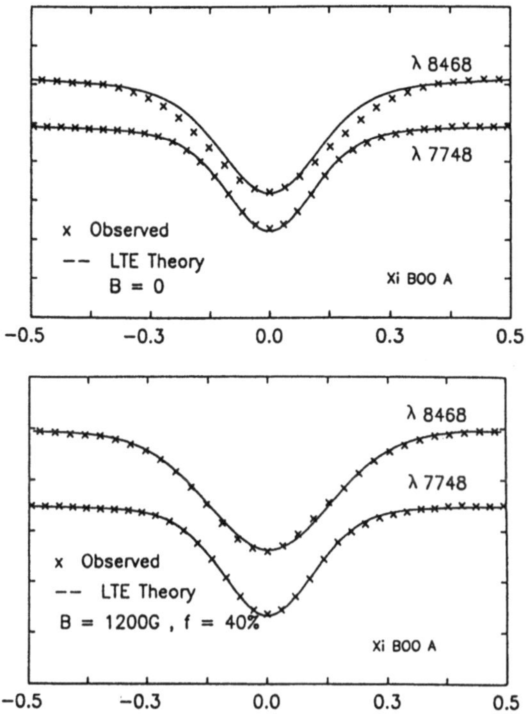

Fig. 4. Comparison of observed (crosses) to synthetic (solid curves) profiles of Fe I 8468Å (Zeeman sensitive line) and Fe I 7748Å (Zeeman insensitive line) for ξ Boo A (G8 V). The upper panel contains overlays of observed on synthetic profiles constructed with no magnetic field. Note the excess width of the observed λ8468Å profile. The lower panel shows the fit obtained when a magnetic field with a strength of 1200 G and a filling factor of 40% is included in the computations. After Basri and Marcy (1988).

compensates for differences in saturation between the two lines. This approach has been taken for most of the newer magnetic field determinations (e.g. Saar 1988a, Basri and Marcy 1988).

3. In general little is known about the geometry of the stellar magnetic field, except that it must be complex (Sect. 6). So far investigators have usually assumed that the field is fragmented into many small regions distributed evenly over the stellar surface. The spherical geometry of the stellar surface and stellar rotation then require the synthetic line profiles to be calculated at many points on the stellar disk and added together before being compared with the data (e.g. Saar 1988b). Time series measurements covering a full rotation period of a star (Saar et al. 1987a) may allow a rough idea of the distribution of the field to be obtained, if the field can be measured with sufficient accuracy. A combination of various direct and indirect magnetic field indicators that re-

act differently to the field distribution (e.g. Ca II K emission, Robinson technique, circular and linear polarization) may also give an indication of the distribution.

4. A major problem is the unknown thermodynamic structure of the magnetic features (Basri et al. 1990). Although B may in principle be determined with reasonable accuracy using current techniques, the derived f values are strongly affected by the unknown line weakening in and continuum intensity of the magnetic features relative to the non-magnetic atmosphere of the star.[3] One promising solution is to use self-consistent models of magnetic flux tubes including a sufficiently sophisticated energy equation (e.g. Grossmann-Doerth et al. 1989a, Steiner and Stenflo 1990, Knölker et al. 1990) to calibrate the measurements. The results of such calculations are getting increasingly realistic for the solar photosphere (Sect. 5.2.). Adapted to other stars the models should be able to predict the thermal structure of the stellar magnetic features and thus help with the interpretation of the observations.

7.3. Survey of Stellar Magnetic Field Measurements

Inspite of the problems discussed in Sect. 7.2., numerous attempts to derive magnetic parameters of cool stars have been undertaken, most of which have met with some degree of success. The following is an incomplete sample of such investigations, listed in roughly chronological order: Robinson et al. (1980), Giampapa et al. (1983), Marcy (1984), Marcy and Bruning (1984), Gray (1984), Gondoin et al. (1985), Saar and Linsky (1985), Saar et al. (1986a,b, 1987a,b), Saar (1987b, 1988a), Basri and Marcy (1988, 1990), Marcy and Basri (1989), Mathys and Solanki (1989), Bopp et al. (1989), Rípodas et al. (1990), Valenti (1990). For a given star the results of the more recent, more sophisticated investigations are in general more reliable and are to be preferred. For example, it is probable that some of the detections in the earlier investigations are spurious, or that some magnetic fields have gone undetected.

Some of the results of these investigations are now summarized with the help of two figures. Fig. 5 summarizes the "reliable" cool star field strengths measured to date (with two exceptions all are on main sequence stars). Plotted is the measured field strength vs. an 'equipartition field strength',

---

[3] Since the determination of f and B is to some extent coupled, in practice the measured B values may also be affected by this uncertainty.

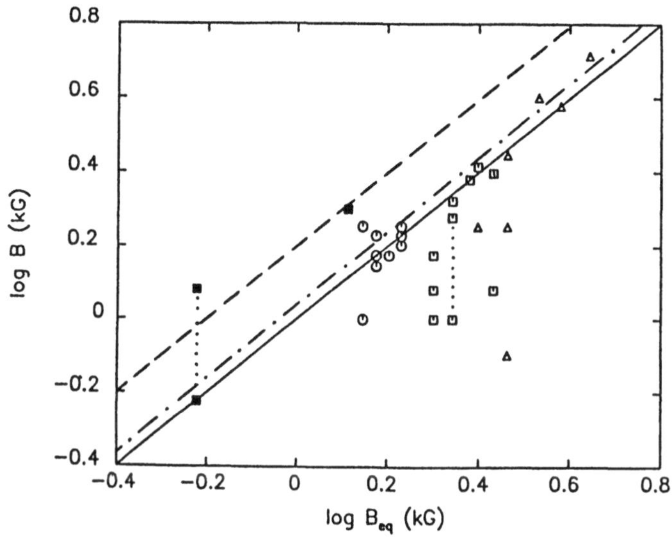

Fig. 5. Measured field strength B plotted vs. an "equipartition field strength" B_{eq}. B_{eq} is the field strength expected if the stellar magnetic field is confined by the gas pressure deficit in the magnetic features and if the Zeeman split lines are formed at the continuum forming level of the non-magnetic atmosphere (B_{eq} is normalized to reproduce the solar case correctly). Circles, squares, triangles and filled squares represent G, K, M dwarfs and RS CVn binaries, respectively. The sun is indicated by $\odot$. The relationship $B = B_{eq}$ is shown as a solid line. Data points to the upper left of the dashed curve refer to magnetic fields in stellar spots, while points to the lower right of the dot-dashed curve correspond to measurements of fields in stellar faculae. Adapted from Saar (1990a).

i.e. the expected strength if the field is mainly confined by gas pressure and if the Zeeman broadened lines in the magnetic features are formed at the $\tau = 1$ level of the non-magnetic atmosphere ($\tau_{nm} = 1$). If inside the magnetic features of a given star the lines are formed above the $\tau_{nm} = 1$ level, then the corresponding data point should lie to the lower right of the solid diagonal, if they are formed deeper down then the point should lie to the diagonal's upper left. Due to the strong dependence of continuum opacity on temperature, lines are formed deeper in cool starspots than in hot faculae.[4] It is thus possible to tentatively separate the measurements into detections of stellar spots and of stellar faculae. The data points to the upper left of the dashed curve refer to stellar spots, those to the lower right of the dot-dashed curve to stellar faculae. The stars lying between the two curves may possess a mixture of spots and faculae, or may simply reflect uncertainties in the field strength determination. Fig. 5 suggests that,

---

4 For very cool stars (M dwarfs) the relation between opacity and temperature may well break down due to the presence of broad molecular bands. For such stars the conclusions drawn here must be considered with care.

with the possible exception of the two giants/subgiants, no star exhibits any reasonable magnetic signal due to stellar spots.

The above conclusion is supported by more direct observational indicators. For example, on ϵ Eri the comparison between the field derived from two independent sets of spectral lines with different temperature sensitivities indicates that the measured field is concentrated in hot, bright regions (Solanki and Mathys 1987). On HR 1099 the region showing the maximum of the Stokes V signal is spatially distinct from the dark spots on that star (Donati et al. 1990). Since many late-type stars show unmistakable signs of cool and dark spots on their surfaces (e.g. Vogt and Penrod 1983, Vogt 1988, Ramsey and Nations 1980, Huenemoerder and Ramsey 1987) does the absence of a magnetic signature from the dark stellar patches imply that they are non-magnetic? Not necessarily, since on a spatially unresolved star any magnetic field concentrated into localized dark structures is very hard to detect due to the low continuum light level in such structures. For example, in the visible solar spectrum the continuum intensity I_c of facular magnetic flux tubes lies in the range

$$I_c^{\rm quiet} \lesssim I_c^{\rm faculae} \lesssim 2 I_c^{\rm quiet},$$

while for sunspots

$$0.1 I_c^{\rm quiet} \lesssim I_c^{\rm spot} \lesssim 0.5 I_c^{\rm quiet}.$$

Here $I_c^{\rm quiet}$ is the continuum intensity of the quiet sun. Therefore, a considerable fraction of the stellar field may be present in the form of spots, yet may still escape detection (cf. Saar et al. 1986a).[5] It may nevertheless be possible to detect fields in stellar spots by considering purely umbral lines with large Zeeman sensitivities. Of particular interest are lines in the infrared, since the continuum contrast decreases with wavelength.

Another often used way of summarizing the results of magnetic field measurements on multiple stars is to plot the magnetic filling factor f, or average magnetic flux density fB vs. stellar rotation frequency Ω, stellar mean convective turnover time τ_c, or some product of the two. Such diagrams may, e.g., be compared to the predictions of different dynamo models and the hope is that they can distinguish between rivalling theories. An example of such a diagram with $\log(fB)$ plotted vs. $\log(\Omega)$ is shown in

[5] A preponderance of faculae may, however, be a real effect on at least some active stars. There is evidence that the average upper photospheric temperature in active cool dwarfs is higher than of their inactive counterparts (Holweger 1988, Basri et al. 1989).

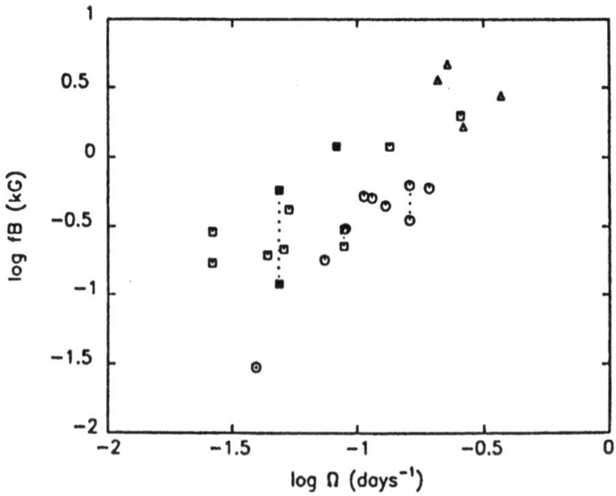

Fig. 6. Magnetic flux density fB vs. stellar rotation frequency Ω. The symbols have the same meanings as in Fig. 5. After Saar (1990a).

Fig. 6. A general trend is visible and a least squares fit to the data gives $fB \sim \Omega^{1.3\pm0.1}$ (Saar 1990). However, a caveat is in order. Even on the sun the magnetic filling factor and flux density are extremely difficult to measure without bias (e.g. Schüssler and Solanki 1988). The f values on other cool stars are yet more unreliable, since their determination does not take into account the influence of the thermodynamic properties of the unresolved magnetic features on the line profiles. This can lead to considerable errors in f (e.g. Basri et al. 1990) and may be responsible for a part of the scatter in Fig. 6. For stars with mutually similar magnetic structures (e.g. with a predominance of faculae) the errors in f may be systematically similar.

8. A Particular Case: ϵ Eri

The active dwarf ϵ Eri (K2 V) is one of the most thoroughly investigated cool stars. In his review Saar (1990) compiled a list of magnetic field measurements of this star (his Table 2). Four different groups have carried out 10 independent magnetic field measurements. The results agree only partially and the total bandwidth of measured B and f values is disturbingly large. Values of B range between 1kG and 3kG, f values between 8% and 67%. Even fB is found to vary by a factor of three between 0.24 kG and 0.78 kG. I see four possible causes of these differences:

1. One possibility is that mutually incompatible measurements simply reflect the magnetic field at different stages of its evolution. Early measurements suggested large temporal variations of the field strength and filling factor — Timothy et al. (1981) and Marcy (1984) found evidence for night-to-night variations. However, these variations have not been confirmed by newer, more sophisticated measuring techniques (Saar et al. 1986b). All in all, we currently can neither confirm nor rule out evolution effects as the cause of the discrepancies described above, but it appears unlikely that they are the sole cause. A systematic study of magnetic field variations with time on individual stars would be welcome.
2. The largest values of f and of fB are due to the two oldest and probably least reliable measurements. If we exclude these, then the scatter decreases significantly. In particular, the measured values of the average magnetic flux density fB are relatively similar for the rest of the investigations.
3. The derived magnetic field parameters depend on the used spectral lines. The choice of spectral lines can influence the results in three ways. *i*) Some lines are formed at greater heights in the atmosphere than others. Since the field strength decreases with height, two lines formed at different heights are also Zeeman split by different amounts. Grossmann-Doerth and Solanki (1990) showed that this effect can explain a large fraction of the approximately factor of 2 difference between the B values derived from lines in the visible (Saar 1988a, Mathys and Solanki 1989) and in the near infrared (Basri and Marcy 1988, Marcy and Basri 1989). The positive aspect of this effect is that it allows the vertical gradient of the field strength to be estimated. Besides being of intrinsic interest, this presently appears to be the most reliable way of testing whether the confinement mechanism of solar magnetic fields (horizontal gas pressure gradients) is also valid for cool stars. However, note that fB should be independent of the formation height. *ii*) The lines used by the different groups also have somewhat different temperature sensitivities, so that they react differently to the temperature in the magnetic features. The derived f value (and partly also the field strength) will therefore depend on the selected lines. This effect can be turned to advantage to derive the temperature in the stellar magnetic features (Solanki and Mathys 1987). *iii*) The strength and influence of weak blends also varies from line to line and may account for some of the discrepancies.

4. Finally, a part of the scatter is certainly due to noise in the data, and probably also to shortcomings of the present techniques. The improvement of these techniques must be given a high priority in the near future.

9. Conclusions

Let me end by summarizing some of the main points of this review and drawing some general and possibly rather obvious conclusions.

1. Magnetic fields have highly visible effects on the radiation from cool stars, mainly through their influence on the thermodynamic structure of the stellar atmospheres. However, the fields themselves are difficult to observe directly.
2. Nevertheless, magnetic fields have been detected on approximately 40 main sequence stars of spectral types between G0 and M4.5, on 2 giant/subgiant members of close binaries and on a pre-main sequence star. So far only facular fields on active stars with small $v \sin i$ (projected rotation velocity) have been definitely detected (exceptions are mentioned under points 3 and 5).
3. The only inactive star with a reliably measured magnetic field is the sun. Its field can be measured in great detail and theoretical models of the magnetic structures (e.g. flux tube models) can be rigorously tested. Such models, once they have been adapted in a physically consistent manner to the conditions on other stars, are expected to form an important part of improved techniques of stellar magnetic field measurement now under development.
4. Recent improvements in infrared detector technology and new efficient high dispersion gratings for the infrared should increase the sensitivity to stellar magnetic fields by a factor of 3–10. Possibly the new instruments will finally allow the reliable measurement of magnetic fields in starspots.
5. For rapid rotators Stokes V polarimetry shows the most promise. In particular, it may be possible to roughly derive the distribution of the field on the stellar surface through the application of Zeeman Doppler imaging (Semel 1989, Donati et al. 1989).

Acknowledgements: Ich möchte mich für die freundliche Einladung zu diesem Vortrag bei den Vorstandsmitgliedern der AG und insbesondere beim Vorsitzenden, Prof. E.H. Schröter, bedanken.

References

Babcock, H.W.: 1947, *Astrophys. J.* **105**, 105

Babcock, H.W.: 1958, *Astrophys. J. Suppl. Ser.* **3**, 141

Babcock, H.W., Babcock, H.D.: 1952, *Publ. Astron. Soc. Pacific* **64**, 282

Basco, M.M., Sunyaev, R.A.: 1975, *Astron. Astrophys.* **42**, 311

Basri, G., Marcy, G.W.: 1988, *Astrophys. J.* **330**, 274

Basri, G., Marcy, G.W.: 1990, in *The Sun and Cool Stars: Activity, Magnetism, Dynamos*, Helsinki, July 17–21, 1990, , *IAU Coll.* **130**, in press

Basri, G., Marcy, G.W., Valenti, J.A.: 1990, *Astrophys. J.* **360**, 650

Basri, G, Wilcotts, E., Stout, N.: 1990, *Publ. Astron. Soc. Pacific* **101**, 528

Boesgaard, A.M.: 1974, *Astrophys. J.* **188**, 567

Boesgaard, A.M., Chesley, D., Preston, G.W.: 1975, *Publ. Astron. Soc. Pacific* **87**, 353

Bonsack, W.K., Simon, T.: 1983, in *Solar and Stellar Magnetic Fields: Origins and Coronal Effects*, J.O. Stenflo (Ed.), *IAU Symp.* **102**, 35

Bopp, B.W., Saar, S.H., Ambruster, C., Feldman, P., Dempsey, R., Allen, M., Barden, S.P.: 1989, *Astrophys. J.* **339**, 1059

Borra, E.F., Edwards, G., Mayor, M.: 1984, *Astrophys. J.* **284**, 211

Brown, D.N., Landstreet, J.D.: 1981, *Astrophys. J.* **246**, 899

Bünte, M., Steiner, O., Solanki, S.K.: 1990, in *Solar Polarimetry*, L. November (ed.), Proc. 11th Sacramento Peak Workshop, National Solar Obs., Sunspot, NM.

Deutsch, A.J.: 1970, *Astrophys. J.* **159**, 985

Donati, J.-F., Semel, M., Rees, D.E., Taylor, K., Robinson, R.D.: 1990, *Astron. Astrophys.* **232**, L1

Donati, J.-F., Semel, M., Praderie, F.: 1989, *Astron. Astrophys.* **225**, 467

Elias, N.N., Dorren, J.D.: 1990, *Astron. J.* **100**, 818

Giampapa, M.S., Golub, L., Worden, S.P.: 1983, *Astrophys. J.* **268**, L121

Gnedin, Yu.N., Sunyaev, R.A.: 1974, *Astron. Astrophys.* **36**, 379

Gondoin, P., Giampapa, M.S., Bookbinder, J.A.: 1985, *Astrophys. J.* **297**, 710

Gray, D.F.: 1984, *Astrophys. J.* **277**, 640

Gray, D.F.: 1985, *Publ. Astron. Soc. Pacific* **97**, 719

Gray, D.F.: 1988, *Lectures on Spectral-Line Analysis: F, G and K Stars*, The Publisher, Arva

Grossmann-Doerth, U., Knölker, M., Schüssler, M., Weisshaar, E.: 1989a, in *Solar and Stellar Granulation*, R.J. Rutten and G. Severino (Eds.), Reidel, Dordrecht, 481

Grossmann-Doerth, U., Schüssler, M., Solanki, S.K.: 1988, *Astron. Astrophys. Letters* **206**, L37

Grossmann-Doerth, U., Schüssler, M., Solanki, S.K.: 1989b, *Astron. Astrophys.* **221**, 338

Grossmann-Doerth, U., Solanki, S.K.: 1990, *Astron. Astrophys.* **238**, 279

Güdel, M., Benz, A.O.: 1989, *Astron. Astrophys.* **211**, L5

Hale, G.E.: 1908, *Astrophys. J.* **28**, 315

Hale, G.E.: 1922, *Monthly Notices Royal Astron. Soc.* **82**, 168

Hartmann, L.: 1987, in *Cool Stars, Stellar Systems, and the Sun,* V., J.L. Linsky, R.E. Stencel (Eds.), Lecture Notes in Physics Vol. 291, Springer-Verlag, Berlin, p. 1

Holweger, H.: 1988, in *The Impact of Very High S/N Spectroscopy on Stellar Physics*, G. Cayrel de Strobel, M. Spite (eds.), Kluwer, Dordrecht, *IAU Symp.* **132**, 411

Huendemoerder, D.P., Ramsey, L.W.: 1987, *Astrophys. J.* **319**, 392

Kemp, J.C.: 1970, *Astrophys. J.* **162**, 169

Kemp, J.C., Henson, G.D., Kraus, D.J., Dunaway, M.H., Hall, D.S., Boyd, L.J., Genet, R.M., Guinan, E.F., Wacker, S.W., McCook, G.P.: 1987b, *Astrophys. J.* **317**, L29

Kemp, J.C., Henson, G.D., Steiner, C.T., Powell, E.R.: 1987a, *Nature* **326**, 270

Kemp, J.C., Swedlund, J.B., Landstreet, J.D., Angel, J.R.P.: 1970, *Astrophys. J.* **161**, L77

Kiepenheuer, K.O.: 1953, *Astrophys. J.* **117**, 447

Knölker, M., Grossmann-Doerth, U., Schüssler, M., Weisshaar, E.: 1990, *Adv. Space Res.* in press.

Kurucz, R.L., Hartmann, L.: 1984, SAO Preprint, No. 2015

Linsky, J.L.: 1985, *Solar Phys.* **100**, 333

Marcy, G.W.: 1983, in *Solar and Stellar Magnetic Fields: Origins and Coronal Effects*, J.O. Stenflo (Ed.), *IAU Symp.* **102**, 3

Marcy, G.W.: 1984, *Astrophys. J.* **276**, 286

Marcy, G.W., Basri,G.: 1989, *Astrophys. J.* **345**, 480

Marcy, G.W., Bruning, D.H.: 1984, *Astrophys. J.* **281**, 286

Martin, S.F.: 1990, in *Solar Photosphere: Structure, Convection and Magnetic Fields*, J.O. Stenflo (Ed.), Kluwer, Dordrecht, *IAU Symp.* **138**, 129

Mathys, G.: 1989, *Fundam. Cosmic Phys.* **168**, 184

Mathys, G., Lanz, T.: 1990, *Astron. Astrophys.* **230**, L21

Mathys, G., Solanki, S.K.: 1989, *Astron. Astrophys.* **208**, 189

Mürset, U., Solanki, S.K., Stenflo, J.O.: 1988, *Astron. Astrophys.* **204**, 279

Nordlund, Å.: 1986, in *Proc. Workshop on Small Magnetic Flux Concentrations in the Solar Photosphere*, W. Deinzer, M. Knölker, H.H. Voigt (Eds.), Vandenhoeck & Ruprecht, Göttingen, p. 83

Pallavicini, R.: 1987, in *Activity in Cool Star Envelopes*, O. Havnes et al. (Eds.), Reidel, Dordrecht, p. 25

Ramsey, L.W., Nations, H.L.: 1980, *Astrophys. J.* **239**, L121

Rípodas, P., Collados, M., García López, R., Sánchez Almeida, J.: 1990, in *The Sun and Cool Stars: Activity, Magnetism, Dynamos*, Helsinki, July 17–21, 1990, , *IAU Coll.* **130**, in press

Robinson, R.D.: 1980, *Astrophys. J.* **239**, 961

Robinson, R.D.: 1986, *Highlights of Astronomy*, **7**, 417

Robinson, R.D., Worden, S.P., Harvey, J.W.: 1980, *Astrophys. J.* **236**, L155

Russell, C.T., Priest,, E.R., Lee, L.C.: 1990, (eds.) *Physics of Magnetic Flux Ropes*, Geophysical Monograph 58, American Geophysical Union, Washington DC

Saar, S.H.: 1987a, in *Cool Stars, Stellar Systems and the Sun. V.* J.L. Linsky, R.E. Stencel (eds.), Springer, New York, p. 10

Saar, S.H.: 1987b, *Ph.D. Thesis*, JILA, Boulder, CO

Saar, S.H.: 1988a, *Astrophys. J.* **324**, 441

Saar, S.H.: 1988b, in *The Impact of Very High S/N Spectroscopy on Stellar Physics*, G. Cayrel de Strobel, M. Spite (eds.), Kluwer, Dordrecht, *IAU Symp.* **132**, 295

Saar, S.H.: 1990, in *Solar Photosphere: Structure, Convection and Magnetic Fields*, J.O. Stenflo (Ed.), Kluwer, Dordrecht, *IAU Symp.* **138**, 427

Saar, S.H., Huovelin, J., Giampapa, M.S., Linsky, J.L., Jordan, C.: 1987, in *Activity in Cool Star Envelopes*, O. Havnes et al. (Eds.), Reidel, Dordrecht, p. 45

Saar, S.H., Linsky, J.L.: 1985, *Astrophys. J.* **299**, L47

Saar, S.H., Linsky, J.L., Beckers, J.M.: 1986, *Astrophys. J.* **302**, 777

Saar, S.H., Linsky, J.L., Duncan, D.K.: 1986, in *Cool Stars, Stellar Systems and the Sun. IV.* M. Zeilik, D.M. Gibson (eds.), Springer, New York, p. 275

Saar, S.H., Linsky, J.L., Giampapa, M.S.: 1987, in *Observational Astrophysics with High Precision Data*, Proc. Liège Colloq. No. 27, Inst. d'Astrophys., Cointe-Ougrée, p. 103

Sánchez Almeida, J., García López, R.: 1990, in *The Sun and Cool Stars: Activity, Magnetism, Dynamos*, Helsinki, July 17–21, 1990, , *IAU Coll.* **130**, in press

Schüssler, M.: 1986, in *Small Scale Magnetic Flux Concentrations in the Solar Photosphere*, W. Deinzer, M. Knölker, H.H. Voigt (Eds.), Vandenhoeck & Ruprecht, Göttingen, p. 103

Schüssler, M.: 1987, in *The Role of Fine-Scale Magnetic Fields on the Structure of the Solar Atmosphere*, E.-H. Schröter, M. Vázquez, A.A. Wyller (Eds.), Cambridge University Press, p. 223

Schüssler, M.: 1990, in *Solar Photosphere: Structure, Convection and Magnetic Fields*, J.O. Stenflo (Ed.), Kluwer, Dordrecht, *IAU Symp.* **138**, 161

Schüssler, M., Solanki, S.K.: 1988, *Astron. Astrophys.* **192**, 338

Semel, M.: 1989, *Astron. Astrophys.* **225**, 456

Skumanich, A., Lites, B.W.: 1987, *Astrophys. J.* **322**, 483

Solanki, S.K.: 1987a, in *The Role of Fine-Scale Magnetic Fields on the Structure of the Solar Atmosphere*, E.-H. Schröter, M. Vázquez, A.A. Wyller (Eds.), Cambridge University Press, p. 67

Solanki, S.K.: 1987b, in *Proc. Tenth European Regional Astronomy Meeting of the IAU. Vol. 1: The Sun*, L. Hejna, M. Sobotka (Eds.), Publ. Astron. Inst. Czechoslovak Acad. Sci., p. 95

Solanki, S.K.: 1989, *Astron. Astrophys.* **224**, 225

Solanki, S.K.: 1990, in *Solar Photosphere: Structure, Convection and Magnetic Fields*, J.O Stenflo (Ed.), *IAU Symp.* **138**, 103

Solanki, S.K., Mathys, G.: 1987, in *Activity in Cool Star Envelopes*, O. Havnes et al. (Eds.), Reidel, Dordrecht, p. 39

Spruit, H.C.: 1981, in *The Sun as a Star*, S. Jordan (Ed.) NASA SP-450, p. 385

Spruit, H.C.: 1983, in *Solar and Stellar Magnetic Fields: Origins and Coronal Effects*, J.O. Stenflo (Ed.), *IAU Symp.* **102**, 41

Spruit, H.C. and Roberts, B.: 1983, *Nature*, **304**, 401

Spruit, H.C., Schüssler, M., Solanki, S.K.: 1990, in *Solar Interior and Atmosphere*, A.N. Cox, W. Livingston, M.S. Matthews (Eds.), in press (1989).

Steiner, O., Stenflo, J.O.: 1990, in *Solar Photosphere: Structure, Convection and Magnetic Fields*, J.O. Stenflo (Ed.), Kluwer, Dordrecht, *IAU Symp.* **138**, 181

Stenflo, J.O.: 1973, *Solar Phys.* **32**, 41

Stenflo, J.O.: 1984, *Adv. Space Res.* **4**, 5

Stenflo, J.O.: 1985, *Solar Phys.* **100**, 189

Stenflo, J.O.: 1986, in *Small Scale Magnetic Flux Concentrations and the Solar Photosphere*, W. Deinzer, M. Knölker, H.H. Voigt (Eds.), Vandenhoek & Ruprecht, Göttingen, p. 59

Stenflo, J.O.: 1989, *Astron. Astrophys. Review* **1**, 3

Stibbs, D.W.N.: 1950, *Monthly Notices Royal Astron. Soc.* **110**, 395

Stix, M.: 1990, *The Sun, an Introduction*, Springer, Berlin

Thiessen, G.: 1952, *Z. Astrophys.* **30**, 185

Timothy, J.G., Joseph, C.L., Linsky, J.L.: 1981, *Bull. American Astron. Soc.* **13**, 828

Trümper, J., Pietsch, W., Reppin, C., Sacco, B., Kendziorra, E., Staubert, R.: 1977, *Mitt. Astron. Gesell.* **42**, 120

Trümper, J., Pietsch, W., Reppin, C., Voges, W., Staubert, R., Kendziorra, E.: 1978, *Astrophys. J.* **219**, L105

Valenti, J.A.: 1990, in *The Sun and Cool Stars: Activity, Magnetism, Dynamos*, Helsinki, July 17–21, 1990, , *IAU Coll.* **130**, in press

Vogt, S.S.: 1980, *Astrophys. J.* **240**, 567

Vogt, S.S.: 1988, in *The Impact of Very High S/N Spectroscopy on Stellar Physics*, G. Cayrel de Strobel, M. Spite (eds.), Kluwer, Dordrecht, *IAU Symp.* **132**, 253

Vogt, S.S., Penrod, G.D.: 1983, *Publ. Astron. Soc. Pacific* **95**, 565

Zeeman, P.: 1897, *Phil. Mag.* **43**, 226

The Expanding Envelope of Supernova 1987A in the Large Magellanic Cloud

R. W. Hanuschik

Astronomisches Institut, Ruhr-Universität Bochum,
Postfach 102148, W-4630 Bochum 1, Fed. Rep. of Germany

1. Introduction

The Ludwig Biermann Award 1990 of the Astronomische Gesellschaft clearly is not only a distinction for me and my collaborators in the Bochum SN 1987A project, but also an excellent opportunity to express my thanks to the Bochum Institute Director, Prof. Theodor Schmidt-Kaler, who invested a lot of time and dedication in this programme. In fact this work would have been impossible without his permanent and inspiring support in the background.

Supernovae indicate the explosive end of a star. With a mean absolute brightness at maximum of M_b = - 19.0 (type Ia), -17.1 (type Ib), and -16.9 (type II; Miller and Branch, 1990), they are the intrinsically brightest single objects in the sky. Type-II SNe are the most energetic (albeit not the most luminous) ones. They release a total energy of ~ 10^{53} erg, most of which is emitted *within seconds* after core collapse (Burrows, 1988) in form of neutrinos. (For comparison, the Sun emits 10^{51} erg in 10^{10} years).

Meanwhile, supernovae are frequently observed objects. The Asiago Supernova Catalogue (Barbon *et al.*, 1989) lists 661 extragalactic objects. Automatic survey programmes like the Berkeley Automated Supernova Search (Kare *et al.*, 1989), are now inflating this list dramatically. By October 29, 1990 (IAU *Circular* 5124), 32 supernovae have been discovered this year. Prospects are good that soon the discovery rate will increase to one SN per night (Perlmutter, 1990).

Most of these supernova detections, however, are only valuable for improving supernova statistics. For understanding the physics of the explosion, a nearby supernova is needed, and these are rare. Only five galactic SNe have been observed this millenium, and none of them by telescope. This is the reason why SN 1987A in the LMC immediately caused restless activity at all southern observatories. It was by far the brightest extragalactic supernova in historical times, and by far the brightest supernova since the invention of the telescope. It is now the probably best investigated (although not best understood) extra-solar system object in astronomy, with more than 1230 papers dealing with it in 1987-89 (as counted from the *A&A Abstracts*).

2. What we have seen: collapse and explosion

Nowadays supernovae are classified according to their spectra, not to their lightcurve which is thought of as being only a secondary criterion for classification (Filippenko, 1990). Type-I's do not show Balmer lines, type-II's do. The physical reason therefore are the different progenitor histories: type-II SNe arise from evolved single massive stars with large hydrogen envelopes, while type-I progenitors supposedly are white dwarfs in evolved binary systems or exploding WR stars, with little, if any, hydrogen. SN 1987A, as a type-II event, was caused by the *collapse* of the $\approx 1.4\ M_\odot$ iron core of its $\approx 20\ M_\odot$ progenitor, releasing the binding energy $GM^2/R \sim 2$–$3\ 10^{53}$ erg mostly in a neutrino/anti-neutrino burst. This event is the only information we have about the physics of the collapse.

The subsequent shock wave running through the envelope finally reached the surface and caused the brightening of the supernova. The energy of this *explosion*, mostly in form of mechanical energy, is only a small fraction of the collapse energy, ~ 1%, and the electromagnetic energy, the primary source of information about the explosion, is again only ~ 1 % of the kinetic energy. Peering at the photons from SN 1987A, we are thus watching a phenomenon subordinate in terms of energy, but highly informative: the *perfume*.

Canonical progenitors of type-II SNe are red supergiants. The progenitor of SN 1987A, the first one ever really observed, was a blue supergiant. This is already implied by the fact that the time required for the shock wave to traverse the extended ($\sim 10^{14}$ cm) envelope of a red supergiant would be days, not hours as observed. Thus the progenitor must have been a comparatively "compact" object, i.e. a blue supergiant.

Serendipitous *observations* of the progenitor started in the early 1970's when Sanduleak (1969) took objective prism spectra of OB-type stars in the LMC, among them Sk -69^0 202. Isserstedt (1975) proved wise foresight when he performed photoelectric photometry on the Sanduleak star in 1971-73. Spectrograms exist from 1972 (Walborn *et al.*, 1989) and 1977 (González *et al.*, 1987). The next two important time points both occurred on February 23, 1987: at $t_0 = 23.316$ UT, a total number of 19 neutrinos from the core collapse of Sk -69° 202 were registrated in the Kamioka and IMB detectors (Hirata *et al.*, 1987; Bionta *et al.*, 1987), and $\Delta t = 2.41 \pm 0.64$ hours later, on $t_1 = 23.390 \ldots 23.443$ UT, photons from the explosion arrived on Earth. This time point is fixed by the non-detection of SN 1987A by the amateur astronomer Jones, and the first positive detection on a photographic plate.

This review, as necessarily all others on SN 1987A, does not intend to be comprehensive, not even concerning the observational facts, and by no means with regard to the theoretical progress in understanding these observations. The diligent reader must instead be referred to the conference proceedings hitherto dedicated to SN 1987A, i.e. the ESO 1987 and 1990 conferences (Danziger, 1987, 1990), the 1987 George Mason Workshop (Kafatos and Michalitsianos, 1988), the 1988 La Thuile Workshop (Greco, 1988), the 1988 White Conference (Proust and Couch, 1988) and the 1989 Santa Cruz Workshop (Woosley, 1990). Recent reviews are, e.g., by Dopita (1988), Arnett *et al.* (1989), McCray (1989) and Hillebrandt and Höflich (1989).

Spectroscopic observations of the explosion started on February 24.8 with the IUE satellite (e.g., Wamsteker *et al.*, 1987; Kirshner *et al.*, 1987) and at the South African Observatory (Menzies *et al.*, 1987). Slightly later the Chilean observatories took over (ESO, CTIO, Bochum telescope). At that time, the supernova was at V $\approx 5^m$, and small telescopes were favoured in the observations. The smallest and least efficient of them could do the best job: the 61cm telescope of the Ruhr-Universität Bochum on La Silla, equipped with a rapid spectrum scanner spectrograph (Haupt *et al.*, 1976) and a photomultiplier. (This telescope has already a good tradition in observing what we now call SN 1987A: Isserstedt [1975] performed his 1971–73 photometry at the Bochum telescope.) With its large entrance slit (20–30"), it is ideally suited for performing precise spectrophotometry of bright (V $< 7^m$) objects. Observers in the Bochum SN 1987A campaign are listed in Table 1.

The scanner spectrograph was used until day 314 (1988 January 3); observations could be continued, albeit at lower accuracy, with a different detector (OMA, kindly provided by Prof. Geyer from Hoher List Observatory) until day 450. At present, data for the first 314 days are fully re-

Tab. 1: List of observers in the Bochum SN 1987A program

Time of observation	Days[a]	Observer
25. 2. – 15. 3. 1987........	1.7–20	J. Dachs
16. 3. – 15. 4. 1987........	21–50	R. Hanuschik
16. 4. – 14. 6. 1987........	51–111	G. Thimm
15. 6. – 30. 7. 1987........	112–157	K. Seidensticker
31. 7. – 10. 9. 1987........	158–198	J. Gochermann
11. 9. – 15.11. 1987........	199–265	S. Kimeswenger
16.11. – 15.12. 1987........	266–293	R. Poetzel
16.12. – 3. 1. 1988........	294–314	G. Schnur
4. 1. – 20. 2. 1988........	315–362	U. Lemmer
21. 2. – 30. 4. 1988........	363–432	J. Stüwe
30. 4. – 17. 5. 1988........	433–449	M. Werger

[a] Days since 23.316 February 1987 (UT)

duced. They usually cover the wavelength range 3 200–8 700 Å at a resolution of 10, later 15 Å. Most spectral regions have been additionally measured at 3–5 Å resolution and fluxed with the spectrophotometry. Flux calibration was performed with flux standard stars from the list of Tüg (1980a, b). Accuracy of the flux scale is 3 %, slightly worse at the corners of the spectra. The Bochum spectrophotometric atlas of SN 1987A therefore represents the *fully resolved, precisely calibrated, quasi-continuous time-record of the optical evolution of an expanding supernova envelope*. The reduced part of this data set is published or will be published soon (Hanuschik and Dachs, 1988; Hanuschik *et al.*, 1989, 1991) and is availabe upon request on magnetic tape.

In the following, I will focus on recent and interesting results about the energetics in the expanding envelope (chapter 3), the evolution of the flux in the lines (chapter 4), mixing and clumping processes (chapter 5) and about the surrounding circumstellar medium (chapter 6). Other aspects, like kinematics, distance determination and the *Bochum event*, are reviewed in more detail in Hanuschik (1989, 1990a).

3. Bolometric luminosity and energetics

3.1 Overview over the spectrum

Figure 1 shows the colour-coded evolution of the optical spectrum of SN 1987A, and of its first derivative. The flux record shows best the brightness evolution, as well as the change from a rather blue, hot spectrum into a red, cool spectrum during the first few days. The flux derivative demonstrates the evolution of the line spectrum and especially emphasizes the kinematics.

Some momentary shots of the spectral evolution of SN 1987A, as seen from the Bochum telescope, are shown in Fig. 2. While essentially a black body with few, extremely broad hydrogen and helium lines on day 2 (Feb. 25), the optical spectrum evolves quickly into one with strong absorption lines, and finally an emission-line dominated nebular-type spectrum around day 300. This transition from a stellar into a nebular spectrum is essentially caused by the different evolution of continuum- and line-forming regions in the expanding envelope. Initially, photosphere and atmosphere are close together, frozen in the expanding matter. After 5 days, the photosphere starts falling back in comoving frame (but still expands for the observer), while the line-forming regions keep on expanding. After day 100, the photosphere shrinks, and most light is emitted in the lines.

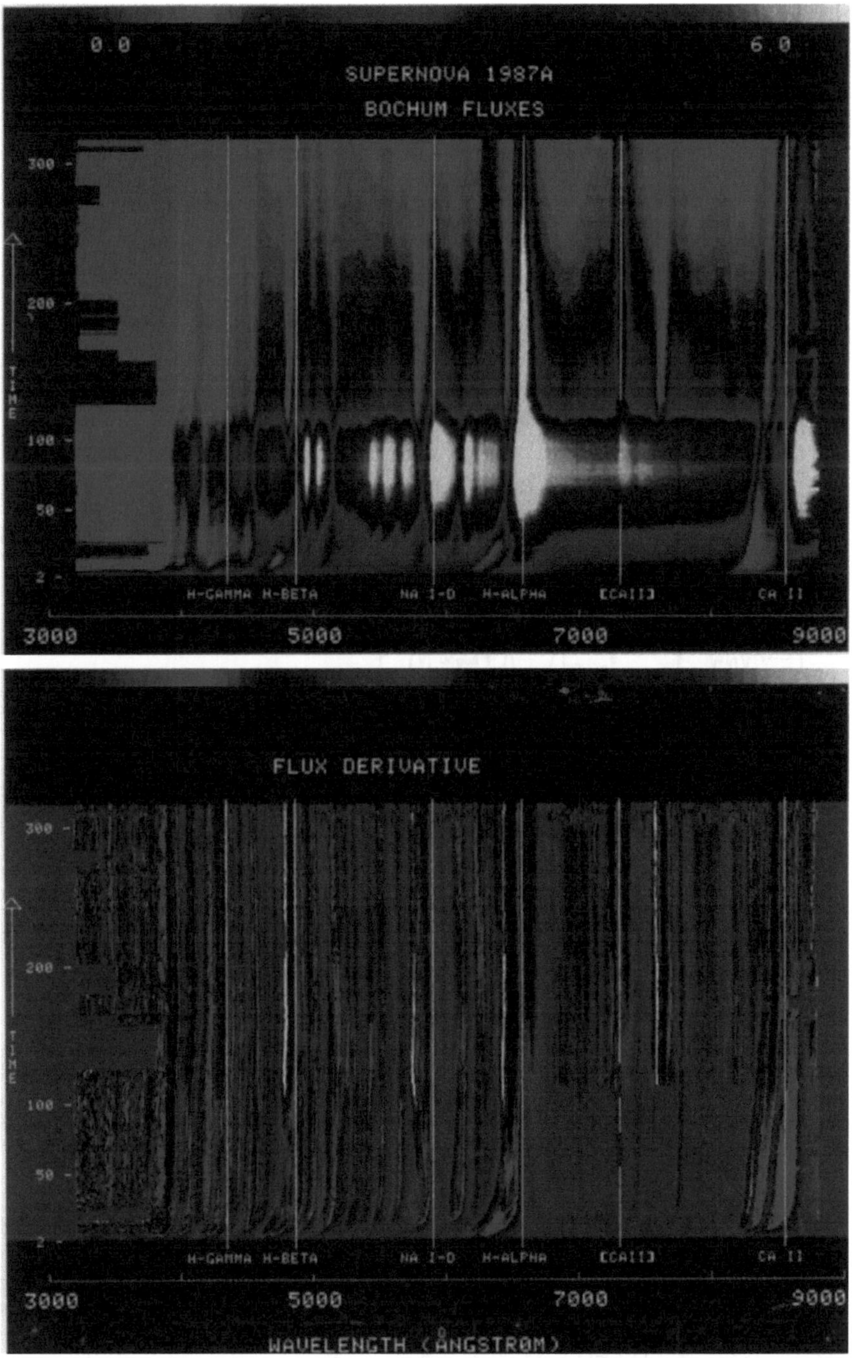
0.0
6.0
SUPERNOVA 1987A
BOCHUM FLUXES
TIME
300
200
100
50
2
H-GAMMA H-BETA
NA I-D
H-ALPHA
[CAII]
CA II
3000
5000
7000
9000
FLUX DERIVATIVE
TIME
300
200
100
50
2
H-GAMMA H-BETA
NA I-D
H-ALPHA
[CAII]
CA II
3000
5000
7000
9000
WAVELENGTH (ÅNGSTROM)

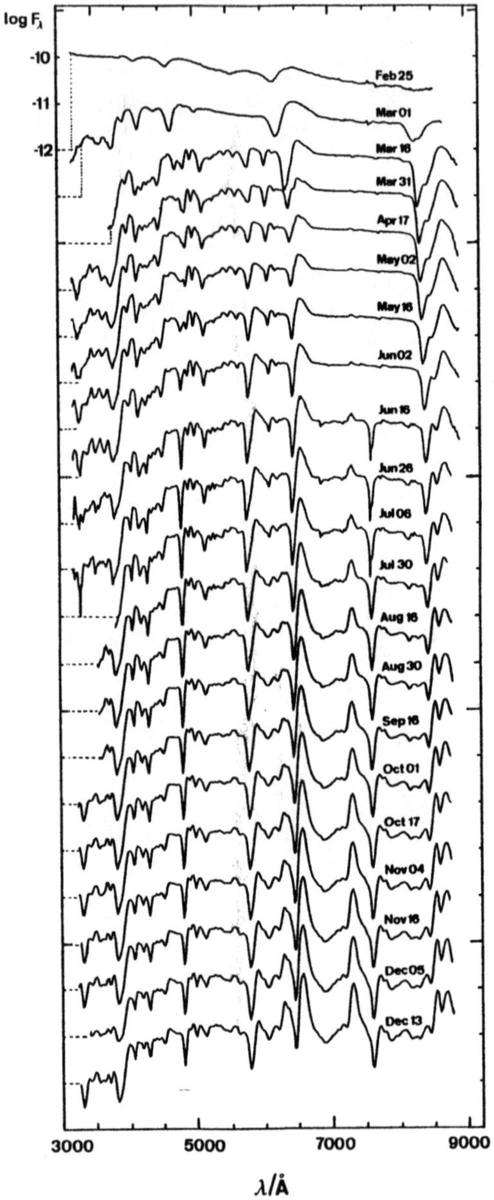

Fig. 2: Selected spectra of SN 1987A. Logarithmic flux scale is in erg $s^{-1}cm^{-2}Å^{-1}$. Spectra are shifted by -1 dex each, the broken line marking log F = -12. The strong absorption lines in the red after June 16 are telluric

Fig. 1 (previous page). Top: Colour-coded fluxes of SN 1987A as measured at the Bochum telescope. The flux scale is logarithmic. The ordinate is in days since core collapse. Rest wavelengths of a few strong lines are marked. This figure particularly emphasizes the brightness evolution, with its maximum occurring on day 82. **Bottom:** Normalized spectral derivative $[F(\lambda + \Delta\lambda) - F(\lambda)]/F(\lambda)$. Negative flux gradients appear blue-white, positive red. Green codes spectral regions with virtually constant flux across $\Delta\lambda \approx 5$ Å. This figure emphasizes the line spectrum, and particularly its kinematical evolution

3.2 The bolometric lightcurve

The bolometric lightcurve (BLC) of a supernova provides the record of electromagnetic energy radiated away. It is defined as

$$L_{bol}(t) = 4\pi D^2 \int_0^\infty F'_\lambda \, d\lambda, \quad F'_\lambda = F_\lambda \cdot 10^{0.4\ A(\lambda)\cdot E(B-V)} \qquad (1)$$

where F_λ are the measured fluxes, D is the distance, $A(\lambda)$ is the extinction law and E_{B-V} is the reddening. SN 1987A is the first supernova for which a BLC could be constructed. The hitherto derived BLCs for SN 1987A are from photometric data (Whitelock *et al.*, 1988; Dopita *et al.*, 1988; Suntzeff and Bouchet, 1990). These all suffer from intrinsic systematic errors due to calculating monochromatic fluxes from broad-band magnitudes, i.e. weighted means of the spectrum. The best way to derive the BLC therefore is to use spectrophotometry. Figure 3 shows the BLC derived from the Bochum data (Hanuschik, 1990b). This lightcurve includes 0.9–20 μ IR photometry (Bouchet *et al.*, 1989b) and therefore is not ideal, but offers a better time coverage than the truly spectrophotometric BLC of Bouchet *et al.* (1990). Its inner precision is 3 %, but the remaining uncertainties are larger: the uncertainties of the extinction towards SN 1987A, E_{B-V} = 0.17 ± 0.02 (Walker and Suntzeff, 1990), results in a 4 % error in L_{bol}, but the

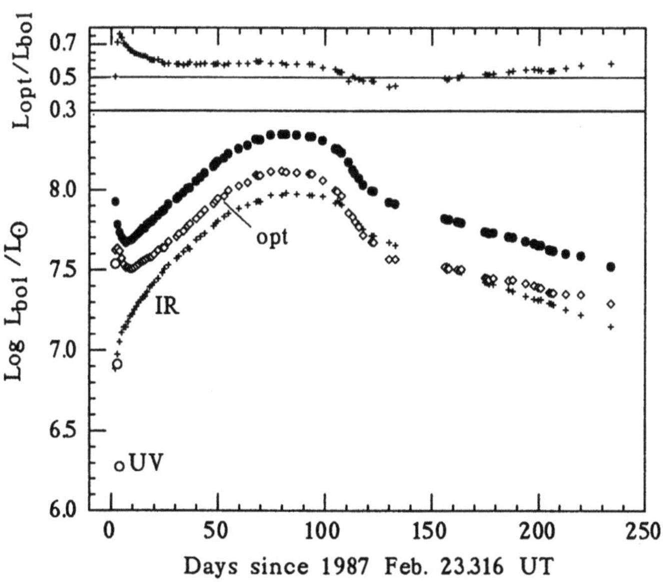

Fig. 3: The Bochum semi-spectrophotometric BLC, for E_{B-V} = 0.16 and D = 50 kpc. Besides the bolometric lightcurve (solid circles), also its contributions from the optical (diamonds), infrared (crosses) and UV (open circles) are shown. Furthermore the fraction L_{opt}/L_{bol} is shown on top

largest error, 15–20 %, comes from the distance which is only known within $\lesssim$ 10 %.

As shown in Fig. 3, the optical range (0.32–0.86 μ) dominates L_{bol} throughout the investigated period except for a short phase during the steep fall-off after maximum. The IR range (0.86–20 μ) contributes only a small fraction in the initial period, later-on it rivals with the optical range. The UV range (0.12–0.32 μ) is very strong on day 2, but drops to an insignificant (< 1 %) level after day 5. This evolution demonstrates the influence of the dramatic temperature fall-off during the first ~10 days (see, e.g., Hanuschik 1989).

3.3 Phases of evolution

The general form of the BLC is essentially determined by the balance between the release of shock-deposited energy (thermal plus nuclear) and diffusion through the opaque envelope (e.g., McCray, 1989; Shigeyama and Nomoto, 1990). During the first 10 days, the dense *bsg* envelope effectively hinders diffusion of most of the internal energy, thereby favouring its transformation into kinetic energy while the envelope cools off adiabatically. If we simply assume that the photosphere moves with the matter, $r_{ph} \sim t$, then from a diffusion ansatz it follows that

$$L(t) \sim t^{-1} \ , \ T_{eff}(t) \sim t^{-3/4} \tag{2}$$

which is quite good an approximation to the observed values. When T has dropped to ≈ 5 000 K, hydrogen recombination starts, and energy is transported more efficiently than by diffusion. The photosphere now starts to recede. As a result, L increases again, at a rate

$$L(t) = \varepsilon_H \cdot \dot{M}_H(t) \qquad (\varepsilon_H = 1.3\ 10^{13}\ \text{erg g}^{-1}). \tag{3}$$

From day 9 until day 21, a mass of 10 $M_\odot$ recombining hydrogen is required to penetrate the photosphere, in order to produce the observed luminosity. As this is the estimated total hydrogen mass in the envelope of SN 1987A, luminosity should drop steeply after that date. Instead another energy source takes over: radioactive decay of nickel-56 (mother) and cobalt-56 (daughter), nucleosynthesized by the shock deep inside the core. As the envelope becomes thinner, diffusion of thermalized energy from this store becomes more and more important, dominating the lightcurve around maximum (day 82). When the store is empty, only the actually produced decay energy is radiated away, and the lightcurve finds its final, exponential decay phase,

$$L(t) = M_0(^{56}\text{Ni}) \cdot \varepsilon_{Co} \cdot \tau_{Co}^{-1} \cdot \exp\,(-t/\tau_{Co})\ , \tag{4}$$

with $\varepsilon_{Co} = 6.4\ 10^{16}$ erg g^{-1} and $\tau_{Co} = 111.3^d$.

A fit to the exponential tail of the BLC yields the initially produced mass of ^{56}Ni,

$$M_0(^{56}\mathrm{Ni}) = (0.078 \pm 0.002)\ (D/50\ \mathrm{kpc})^2\ M_\odot\ . \qquad (5)$$

During this exponential tail phase, the SN 1987A lightcurve and those of ordinary type-II SNe look similar (Phillips, 1988). They are dominated by the initial ^{56}Ni mass which is a function of the explosion energy and of processes in the interior, and therefore is not different for blue and red supergiants. In the initial phase, however, this difference is very important and caused what was called an unusual lightcurve in the case of SN 1987A: the initial shape of the lightcurve (and the maximum brightness!) is a function of how easy photons can diffuse through the envelope. A red supergiant with its tenuous envelope enables a high initial luminosity, while a blue supergiant with its dense and rather compact envelope hinders diffusion efficiently. Popping up the progenitor's radius from 3 10^{12} cm (*bsg*) to 4 10^{14} cm (*rsg*), hydrodynamic model calculations can smoothly transform a type-1987A lightcurve into an ordinary type-II lightcurve (Arnett, 1989).

By now, the brightness of SN 1987A has dropped by a factor of 2 10^4 since maximum ($L_{bol} \approx 9\ 10^3\ L_\odot$) and it becomes more and more difficult to measure, because contributions from background stars or light-echoes become increasingly more important. At day 1000, 80 % of the total light come from the far-IR, dominated by a ~100 K black body, the dust in the envelope. Bouchet *et al.* (1989a) detected a levelling-off in the ligthcurve around that date, but this is not confirmed by CTIO measurements (Bouchet, 1990). Such a levelling-off would be an evidence for energy input by a central pulsar.

3.4 Energetics

The total amount of electromagnetic energy radiated away from SN 1987A is

$$E_{rad} = \int_0^\infty L_{bol}(t)\, dt = 9.4\ 10^{48}\ (D/50\ \mathrm{kpc})^2\ \mathrm{erg} \qquad (6)$$

(Hanuschik, 1990b), including contributions from Woosley's (1988) model 10H for the unobserved period day 0—1.7, and from the extrapolation of the radioactive tail from day 234 to infinity. This energy has to be compared to the total energy from the decay of 0.078 $M_\odot$ nickel-56, 1.47 10^{49} erg. Allowing for the fact that part of the radiated energy is due to hydrogen recombination (2.7 10^{47} erg), we find that

$$\Delta E \approx 5.6\ 10^{48}\ \text{erg} \tag{7}$$

from the ^{56}Ni decay has not been radiated away. Instead, this energy went presumably into thermal, kinetic and turbulent energy of the nickel bubble (e.g., Müller *et al.*, 1989; Arnett *et al.*, 1989), driving acceleration and turbulent mixing there. This energy is sufficient to accelerate the whole nickel bubble from its initial ~1000 km s^{-1} to 3 000 km s^{-1}.

4. Troughs and humps: absorption and emission line lightcurves

4.1 Absorption lines

At first view, a supernova spectrum is completely different from a stellar one. Absorption and emission features overlap, both broadened to several 10^3 km s^{-1}, and there are wide spectral regions without any continuum. Only few features can be reliably assigned to a single or dominating transition.

The equivalent width of absorption lines, W_λ, is a measure of the integral opacity within the envelope. In general, it is a complex function of density and excitation conditions and can only be fully understood by careful modelling. Its temporal evolution is expected to follow the following trends: initially, W_λ for most optical lines is expected to increase as a result of cooling in the envelope and increasing size of the line absorbing volume above the photosphere. Subsequently, W_λ will eventually fall off again, because expansive dilution ($n \sim t^{-3}$) causes a decrease of the Sobolev optical depth, $\tau_{\text{Sob}} \sim t \cdot n \sim t^{-2}$.

Only few lines are embedded in pseudo-continuum such that their equivalent width is defined in a sufficiently reliable way. From among them, the Balmer lines Hα and Hβ, and furthermore Ca II-2 λ8500, Ba II-2 λ6141 and the ground-state line Na I-D were selected for measuring W_λ. Figure 4 shows the results for days 2—314.

The general evolutionary scenario described above is best followed by the Ba II and the Ca II line: their absorption strength indeed shows a steep increase and a subsequent slow decrease. Hα, Hβ and Na I-D troughs exhibit a more complex evolution after the initial increase of W_λ: absorption in these lines drops after day 20 (especially in the Balmer lines), and a transient minimum is obtained around day 40—50; Hβ disappears completely by that time. Afterwards absorption increases again. This purely *spectroscopic* behaviour between days 20 and 100 has no correspondance in the lightcurve, but in another spectroscopic phenomenon, the *Bochum event*: the fading of absorption in these lines correlates with spectral finestructure.

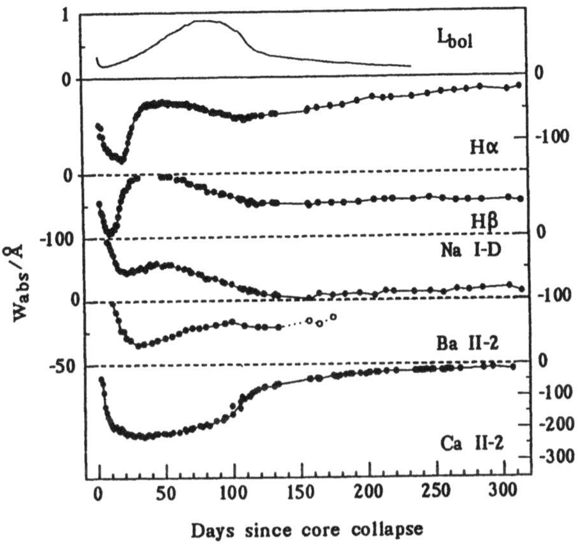

Fig. 4: Equivalent width W_{abs} for optical absorption troughs of the Balmer lines Hα and Hβ, of Na I-D λ5890, Ba II-2 λ6141 and Ca II-2 λ8500. Shown on top is the bolometric lightcurve. The Ba II-2 line fades away after day ~170, open circles denote uncertain measurements. Hβ disappears completely between days 35 and 50

In Na I-D absorption remains strong even beyond day 314 (cf. Spyromilio *et al.*, 1990). The same is true for Hβ after its revival. While this behaviour is not surprising at $T \approx 5\,000$ K for Na I-D as a resonance line exhibiting a very large optical depth, the run of the Hβ absorption is surprising as Hα absorption fades away at the same time! By day 300, W_α is only half as large (~ 25 Å) as W_β (50 Å). However, the definition of Hα absorption becomes worse with time due to the increasing strength of [O I] emission, veiling part of the Hα absorption. The measured values in Fig. 4 should therefore be regarded only as a lower limit. Furthermore, weak blends in Hβ could mimic too high an absorption strength at late times. However, no clear evidence for line-blending in the post-maximum phase is visible in the flux vs.time map of Hβ (Hanuschik and Thimm, 1990). It is therefore not yet clear what the cause for the rather strange Hα/Hβ ratio is.

4.2 Emission-line lightcurves

Emission-line lightcurves are even more complex than curves of absorption-line equivalent width because of the explicit dependence on the powering light source. Figure 5 shows the evolution of Hα emission, with three maxima of emission: on day 5, around day 95 and around day 180, well in the

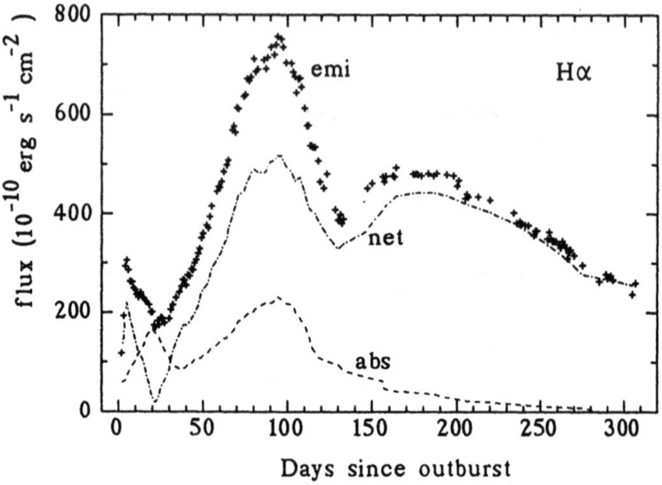

Fig. 5: Evolution of fluxes in Hα. Crosses denote the emission part of the P Cygni profile; the absorbed flux is marked by a broken, the net flux by a dash-dotted line. Fluxes are dereddened with E_{B-V} = 0.16 and can easily be transformed into luminosity by multiplying with 47.476 dex

exponential tail. As proposed by Thimm et al. (1989) and Phillips and Heathcote (1989), the early-time emission of Hα can be understood in terms of powering by recombination (first maximum) and thermalized cobalt decay (second maximum). This is supported by the fact that both phases are separated by a short drop of the Hα net emission to zero (Fig. 5). The third maximum could be due to the effect of an optically thick emitting Hα disk ($\sim t^2$), combined with the exponential fall-off in the powering continuum ($\sim \exp[-t/111^d]$).

A similar maximum in the exponential phase, around day 200, occurs in the Ca II-2 λ8600 emission (cf. Fig. 6) which runs parallel to the Ha emission. In contrast, Na I-D emission fades away after the overall brightness maximum. The forbidden lines evolve differently: [Ca II] and [O I] emission approach their maximum well in the exponential tail, around day 250. There is no correlation with the lightcurve what is not surprising if theses lines are collisionally excited. The increase in [Ca II] emission strength is roughly proportional to t^{-2}, as expected for an optically thick sphere.

4.3 Optical depth in forbidden lines

Even forbidden lines can become optically thick in expanding SN envelopes with their huge optical pathlengths. In a differentially expanding envelope, the emergent flux in a line is given, according to the Sobolev theory, by

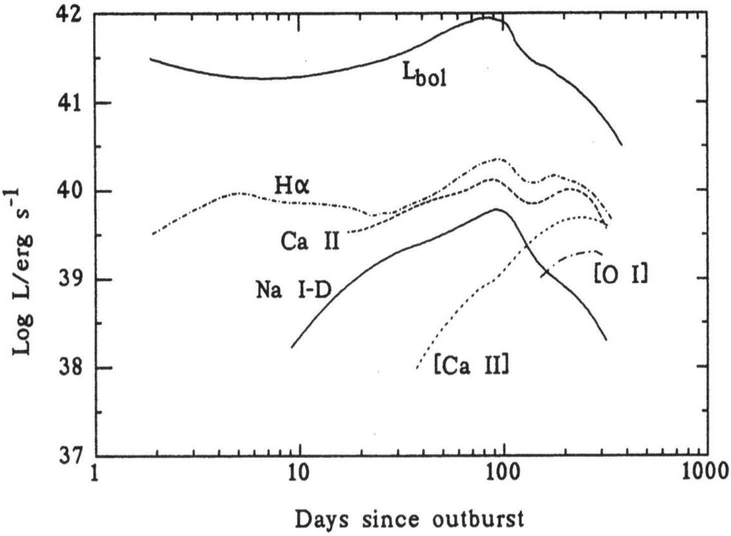

Fig. 6: Comparison of bolometric and emission-line luminosities in SN 1987A. For the evolution after day 314, compare, e.g., Phillips and Williams (1990) or Spyromilio *et al.* (1990). Here, the early-time evolution is emphasized by using a logarithmic timescale. Lines shown are Hα, Ca II λ8600, Na I-D, [Ca II] λλ7291/7323 and [O I] λλ6300/6363

$I \sim \beta A_{ul} n_u$, with the escape probability $\beta = \tau^{-1}$ [1 - exp(-τ)], the number density in the upper state n_u and the Einstein coefficient for the transition investigated, A_{ul}. The optical depth τ_{Sob} is, for homologous expansion, proportional to $t \cdot n_1 \sim t^{-2}$. Then, the intensity ratio of a fine-structure doublet, such as [O I] λλ6300/6363 and [Ca II] λλ7291/7323, can be used to measure the optical depth in the line (see Phillips and Williams, 1990; Spyromilio *et al.*, 1990). The line ratio $I_1/I_2 = \beta_1 A_1/\beta_2 A_2$ becomes

$$I_1/I_2 = \left[1 - \exp(-\tau_1)\right] / \left[1 - \exp(-\tau_2)\right] \tag{8}$$

which approaches unity in the optically thick limit, and A_1/A_2 in the optically thin limit.

Figure 7a shows that the optical depth in the primary [O I] component, calculated from (8), starts at $\tau_1 > 10$ and is still larger than 1 (≈ 2.5) on day 314. It falls off according to t^{-2}, as expected for expansive dilution. Evolution in the [Ca II] doublet is more puzzling (Fig. 7b). While this line seems to be optically thin when it first shows up, its optical depth *increases* to a value close to 2 around day 260. However, measuring the [Ca II] line ratio is more difficult than in [O I] as the separation is only 32 Å or 1300 km s^{-1}, less than the typical width of the main component, 2500 km s^{-1} (cf. Hanuschik *et al.*, 1989). As optical depth effects play a role, the simple

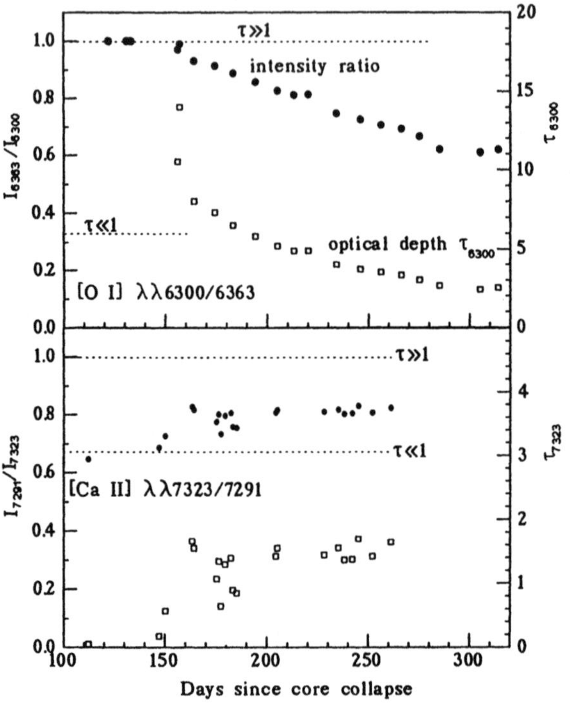

Fig. 7: Intensity ratio I_1/I_2 (full circles, left ordinate) and Sobolev optical depth τ_1 [open squares, right ordinate; cf. (8)] for the [O I] (top) and the [Ca II] (bottom) doublet in SN 1987A. As the [Ca II] doublet is rather narrow, only higher-resolution measurements at 3 Å from the Bochum telescope have been investigated, covering days 112-260. The line shows up, however, as early as on day 36 (Hanuschik *et al.*, 1989)

deblending technique described by Hanuschik *et al.* is probably not appropriate, and more sophisticated modelling is needed to treat the [Ca II] line profile properly.

5. Clumping and mixing

The simplest model for the kinematics of an expanding supernova envelope is the solution of the equation of continuity, i.e. homologous expansion:

$$v(r,t) = (r - r_0)/(t - t_0) \ , \tag{9}$$

with r_0 the radius of the progenitor and t_0 the onset time of the expansion. However, there is evidence, e.g., for subsequent acceleration of the core (see above). Furthermore, it is well known that filaments in the youngest galactic SN remnant, Cassiopeia A, move in a rather chaotic manner (Kirshner and Chevalier, 1977). Thus, turbulent mixing and clumping is expected to play an important role in the expansion of a supernova envelope (see Woosley *et al.*, 1988, and references therein). SN 1987A offers a first-row opportunity to study such mixing. Both indirect and direct evidence for mixing in SN 1987A exists.

5.1 Indirect evidence for mixing

The first evidence is the smooth transition in the bolometric lightcurve from powering by recombination into powering by radioactive decay, as well as the rather broad maximum (Fig. 3). Numerical models without mixing (e.g., Woosley, 1988, Fig. 15) fail to reproduce this smooth transition, but show a steep bump when all hydrogen has passed the photosphere and photons from the radioactive decay cannot yet diffuse into upper layers. Furthermore, their lightcurve maximum is rather steep and narrow. To better fit the observations, mixing-up of nickle bubbles, as well as mixing-down of hydrogen must be inferred (Pinto and Woosley, 1988).

The same kind of argumentation must be inferred to explain the early emergence of γ-rays from the ^{56}Co decay (Matz *et al.*, 1987). Contrary to the above described scenario that all decay energy was thermalized, a small fraction of γ-rays could obviously escape directly from the envelope, without being down-scattered and thermalized to optical photons, although the envelope was still optically thick at that time. Again, some ^{56}Co mixing-up must be assumed.

Thirdly, the high velocity ($\approx$ 4 000 km s^{-1}) observed in the wings of Co II and Ni II infrared lines (Aitken, 1988; Witteborn *et al.*, 1989) is an indication for mixing because unmixed core material moves at less than 1 000 km s^{-1} according to model calculations (Woosley, 1988). Although electron scattering plays a role in broadening especially the red wing of the line profiles (see Woosley *et al.*, 1988, Fig. 12), the basic conclusion is valid that we are observing material freshly nucleosynthesized in the explosion at much higher velocity than expected from unmixed explosion models.

A fourth indirect evidence for mixing is the unusual strength of lines of Ba II ($\lambda\lambda$612, 4924, 4554) and Sc II ($\lambda\lambda$6245, 5658, 5527). These elements are produced by slow neutron capture (*s*-process) deep in the neutron-rich inner layers, not in the hydrogen envelope. Since they became visible in the optical spectrum as early as on day 10–20 (Hanuschik and Dachs, 1988), long before helium-rich layers appeared at the photosphere, again mixing-up of heavier elements is required.

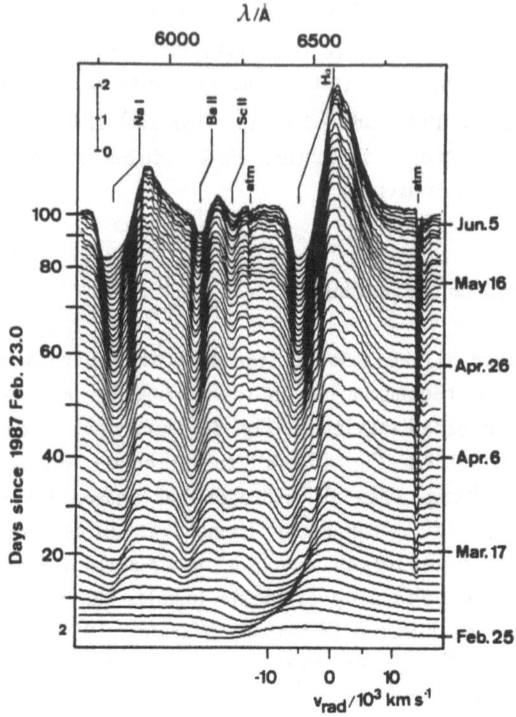

Fig. 8: The Bochum event in Hα and Na I-D (Hanuschik *et al.*, 1988). In addition, the Ba II-2 λ6142 and Sc II-28 λ6245 lines are visible. To avoid crowding, only scans for every second day are shown

5.2 Direct evidence for mixing

The *Bochum event* is a transient spectroscopic finestructure feature in hydrogen and sodium lines, as documented by Hanuschik and Dachs (1987), Hanuschik *et al.* (1988), Phillips and Heathcote (1989) and Hanuschik and Thimm (1990). It was first detected as a satellite feature in the absorption trough of Hα (Fig. 8), on day 19 = March 14, 1987. Slightly later, a similar structure evolved in the emission hump of this line, at the same (absolute) radial velocity. Related finestructure could be observed in Hβ and Na I-D, as well as in infrared hydrogen lines (Phillips and Heathcote, 1989). The radial velocity of the finestructure in the Hα trough was initially -5 800 km s^{-1}, close to the photospheric value, and subsequently fell off to -2 000 km s^{-1}. It disappeared during the transition to the exponential tail, around day ~120.

The kinematic behaviour has been interpreted as evidence for a change of the excitation conditions of the matter penetrating the photosphere around

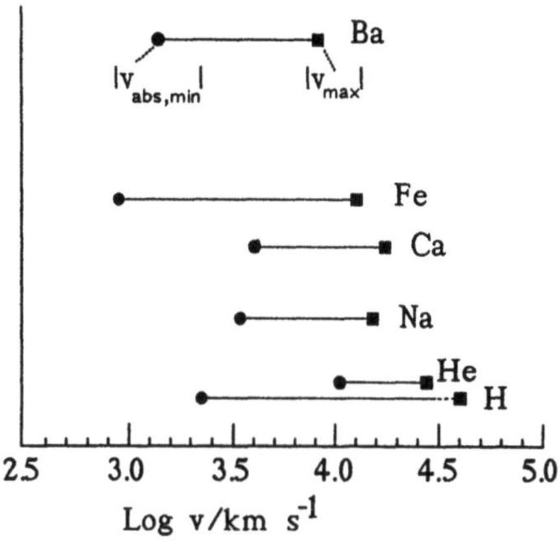

Fig. 9: Chemical stratification in the envelope. Indicated is the observed velocity spread of several species with prominent lines in the optical. See text for the definition of $v_{abs,\,min}$ and v_{max}

day 20 (Lucy, 1988; Hanuschik *et al.*, 1988); Phillips and Heathcote, 1989; Hanuschik and Thimm, 1990). This change is likely to be caused by mixed-up ^{56}Co bubbles or fingers. (Remember that direct evidence for cobalt at ≈4 000 km s^{-1} exists at later times, see above.) This suggestion is supported by recent detailed lightcurve modelling with ^{56}Co mixing (Shigeyama and Nomoto, 1990). The *Bochum event* thus is the earliest evidence for mixing in SN 1987A.

Another interesting diagnostic for the investigation of mixing processes is the search for chemical stratification from the kinematics of spectral lines. Such stratification should be rather pronounced in absence of mixing. In general, relating observed *radial* velocities to *true* velocities and radii is model dependent, except for two cases:

(1) The maximum blueshifted or redshifted (whatever is larger) radial velocity observed in a certain line, $|v_{max}|$, corresponds to (or sets a lower limit to) the highest outflow velocity of the species investigated, and hence of the *outer radius* of the corresponding shell.

(2) The velocity of the *inner boundary* of the shell is expected to become visible, in a trough velocity *vs.* time diagram (as, e.g., shown in Hanuschik and Schmidt-Kaler, 1989), as an asymptotic lower value, $|v_{abs,\,min}|$. Trough velocities are radial velocities of the flux minium in an absorption trough.

For hydrogen, the maximum observed velocity is $|v_{max}|$ = 31 000 km s^{-1}, as observed in Hα on day 1.8 (Hanuschik and Dachs, 1987). The lowest hydrogen velocity is seen in Hγ, $|v_{abs,\,min}|$ = 2 240 km s^{-1}, reached asymptotically around day 170 (Hanuschik *et al.*, 1991). No lower trough velocity has been observed in hydrogen lines so far. It is therefore reasonable to conclude that hydrogen is present in the envelope from at least 2 240 up to at least 31 000 km s^{-1}.

In a similar way, upper and lower boundaries for He, Na, Ca, Fe and Ba zones can be derived from optical lines (Fig. 9). Obviously, heavier elements, even iron and barium, are found far out in what in unmixed

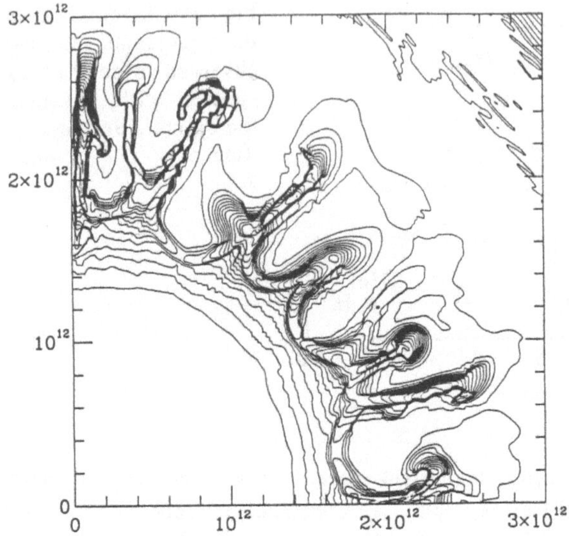

Fig. 10: Model calculation showing Rayleigh-Taylor instabilities and mixing 2.7 hrs after explosion (Müller *et al.*, 1989). Density contours are shown at 5 % spacing

models is the pure hydrogen envelope, i.e. at velocities of up to 10^4 km s^{-1}. For comparison, in unmixed models (Woosley, 1988) helium and calcium should mainly show up at ≈ 10^3 km s^{-1}.

A number of model calculations exist (e.g., Müller *et al.*, 1989, Arnett *et al.*, 1989, and references therein) that predict turbulent mixing by Rayleigh-Taylor instabilities at the He-H boundary, a few hours after shock-breakout (Fig. 10). A few mushroom-like bubbles may have reached 4 000 km s^{-1} in the course of this early-time mixing. A second phase with mixing could have occurred later, after ≈ 10 days, when a substantial amount of energy had been deposited in the inner layers by the onset of the Ni-Co decay.

5.3 Clumping

Closely related to mixing processes is clumping, i.e. small-scale density, not necessarily velocity gradients. These clumps are directly visible in the line profiles as bumps or wiggles at sufficiently high resolution and S/N. Because their radial velocity is expected to be constant with time, they can easily be discerned from the *Bochum event* which is variable in radial velocity.

Clumping in SN 1987A line profiles was observed by Stathakis and Cannon (1988) in December 1987 and thereafter, and by Thimm (1989,

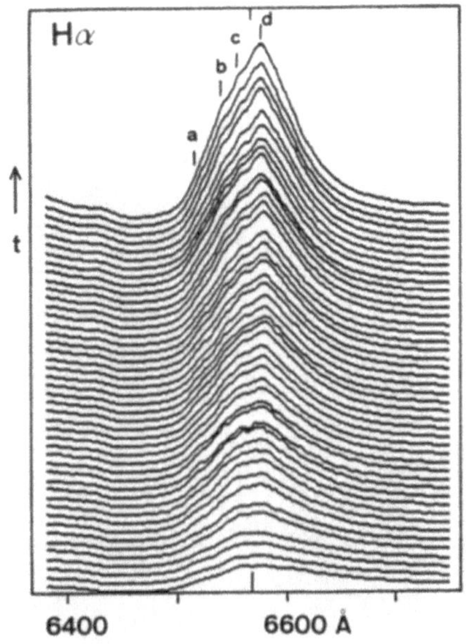

Fig. 11: Bumps and clumping. All Hα profiles measured between days 111 and 260 at 3 Å resolution are plotted, with the fluxes being normalized to their value at 6875 Å. Four bumps (a-d) appear around day 160 (cf. Fig. 12)

private communication). It appears as finestructure humps on top of the envelope emission in [O I] λλ6300/63, [Ca II] λλ7291/7324 and Hα. These humps have a typical width of ≈ 100 km s^{-1} (FWHM).

At lower resolution (3 Å ≙ 140 km s^{-1}), the high-S/N profiles of Hα obtained with the Bochum telescope show the onset of clumpy finestructure around day 160 ± 10 (Figs. 11, 12). There are four bumps, constant in radial velocity at -2 000, -900, -200, and +500 km s^{-1}, persistent through day 300. Their width is 300–400 km s^{-1}. These bumps are quite pronounced and stronger than those visible on high-resolution spectra, but they appear on the blue side of the emission profile, in the steep flank, and therefore are barely visible on single spectra. Only an extensive time record, together with contrast enhancement (Fig. 12), makes them stand out clearly.

We can exclude the possibility that these clumps were existing before – hidden below the photosphere –, because the photospheric velocity on day 160 is only 660 km s^{-1}, while the lower boundary of the hydrogen shell is ≈ 2 200 km s^{-1}, so that the bulk of the hydrogen had already passed the photosphere around day 50. It is therefore likely that *we observe, on day 160, the onset of clumping in the hydrogen envelope.*

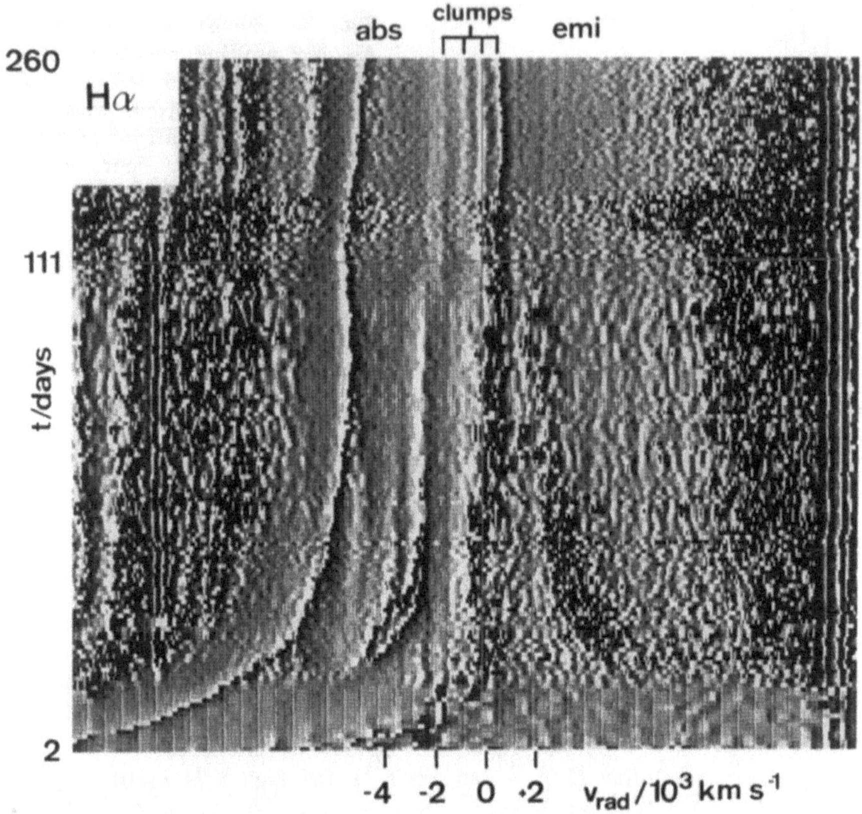

Fig. 12: Grey-scaled plot of the Hα fine-structure. Shown in this plot is essentially the second derivative of fluxes, resulting in considerable contrast enhancement of faint flux gradient changes. Note that scaling of the time axis is compressed after day 111 (one spectrum per ≈ 3 days on the average). The four bumps indicating clumping appear as straight lines after day 160, while the Bochum event shows up as complex features between day 20 and 120, variable in radial velocity

6. History of the progenitor: the circumstellar medium

6.1 The CSM shell

According to evolutionary models (e.g., Saio *et al.*, 1988; Woosley *et al.*, 1988), a 20 $M_\odot$ star looses a significant amount of mass during its red supergiant phase in form of a dense, slow (≈ 10 km s^{-1}) *rsg* stellar wind. When Sk -69°202 turned into its blue supergiant phase, it started to develop a fast (10^3 km s^{-1}), low density *bsg* stellar wind. This *bsg* wind blew a cavity into the preexisting *rsg* wind and developed a shock front marking the *rsg-bsg* wind boundary. Remnants of these fossil stellar winds, and especially the density-enhanced shock zone, are expected to constitute the circumstellar medium (CSM) surrounding SN 1987A.

The CSM shock zone was struck by the EUV flash from shock break-out in SN 1987A and began to fluorescence in nebular emission lines which became visible as soon as the envelope emission faded. It is important to realize that the signal in these lines comes from the shock zone itself and not from the expanding explosion light, like in the case of the famous light echoes. It is therefore geometrically stationary, apart from light-travel time and recombination effects within the shock zone itself. The narrow emission lines were first detected in the UV (Fransson *et al.*, 1989), later also in the optical (Wampler and Richichi, 1989). From the [O III] lines, a temperature of 55 000 K is derived for the excited shock zone in December 1989. Their peak radial velocity, 286.3 ± 0.8 km s^{-1}, is at present the best estimate we have for the system velocity of Sk -69°202/SN 1987A. The resolved width at Hα (thermal plus kinematic) is 25.9 ± 0.3 km s^{-1}, the velocity gradient, measured across the spatially resolved Hα line, is only 3–6 km s^{-1} (Wood and Faulkner, 1989; Hanuschik, 1990c). The expansion velocity of the shell cannot be larger than $\sim$ 10 km s^{-1}, otherwise the Hα line would be broadened stronger.

The spatial extent of the shock zone is 1.6" by 1.1" with a ring-like geometry, as seen on NTT images taken at 0.5" seeing in December 1989 (Wampler *et al.*, 1990), These parameters are confirmed by the recent HST image. We thus see a flattened ellipsoid, or a flattened torus, or a circular ring, centred on SN 1987A at distance 7 10^{17} cm or 0.7 lightyears. However, one has to be careful in translating these observations into the geometry of the shock zone because apart from aspect effects both light travel time and recombination time play a role (Fransson *et al.* 1989; Fransson and Lundqvist, 1989).

The apparent asphericity of the CSM shell is remarkable, no matter whether it is, in reality, a flattened ellipsoid (then the ring-like appearance is due to limb-brightening as in the case of a planetary nebula) or a symmetrical, but inclined torus (then inclination must be 45°): in either case, this asphericity hints towards a *non-spherical outflow*, most likely due to rapid rotation of the blue supergiant. One should expect that this asymmetry might also be also visible in the *ejecta* of SN 1987A. Indeed, there is multiple evidence for a non-spherically symmetric explosion: Polarimetric observations (Schwarz and Mundt, 1987; Cropper *et al.*, 1988) reveal intrinsic continuum polarization of the envelope of 0.28 %, as early as on day 12. This value of polarization would result, according to models of Shapiro and Sutherland (1982), either from an oblate spheroid with axis ratio μ = 0.77, or from an prolate spheroid with μ = 0.70. Furthermore, direct speckle imaging (Karovska *et al.*, 1989) indicates an axis ratio of the supernova envelope of 0.6 at Hα on days 95–98, and less elongation at other wavelengths. The elongated axis is at position angle 20–30°. For comparison, the NTT and HST images show the major axis of the CSM shell at ~ 80°.

We can conclude that SN 1987A is the first one to prove that SNe do not necessarily explode in a nice, easy-to-model spherical symmetric way. Furthermore, as the observations in the CSM also suggest asphericity in the fossil outflow from the progenitor star, one can conclude that both asymmetries, of the outflow and of the explosion, arise from rotation of the star.

6.2 Other CSM components

The 1.6" by 1.1" ring is the most prominent, but not the only CSM feature yet seen. On high-resolution spectra taken in September 1989, substructure of the CSM Hα feature was detected (Hanuschik, 1990c), in form of a very narrow (unresolved at 10 km s^{-1} resolution) source at 2.1" (1.7 lightyears) projected distance NW of the supernova (Fig. 13). As finestructure broadening of Hα contributes about 6 km s^{-1} (Dyson and Meaburn, 1971) to any observed width, thermal broadening must be very small, corresponding to less than 1 400 K. This Hα feature could arise from a condensation of gas and dust, visible as an arc on the NTT images (Wampler *et al.*, 1990) extending to the NW of the supernova. Recent images of the area (Crotts, 1990; Elias, 1990) show several knots in Hα and He I 1.083 μ, one of them coincident with the spectroscopic Hα blob. These condensations probably are remnants of dust shells expelled by the progenitor star in its *rsg* phase and now are excited to fluorescence, like the shock zone, by the EUV flash.

Finally, the recent detection of the onset of radio emission in July 1990 (Turtle *et al.*, 1990) seems to herald a revival of the fading envelope of SN 1987A by interaction of the outermost debris of the explosion, moving at 40 000 km s^{-1}, with the fossil CSM. These layers now are 0.5 lightyears away from the core of the explosion, still well within the *bsg* stellar wind zone. Therefore their brightening is most likely due to the collision with a cloud within this innermost CSM region. From the rising time of the signal, a size of 5 10^{16} cm or 20 lightdays can be inferred for this cloud (Chevalier, 1990). Alternatively, the radio signal could mark the onset of pulsar activity.

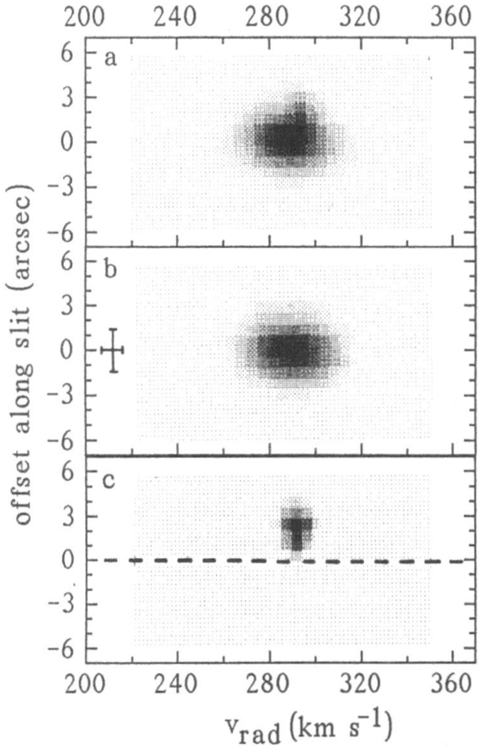

Fig. 13: Spectrum of the CSM Hα component from the vicinity of SN 1987A (cf. Hanuschik, 1990c). **13a**: The Hα line after subtraction of supernova and H II-region emission. **13b**: Primary component arising from the CSM shell. **13c**: The Hα blob, obtained by subtraction of Fig. 13b from Fig. 13a. The spectral narrowness, as well as the offset from the supernova position (indicated by a broken line) is clearly visible. Intensity scaling is the same in all three images. Spectral as well as spatial resolution is indicated in Fig. 13b

7. Conclusions: old and new questions

Supernova 1987A in the LMC has answered many old questions and risen many new ones. The pioneering idea of Baade and Zwicky (1934) that supernovae (of type II) gain their energy from the gravitative collapse of a massive star has been proven. But the first supernova ever with a certain progenitor identification demonstrated that these stars can finish their ordinary life not only as red, but also as blue supergiants. The exploded blue supergiant Sk -69°202 proved, in a certain sense, the suggestion of Shklovskii (1984) that in irregular galaxies, like the LMC, type-II SNe are missed because they are fainter. In fact, the lower metallicity, together with the specific ZAMS mass, mass loss and semi-convection, seem to be the determining parameters for the final state of the massive star at explosion (e.g., Woosley *et al.*, 1988). Thus evolutionary calculations for this part of the HRD have become attractive again.

The same is true for the field of supernova statistics: only 1 type-1987A event for every 50 normal type-II's would be observed if intrinsically both types are equally frequent.

The neutrino signal also confirmed another prediction, viz. that the energy gained in the collapse must be radiated away mainly in form of neutrinos. But no satisfying and consistent answer exists to the question: what happened 4.6 hrs before t_0 in the Mt. Blanc neutrino detector (Aglietta *et al.*, 1987)? Also, because the information contained in 19 officially accepted neutrinos is quite limited, we have virtually nothing learnt about how a supernova explodes, i.e. how a small fraction of the collapse energy (1 %) is transferred into explosion energy of the envelope.

Another important result from SN 1987A is the confirmation that SNe indeed produce heavy elements. Specifically, the exact amount of ^{56}Ni produced in the shock could be measured by comparison to the lightcurve and by analyzing the IR lines of this species.

SN 1987A has also demonstrated that mixing and clumping are important processes in the very early evolution of the expanding envelope, and that asphericity of the envelope is an observational fact that has to be considered in future model calculations.

This supernova will remain a highly interesting object in the future. Some prospective highlights are the manifestation of a central pulsar and the radio and X-rays from the interaction of the envelope with the fossil CSM. If the young remnant of SN 1987A evolves similarly to SN 1054, the remnant of which we observe in the Crab nebula, it will possess a brightness of $\approx 15^m$ and a diameter of 12 arcseconds in the year 2926, then being even more impressive than presently.

References

Aglietta, M. et al.: 1987, *Europhys. Lett.* **3**, 1315/1321

Aitken, D. K.: 1988, *Proc. Astron. Soc. Aust.* **7**, 462

Arnett, W. D.: 1989, *Astrophys. J.* **343**, 834

Arnett, W. D., Bahcall, J.N., Kirshner, R.P., Woosley, S.E.: 1989, *Annu. Rev. Astron. Astrophys.* **27**, 629

Baade, W., Zwicky, F.: 1934, *Proc. Nat. Acad. Sci. US* **20**, 254

Barbon, R., Capellaro, E., Turatto, M.: 1989, *Astron. Astrophys. Suppl.* **81**, 421

Bionta, R.M., Blewitt, G., Bratton, C., Caspar, D.: 1987, *Phys. Rev. Lett.* **58**, 1494

Bouchet, P.: 1990, in *Proc. ESO/EIPC Workshop "SN1987A and other supernovae"*, I. J. Danziger (ed.), ESO, Garching, 1990 *(in press)*

Bouchet, P., Danziger, I.J., Lucy, L.B.: 1989a, IAU *Circ.* 4933

Bouchet, P., Moneti, A., Slezak, E., Le Bertre, T., Manfroid, J.: 1989b, *Astron. Astrophys. Suppl.* **80**, 379

Bouchet, P., Phillips, M.M., Suntzeff, N.B., Gouiffes, C., Hanuschik, R.W., Wooden, D.: 1990, *Astron. Astrophys. (submitted)*

Burrows, A.: 1988, *Astrophys. J.* **334**, 891

Chevalier, R.A.: 1990, in *Proc. ESO/EIPC Workshop "SN1987A and other supernovae"*, I.J. Danziger (ed.), ESO, Garching, 1990 *(in press)*

Cropper, M., Bailey, J., McCowage, J., Cannon, R.D., Couch, W.J., Walsh, J.R., Strade, J.O., Freeman, F.: 1988, *Mon. Not. R. Astron. Soc.* **231**, 695

Crotts, A.P.S.: 1990, in *Proc. ESO/EIPC Workshop "SN1987A and other supernovae"*, I.J. Danziger (ed.), ESO, Garching, 1990 *(in press)*

Danziger, I.J. (ed.): 1987, *Proc. ESO Workshop on the SN1987A*, ESO, Garching

Danziger, I.J. (ed.): 1990, *Proc. ESO/EIPC Workshop "SN1987A and other supernovae"*, ESO, Garching, 1990 *(in press)*

Dopita, M.: 1988, *Space Sci. Rev.* **46**, 225

Dopita, M. et al.: 1988, *Astron. J.* **95**, 1717

Dyson, J.E., Meaburn, J.: 1971, *Astron. Astrophys.* **12**, 219

Elias, J.H.: 1990, in *Proc. ESO/EIPC Workshop "SN1987A and other supernovae"*, I.J. Daziger (ed.), ESO, Garching, 1990 *(in press)*

Filippenko, A.: 1990, in *Proc. ESO/EIPC Workshop "SN1987A and other supernovae"*, I.J. Danziger (ed.), ESO, Garching, 1990 *(in press)*

Fransson, C., Cassatella, A., Gilmozzi, R., Kirshner, R.P., Panagia, N., Sonneborn, G., Wamsteker, W.: 1989, *Astrophys. J.* **336**, 449

Fransson, C., Lundqvist, P.: 1989, *Astrophys. J.* **341**, L59

González, R., Wamsteker, W., Gilmozzi, R., Walborn, N., Lauberts, A.: 1987, in ESO *Workshop on the SN1987A*, I.J. Danziger (ed.), ESO, Garching, p. 33

Greco, M. (ed.): 1988, *Proc. Workshop "SN 1987A, One Year Later"*, Edition Frontières, Gif-sur-Yvette

Hanuschik, R.W.: 1989, *Rev. Mod. Astron.* **2**, 148

Hanuschik, R.W.: 1990a, in *Proc. X^{th} Santa Cruz Summer Workshop in Astronomy and Astrophysics "Supernovae"*, S. Woosley (ed.), Springer, Heidelberg *(in press)*

Hanuschik, R.W.: 1990b, in *Proc. ESO/EIPC Workshop "SN1987A and other supernovae"*, I.J. Danziger (ed.), ESO, Garching, 1990 *(in press)*

Hanuschik, R.W.: 1990c, *Astron. Astrophys.* **237**, 12

Hanuschik, R.W., Dachs, J.: 1987, *Astron. Astrophys.* **182**, L29

Hanuschik, R.W., Dachs, J.: 1988, *Astron. Astrophys.* **205**, 135

Hanuschik, R.W., Schmidt-Kaler, Th.: 1989, *Mon. Not. R. Astron. Soc.* **241**, 347

Hanuschik, R.W., Thimm, G.J.: 1990, *Astron. Astrophys.* **231**, 77

Hanuschik, R.W., Thimm, G.J., Dachs, J.: 1988, *Mon. Not. R. Astron. Soc.* **234**, 41P

Hanuschik, R.W., Thimm, G.J., Seidensticker, K.J.: 1989, *Astron. Astrophys.* **220**, 153

Hanuschik, R.W., Gochermann, J., Kimeswenger, S., Poetzel, R., Schnur, G.F.O., Werger, M.: 1991, *in preparation*

Haupt, W., Desjardins, R., Maitzen, H.M., Rudolph, R., Schlosser, W., Schmidt-Kaler, Th., Tüg, H.: 1976, *Astron. Astrophys.* **50**, 85

Hillebrandt, W., Höflich, P.: 1989, *Rep. Prog. Phys.* **52**, 1421

Hirata, K. et al.: 1987, *Phys. Rev. Lett.* **58**, 1490

Isserstedt, J.: 1975, *Astron. Astrophys. Suppl.* **19**, 259
Kafatos, M., Michalitsianos, A. (eds.): 1988, *Proc. 4th George Mason University Workshop in Astrophysics "Supernova 1987A in the LMC"*, Cambridge University Press, Cambridge
Kare, J.T. et al.: 1989, *Rev. Sci. Instrum.* **59**, 1021
Karovska, M., Koechlin, L., Nisenson, P., Papaliolios, C., Standley, C.: 1989, *Astrophys. J.* **340**, 435
Kirshner, R.P., Chevalier, R.A.: 1977, *Astrophys. J.* **218**, 142
Kirshner, R.P., Sonneborn, G., Grenshaw, D.M., Nassiopoulos, G.E.: 1987, *Astrophys. J.* **320**, 602
Lucy, L.: 1988, in *4th George Mason University Workshop in Astrophysics "Supernova 1987A in the LMC"*, M. Kafatos and A. Michalitsianos (eds.), Cambridge University Press, Cambridge, p. 323
Matz, S.M., Share, G.H., Leising, M.D., Chupp, E.L., Vestrand, W.T., Purcell, W.R., Strickman, M.S., Reppin, C.: 1988, *Nature* **331**, 416
McCray, R.: 1989, in *Proc. Yellow Mountain Summer School "Structure and Evolution of Galaxies"*, L.-Z. Fang (ed.), World Science, Singapur, p. 8
Menzies, J. et al.: 1987, *Mon. Not. R. Astron. Soc.* **227**, 39P
Miller, D.L., Branch, D.: 1990, *Astron. J.* **100**, 530
Müller, E., Hillebrandt, W., Orio, M., Höflich, P., Mönchmeyer, R., Arnett, D., Fryxell, B.: 1989, in *Proc. 5th Workshop on Nuclear Astrophysics (Schloß Ringberg)*, W. Hillebrandt und E. Müller (eds.), MPA/P1, p. 103
Perlmutter, S.: 1990, in *Proc. ESO/EIPC Workshop "SN1987A and other supernovae"*, I. J. Danziger (ed.), ESO, Garching, 1990 *(in press)*
Phillips, M.M.: 1988, *Proc. Astron. Soc. Aust.* **7**, 11
Phillips, M.M., Heathcote, S.R.: 1989, *Publ. Astron. Soc. Pacific* **101**, 137
Phillips, M.M., Williams, R.E.: 1990, in *Proc. Xth Santa Cruz Summer Workshop in Astronomy and Astrophysics "Supernovae"*, S. Woosley (ed.), Springer, Heidelberg *(in press)*
Pinto, P.A., Woosley, S.E.: 1988, *Astrophys. J.* **329**, 820
Saio, H., Kato, M., Nomoto, K.: 1988, *Astrophys. J.* **331**, 388
Sanduleak, N.: 1969, CTIO *contribution No. 89*, Cerro Tololo Interamerican Observatory, La Serena, Chile
Schwarz, H.E., Mundt, R.: 1987, *Astron. Astrophys.* **177**, L4
Shapiro, P.R., Sutherland, P.G.: 1982, *Astrophys. J.* **263**, 902
Shigeyama, T., Nomoto, K.: 1990, *Astrophys. J.* **360**, 242
Shklovskii, I.S.: 1984, *Pis'ma Astron. Zh.* **10**, 723 (english translation: *Sov. Astron. Lett.* **10**, 302)
Spyromilio, J., Stathakis, R.A., Cannon, R.D., Waterman, L., Couch, W.J., Dopita, M. A.: 1990, *submitted to Mon. Not. R. Astron. Soc.*
Stathakis, R., Cannon, R.: 1988, AAO *Newsl.* no. 45
Suntzeff, N., Bouchet, P.: 1990, *Astron. J.* **99**, 650
Thimm, G.J., Hanuschik, R.W., Schmidt-Kaler, Th.: 1989, *Mon. Not. R. Astron. Soc.* **238**, 15P
Tüg, H.: 1980a, *Astron. Astrophys. Suppl.* **39**, 67
Tüg, H.: 1980b, *Astron. Astrophys.* **82**, 195
Turtle, A.J., Campbell-Wilson, D., Manchester, R.N., Staveley-Smith, L., Kesteven, M.J.: 1990, IAU *Circ.* 5086
Walborn, N.R., Prévot, M.L., Prévot, L., Wamsteker, W., González, R., Gilmozzi, R., Fitzpatrick, E.L.: 1989, *Astron. Astrophys.* **219**, 229
Walker, A.R., Suntzeff, N.B.: 1990, *Publ. Astron. Soc. Pac.* **102**, 131

Wampler, E.J., Richichi, A.: 1989, *Astron. Astrophys.* **217**, 31
Wampler, J., Wang, L., Baade, D., Banse, K., D'Odorico, S., Gouiffes, C., Tarenghi, M.: 1990 *(preprint)*
Wamsteker, W., Panagia, N., Barylak, M., Cassatella, A., Clavel, J., Gilmozzi, R., Gry, C., Lloyd, C., van Santvoort, J., Talavera, A.: 1987, *Astron. Astrophys.* **177**, L21
Whitelock, P.A. et al.: 1988, *Mon. Not. R. Astron. Soc.* **234**, 50P
Witteborn, F.C. et al.: 1989, *Astrophys. J. (Letters)* **338**, L9
Wood, P.R., Faulkner, D.J.: 1989, IAU *Circ.* 4739
Woosley, S.E.: 1988, *Astrophys. J.* **330**, 218
Woosley, S.E., Pinto, P.A., Weaver, T.A.: 1988, *Proc. Astron. Soc. Aust.* **7**, p. 355
Woosley, S.: 1990, *Proc. Xth Santa Cruz Summer Workshop in Astronomy and Astrophysics "Supernovae"*, Springer, Heidelberg, in press

Dynamos in Galaxies

F. Krause[1] *and R. Wielebinski*[2]

[1]Astrophysikalisches Observatorium Potsdam,
O-1561 Potsdam, Fed. Rep. of Germany
[2]Max-Planck-Institut für Radioastronomie,
W-5300 Bonn, Fed. Rep. of Germany

1 Introduction

Theoretical arguments provided the first definitive prediction for the existence of an interstellar magnetic field in our Galaxy: There was no other way to explain the observed isotropy of the cosmic radiation. Alfvén, Richtmeyer and Teller (1949), and Fermi (1949) stated that the particles of the cosmic radiation have to be reflected often and this can hardly be explained other than by an interaction with a magnetic field.

In order to prevent the particles from escaping out of the galactic disc the observed maximal energy of $10^{16}eV$ allowed the conclusion that this magnetic field has to be stronger than $10^{-8}G$. There is a different possibility of an estimate based on the requirement that the magnetic energy density cannot be smaller than the observed density of the cosmic radiation, otherwise the magnetic field is incapable to resist the pressure of the cosmic ray particles. In this way a minimal field strength of some $10^{-6}G$ is found, a value which is still representative for the interstellar magnetic fields in our Galaxy.

The above statements characterize the situation around 1950. There were no direct observations; only in papers reporting and interpreting the polarization of star light it was suggested that this polarization may be due to scattering by elongated dust grains that are at least partially aligned by a magnetic field (Davis and Greenstein (1951)). Direct observational evidence for a magnetic field in our Galaxy was then provided by radio observations of:

- the polarization of radio sources and their Faraday rotation (Mayer, Mc Cullough and Sloaneker (1962), Bracewell, Cooper and Cousins (1962), Cooper and Price (1962), Gardner and Whiteoak (1963), Simard-Normandin and Kronberg (1981)),

- the polarized non-thermal radio emission (synchrotron radiation) of the Galaxy (Wielebinski, Shakeshaft and Pauliny-Toth (1962), Westerhout, Seeger, Brouw and Tinbergen (1962), Wielebinski and Shakeshaft (1962, 1964)),

- the Zeeman-splitting of the 21cm-line in HI- and OH-regions (Verschuur (1968), Crutcher and Kazes (1983)).

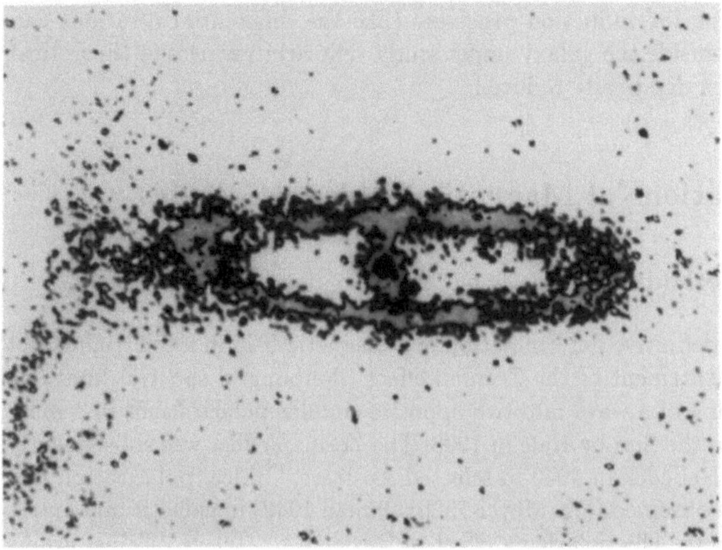

Fig. 1. Synchrotron radiation of the Andromeda nebula M31 at 327 MHZ (courtesy of G. Golla, R.A.M. Walterbos and A.G. de Bruyn (unpublished)). The transparency of a galaxy makes possible to see the toroidal ring field, which is postulated for other cosmical dynamos like the Earth and the Sun. For the Earth the toroidal magnetic field is hidden in the core; for the Sun it is only indirectly observable by the sunspot phenomenon

The radio polarization observations gave evidence for magnetic fields in various parts of our Galaxy. A new chapter in the exploration of these phenomena was opened by the first detection of linear polarization in the nearby galaxy M51 by Mathewson et al. (1972), which indicated synchrotron radiation and consequently a magnetic field in a spiral galaxy. The situation became even more exciting as the first complete maps of linear polarization in the spiral galaxies M31 and M33 obtained with the Effelsberg 100-m telescope revealed global structures of the magnetic field on scales comparable with the scale of the galaxy (e.g. Beck et al. (1987), Beck et al. (1990)).

The latter observations are crucial for the theoretical explanation of these magnetic fields. Whereas for local fields a number of mechanisms may be offered, well ordered magnetic structures which extend over the whole galaxy indicate a self-organization mechanism, the self-excited dynamo.

These observational findings were a great challenge to the dynamo theory as they apparently revealed two different basic types of magnetic structures: The axisymmetric (ASS) ones and the bisymmetric (BSS). Examples of the former are the Andromeda Nebulae (M31) and IC 342, and of the latter M81 and M51.

And there is an additional point of interest for theoreticians: In contrast to other cosmic dynamos like the Sun or the planets the galaxies are in some respect transparent dynamos, that is physical quantities which are responsible for the induction processes (like the differential rotation) can be measured inside the galaxy under study. The arbitrariness of theoretical models is thus drastically reduced.

2 Observations of Magnetic Fields in Galaxies

2.1 The Methods

The earliest definitive determination of a magnetic field in a celestial object was the measurement of the Zeeman effect (doubling of spectral lines due to the splitting of a wave into two opposite circular polarizations in a magnetic field) in the Sun by Hale in 1908. The Zeeman effect was subsequently detected by Babcock in 1946 in selected Ap-stars. Optical polarization had already been observed (e.g. Meyer 1920; Öhman 1942) in galactic and extragalactic objects. The polarizing effect of Rayleigh scattering off dust grains was well known and widely used to interpret the polarization phenomena. Davis and Greenstein (1951) suggested an alternative interpretation: the scattering of light from oblate dust grains aligned in a magnetic field. This latter interpretation can be substantiated today in view of the radio polarization results.

Radio waves (synchrotron emission) which are generated by relativistic electrons can be intrinsically linearly polarized up to 75% in a perfectly aligned magnetic field. Numerous effects combine to reduce this degree of polarization. Firstly beam depolarization must be considered, since radio observations often have only low angular resolution. Secondly Faraday depolarization effects, which are caused by magnetic fields in the line of sight in radio measurements, are important. The depolarization can take place in the disc of the galaxy, in its halo and finally in the intervening space. We certainly know that our Galaxy can provide substantial Faraday rotation (e.g. Simard-Normandin et al. 1981) and hence depolarize radio waves. The Zeeman effect can also be measured at radio wavelengths. In the past few decades the Zeeman effect was detected in the HI, OH and H_2O lines in molecular clouds in the Galaxy. Synchrotron emission is of course a definite signature of magnetic fields and can be used to determine field intensity values based on equipartition arguments (e.g. Moffatt (1973); Beck (1982)).

2.2 Optical Polarization Observations of Galaxies

The first observations of Öhman (1942) used 'surface polarimetry' in the Andromeda Nebula, M31. These observations were extended by Muliarchik (1957) and improved by Hiltner (1958) who studied the polarization of globular clusters beyond M31. All the results could be interpreted in the frame of the Davis-Greenstein effect suggesting that the magnetic fields in M31 follow the spiral arms. Sensitivity, a general problem in all polarization observations, was overcome only recently by CCD detectors. A few observers persevered with this difficult observing technique, notably Elvius (1978); Appenzeller (1967); Bingham et al. (1976); Scarrott et al. (1977, 1987). A number of northern galaxies have been studied, notably M51, M81, NGC7331, M104 and M82. A major polarization study was made of the Magellanic Clouds. Both Schmidt (1970; 1976) and Mathewson and Ford (1970a) made extensive optical polarization observations of the Magellanic Clouds which are in good agreement with recent radio studies.

The vast collection of optical polarization observations for our Galaxy was summarized by Mathewson and Ford (1970b). Basically an azimuthal field is observed with several pronounced local maxima. Again agreement between this data set and the radio polarization studies of the Galaxy (e.g. Vallee (1988)) is evident. In the centre of our Galaxy the magnetic fields are mainly aligned in the vertical direction (with respect to the galactic plane). However, as is well known, it is difficult for an 'inside' observer to delineate the general structure of our Galaxy.

2.3 Radio Polarization Observations of Galaxies

The fact that magnetic fields are responsible for the radio emission was clear since the earliest low frequency total intensity radio observations of M31 (Pooley (1969)) and NGC891 (Baldwin and Pooley (1973)). In statistical studies of a larger sample of galaxies by Gioia et al. (1982) it was shown that thermal emission (free-free) is less then 40% of the total intensity at 10.6 GHz, which means less then 10% at 408 MHz. However, no polarized intensity was detected from galaxies at these low frequencies. This is surely due to the combined effects of beam depolarization and Faraday rotation. Examples of low frequency results in figures 1 and 2 show the 327 MHz maps of M31 and NGC891 respectively.

The first detection of polarized radio emission in an external galaxy was reported by Mathewson et al. (1972) for M51 at 1415 MHz. By going to higher frequencies the dramatic new information about the morphology of the global magnetic fields in galaxies was obtained (e.g. Beck and Gräve (1987), Wielebinski (1990), Beck et al. (1990)). By now most of the nearby galaxies (including the Magellanic Clouds) were mapped at several radio frequencies with a satisfactory angular resolution to enable the determination of the aligned magnetic fields. All galaxies have shown some polarized emission implying aligned magnetic fields. However, in all galaxies the aligned component is only a small fraction of the total so that turbulence seems to play an important role.

The type of information that is available can best be illustrated by looking at the published maps of the galaxy IC342 (M. Krause et al. (1989a))

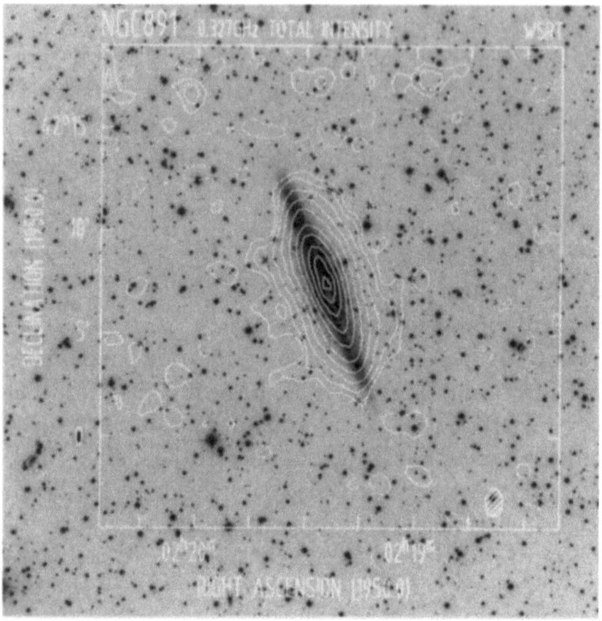

Fig. 2. 327 MHz map of NGC 891 (courtesy of E. Hummel, M. Dahlem, J.M. van der Hulst and S. Sukumar)

shown in figures 3a and 3b. The 1.4 GHz total intensity map (fig 3a) shows a peak in the nucleus (marked by a cross) and radio contours which clearly follow the spiral arms. An unrelated extragalactic source marked by 'E' is also seen in the map. Also an extended halo can be discerned. The map of polarized intensity (fig 3b) shows even more clearly the spiral arms with considerable asymmetry. The peak of polarized intensity in the South-East has no special optical counterpart. It is just somewhat on the inside of the optical arm. There is a second maximum of polarized intensity to the North-West. In the south two distinct filamentary polarized features can be followed. Unpublished 4.8 GHz data (M. Krause and Hummel) confirm that the orientation of the magnetic field is in general along the spiral arms and that the structure is very filamentary. The existence of two maxima of magnetic field alignment, symmetrical about the nucleus seems to be a general observational feature.

The magnetic fields in halos of two galaxies have been recently determined (e.g. Hummel et al. (1988, 1991)). It seems that there are mainly vertical structures, again with fine filaments. This is shown in figure 4 for the galaxy NGC891. A similar result exists for the galaxy NGC4631. The magnetic field strength is estimated to be fairly similar to that found in discs. This has important implications for the Faraday effect. Considerable rotation can be caused in the halo making the previous studies of the morphology of the disc fields questionable.

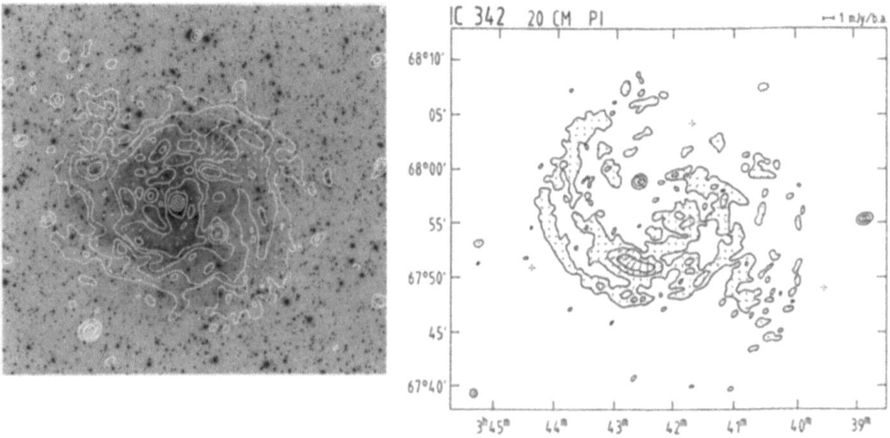

Fig. 3. Radio maps of the galaxy IC 342. a) Total intensity map at 1.4 GHz and radio contours which more or less are aligned to the spiral arms. b) Polarized intensity which clearly shows the spiral arms (courtesy of M. Krause, E. Hummel and R. Beck)

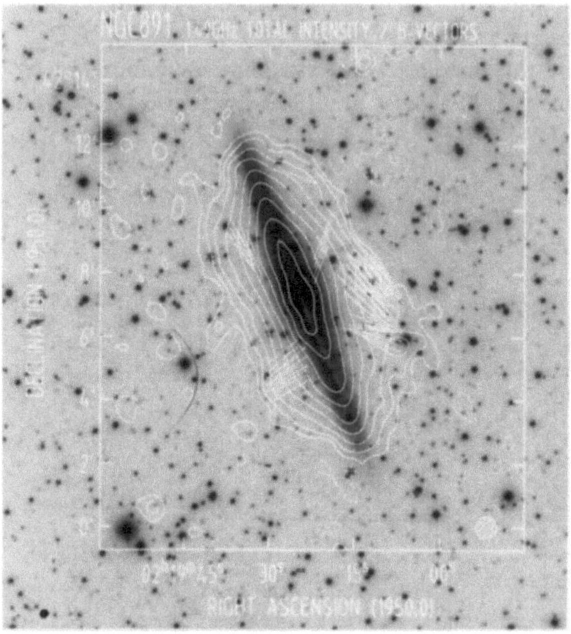

Fig. 4. Magnetic field observations in the halo of the galaxy NGC 891. The structure clearly indicates components which are vertical with respect to the plane of this galaxy (courtesy of E. Hummel, R. Beck and M. Dahlem)

2.4 The Rotation Measure

The rotation measure RM of polarized radiation provides information concerning the magnetic field component parallel to the line of sight, especially, the sign of RM indicates whether the magnetic field is directed towards the observer or in the opposite direction.

If one considers in a galactic disc with vanishing thickness a magnetic field with no vertical component (no component parallel to the rotation axis of the galaxy) the distribution of RM as a function of the azimuth φ seems to provide a clear decision between axial symmetric fields and bisymmetric ones (fig 5).

We determine RM on a circle $r = const.$ and consider it as a function of φ_0, where φ_0 is the azimuth projected on the sky. With φ we denote the azimuth in the central plane of the galaxy.

In case of an axisymmetric field, i.e. B_r, B_φ do not depend on φ, RM varies with $e^{i\varphi_0}$.

If a bisymmetric field is considered, i.e. B_r, B_φ depend on $e^{i\varphi}$, RM varies with $\epsilon^{2i\varphi_0}$ (Sawa and Fujimoto (1986)).

Apparently these simple results are fairly well confirmed by observational findings. In fig 6 are represented the measurements for the galaxy IC 342. The conclusion is here that the magnetic field structure is dominated by an axialsymmetric spiral one (M. Krause et al. (1989a), see also Gräve and Beck (1988)).

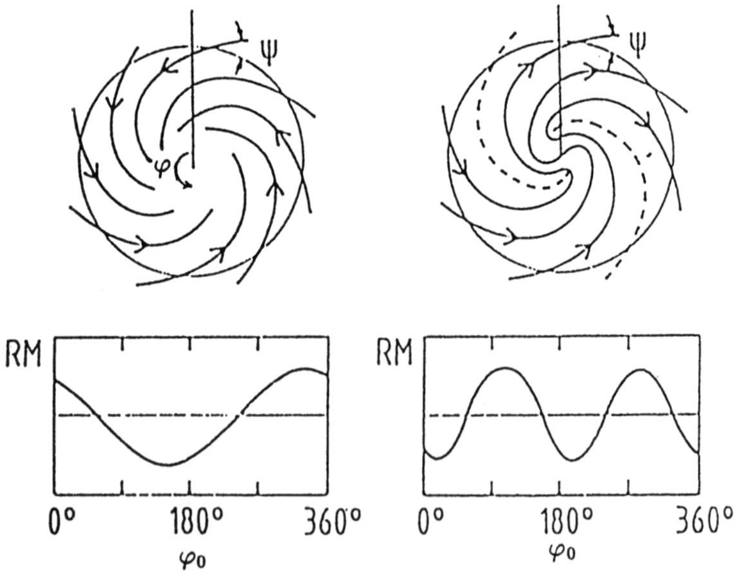

Fig. 5. Draft of an axisymmetric (type S0) and a bisymmetric (type S1) magnetic field in a galactic disc of vanishing thickness and the corresponding RM-profiles if the fields are observed under a certain inclination angle between 0 and $\pi/2$

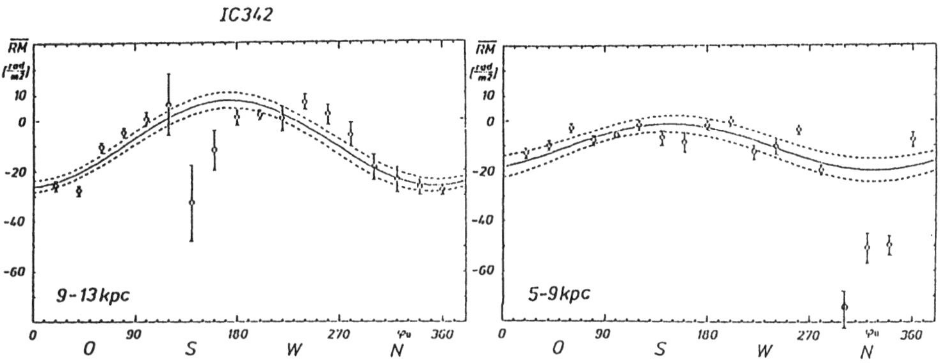

Fig. 6. RM-profiles of the galaxy IC 342 in dependence of φ_0, the azimuth projected on the sky (after M. Krause et al. 1989a). The strong deviations from the harmonic profile may find an explanation by the synthetic RM-profiles in fig 8

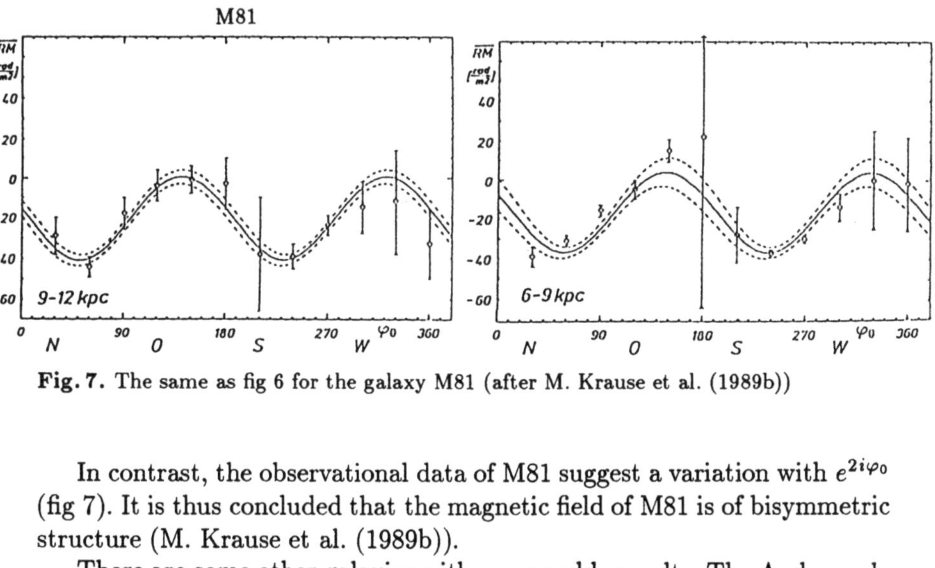

Fig. 7. The same as fig 6 for the galaxy M81 (after M. Krause et al. (1989b))

In contrast, the observational data of M81 suggest a variation with $e^{2i\varphi_0}$ (fig 7). It is thus concluded that the magnetic field of M81 is of bisymmetric structure (M. Krause et al. (1989b)).

There are some other galaxies with comparable results. The Andromeda nebulae, M31, rather clearly show an axialsymmetric ring structure (Beck (1982)). M51, M33 are further examples and candidates for one or the other basic type (Horellou et al. (1990), Buczilowski et al. (1991)).

This analysis looks rather conclusive; however, it has to be noted that a closer inspection reveals certain difficulties. In order to illustrate the situation we present in fig 8 and fig 9 synthetic rotation measures derived from magnetic fields in a disc, which have been calculated from disc dynamo models (Elstner et al. (1991)). In contrast to the simple models discussed above here the curves show clear anharmonic behaviour. Similiar results have been found by Donner and Brandenburg (1990).

The strong deviation from the harmonic character is due to the existence of non-zero components parallel to the axis of rotation and, in general, to the

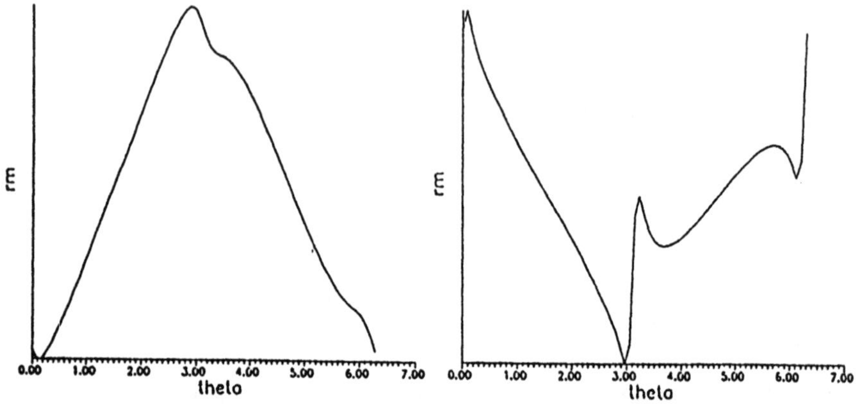

S0−mode rom=20 com=20 rad=23.0 incl=70   S0−mode rom=5 com=50 radius=6.1 incl=70

Fig. 8. Synthetic RM-profile derived from a (type S0) dynamo excited axisymmetric magnetic field in a disc of non-vanishing thickness (courtesy of D. Elstner, R. Meinel and R. Beck (1991))

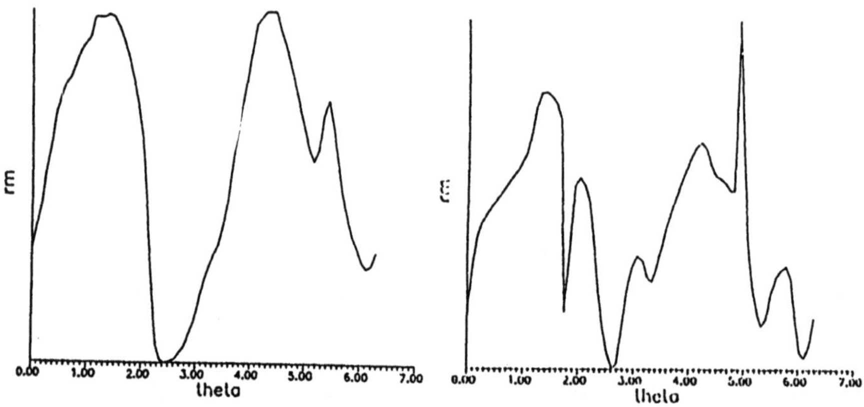

S1−mode rom=2 com=200 rad=2.3 incl=25   S1−mode rom=2 com=200 rad=2.3 incl=45

Fig. 9. The same as fig 8 for a bisymmetric field (type S1), i.e. a field with an $e^{i\varphi}$-dependence on the azimuth (courtesy of D. Elstner, R. Meinel and R. Beck (1991))

spatial distribution of the magnetic field in a disc of non-vanishing thickness. Even the statement that four zeros indicate a non-axisymmetric field is not correct.

We must notice that the simple statement of Sawa and Fujimoto does not resist a closer inspection. Although on its basis an important development in the exploration of the magnetic fields of galaxies was initiated, future investigations have to be based on more refined theoretical models.

3 The origin of cosmic magnetic fields

By the Maxwell equation

$$\operatorname{curl} \mathbf{H} = \mathbf{j} \tag{1}$$

a magnetic field $\mathbf{H}$ is always bound to appear in combination with an electric current $\mathbf{j}$. We use here from the beginning this Maxwell equation in the form (1), i.e. the displacement current is neglected, since we consider electrically sufficiently well conducting media and not too rapid variations with time.

The current obeys Ohms law which we write in the form

$$\mathbf{j} = \sigma(\mathbf{E} + \mathbf{u} \times \mathbf{B} + \mathbf{E}^{(i)}), \tag{2}$$

σ denotes the electric conductivity, $\mathbf{E}$ the electric field strength, $\mathbf{u}$ the velocity field of the conducting medium, $\mathbf{B}$ the magnetic field and $\mathbf{E}^{(i)}$ the impressed electromotive force.

There are several reasons for having an impressed electromotive force, e.g. thermo-electricity, or chemical composition etc. In our context a mechanism is of interest which was first proposed by Schlüter and Biermann (1950) and explicitely applied to the interstellar magnetic field by Biermann (1952). This mechanism rests on the difference of the masses of the electrons, which carry the negative charges, and the ions with the positive charges. A charge separation will appear if the interstellar gas undergoes the influence of an accelerated motion and/or a gravitational field, and consequently an impressed electromotive force

$$\mathbf{E}^{(i)} = \gamma(\frac{d\mathbf{u}}{dt} - \mathbf{g}). \tag{3}$$

With plausible assumptions concerning the scales of the turbulent motions Biermann concluded that in our Galaxy a magnetic background field of about 10^{-16} G must be due to this effect. In addition, the turbulent motions will amplify this background field via the term $\mathbf{u} \times \mathbf{B}$ in (2) until an equipartition of magnetic and kinetic energy is reached. A magnetic field of about 10^{-6} G is thus possible.

In this way one plausible explanation for the interstellar magnetic field in our Galaxy is presented. A characteristic feature of this field has to be mentioned: Its characteristic spatial scales must be the same as those of the turbulent motions of the interstellar gas. It is not possible to explain in this way ordered magnetic fields with a global structure characterized by scales of the whole galaxy. Consequently, the discovery of such ordered magnetic fields in M31, IC 342, M81,... is crucial: The Biermann mechanism cannot explain these phenomena.

The situation reminds one of that in 1919, when J. Larmor at a meeting of the Royal Society presented his famous paper with the title: "How could a rotating body such as the Sun become a magnet?" Larmor analyzed several effects which provide for an impressed electromotive force $\mathbf{E}^{(i)}$, but at the end he concluded that all but one fail to explain the origin of the magnetic field of the Sun (or the Earth). The only possibility which is left is that of a self-excited dynamo, where the Lorentz electromotive force $\mathbf{u} \times \mathbf{B}$ in (2) drive currents which reproduce $\mathbf{B}$. Thus maintenance or even enhancement of the magnetic field is given.

The first dynamo models which have been applied to galaxies were elaborated by Parker (1971) and Vainshtein and Ruzmaikin (1972). Here dynamo generation by a constant shear and a constant α-effect in an infinite slab embedded in insulating space was considered. The constant shear simulates the differential rotation in a galaxy. By appropriate boundary conditions the antisymmetry of the α-effect with respect to the central plane was guaranteed. Let be mentioned here that the α-effect describes an electromotive force parallel to the magnetic field which is due to the concerted action of turbulent motions undergoing the influence of the overall rotation.

Although the models are rather primitive in comparison with the real object, a galaxy, one result was already deduced which proved to be of persistance: A magnetic field showing symmetry with respect to the central plane of the slab is excited with preference.

Models more close to the real objects have been treated by Stix (1975). Starting from spherical dynamos developped for the Sun Stix investigated the conditions for excitation in oblate spheroids, which are embedded in the insulating space. It is interesting to note that the symmetry type of the magnetic field, which is excited with most preference, changes in dependence on the aspect ratio of the spheroid: In a sphere (aspect ratio = 1) a magnetic field showing antisymmetry with respect to the equatorial plane (dipol-type) is excited; however, below a certain aspect ratio smaller than one a symmetric field (quadrupole-type) is preferrently excited. These results have been confirmed by White (1978). With a certain asymptotics Soward (1978) has found some correspondence between the results of Parker and Stix.

4 The Dynamo: A Universal Process in Rotating Systems

As already mentioned J. Larmor was the first who argued that the dynamo process may work in the universe. The dynamo principle was first discovered by W. v. Siemens and in December 1866 he presented it to the Berlin Academy of Science. In short it expresses the possibility that by motions of an electrically conducting matter in a magnetic field there may be induced currents which create the original magnetic field. The dynamo in this way proves to be a self-organization process, where a new quality, the magnetic field, is generated from a minor background.

Most inventions of mankind have been invented before by nature. There is one famous exception: the wheel. For nearly one century it looked like being the same for the dynamo, until it became clear that rotating cosmical objects generally represent dynamos.

In order to illuminate the related induction processes we quote one result of magnetohydrodynamics which is easily to comprehend: The magnetic field lines behave in a well conducting streaming fluid like elastic fibres which are "frozen-in".

With this simple principle in mind we can easily explain the two basic induction processes which in rotating systems are responsible for the dynamo process.

First the differential rotation may be considered, which is in galaxies an observational fact. From the above mentioned principle it is clear that from an originally radial directed magnetic field components in the azimuthal direction are generated because the field lines are wound up. And these azimuthal components become in course of the rotation stronger and stronger, especially, comparably large to the original radial field if a decay is excluded. We thus may understand the ring field in M31.

The strength of the induction action of the differential rotation will be described by the dimensionless parameter

$$C_{\Omega} = \mu\sigma\Omega_0 L_{\alpha}^2, \tag{4}$$

with the characteristic scales Ω_0 for the rotation and L for a spatial dimension which may be either the diameter R or the thickness z_{α} of the disc. The radius r_{Ω} characterizing the transition from rigid to flat rotation has also been used.

However, differential rotation cannot provide any contribution to the original radial field. Consequently, this radial field will decay and with it the induced ring field. An additional induction process is needed which feeds energy back.

This is given by the concerted action of internal turbulent motions which undergo the influence of the overall rotation. The name "cyclonic turbulence" is frequently used to remind us that the important property, the spiral structure, is already well-known from the motions around a high or a low in the Earth's atmosphere.

The behaviour of a magnetic field in a local spiral motion may be easily derived from the above mentioned principle of the "frozen-in" field lines (fig 10). If a flux tube undergoes this motion a loop is formed which looks like a twisted Ω. According to the Maxwell equation (1) the loop is connected with a current, and because of the twist this current has a component parallel to the magnetic field. In fig 10 with a righthanded spiral motion this component is antiparallel to the magnetic field, the opposite will be the case for a lefthanded spiral motions.

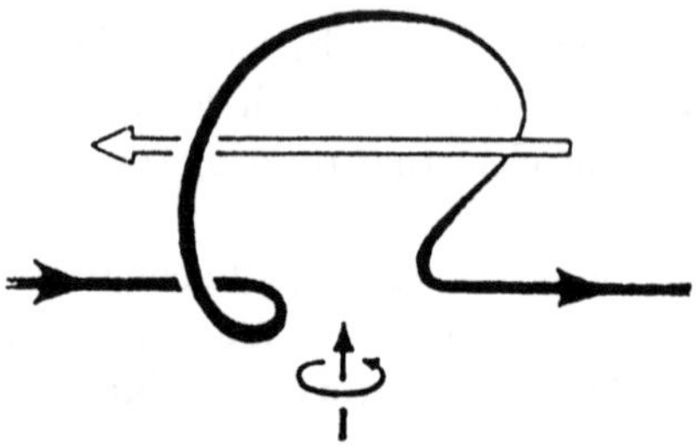

Fig. 10. A magnetic flux rope undergoing the influence of a helical motion is shaped into a twisted Ω. The loop is accompanied by a current which, in case of right-handed helical motions, has a component antiparallel to the magnetic field

In the denotation of mean-field magnetohydrodynamics (F. Krause and Rädler (1980)) one has a turbulent electromotive force $\mathcal{E}$ parallel to the mean magnetic field

$$\mathcal{E} = \alpha \overline{\mathbf{B}}, \tag{5}$$

where α (on resonable simplifying assumptions) is given by

$$\alpha = -\frac{\tau_{cor}}{3} \overline{\mathbf{u}' \cdot \operatorname{curl} \mathbf{u}'}. \tag{6}$$

The bar denotes the average operation, the dash indicates fluctuations. τ_{cor} is the correlation time of the turbulent motions.

From (6) it is clear that a non-zero α-effect requires an unbalance of righthanded and lefthanded spiral motions. This needs in addition to the rotation a radial stratification.

In rotating electrically conducting medium on cosmic conditions one finds (F. Krause, 1967)

$$\alpha = -\tau_{cor}^2 u'^2 \cdot (\Omega \cdot \nabla log \rho u'). \tag{7}$$

Ω denotes the angular velocity, ρ the density and u' the r.m.s. velocity of the turbulent motions. So it is clear that α is zero unless there is a stratification given either by density, or by turbulence intensity (in the case here considered), or by other, for example, topological reasons.

By scaling considerations formula (7) has been strongly simplified: Let us write

$$\alpha \approx -u'^2 (\Omega \tau_{cor}) \frac{\tau_{cor}}{H} \tag{8}$$

where H denotes the scale height connected with the stratification.

If now the (generally accepted) relation

$$u' \approx \frac{\lambda_{cor}}{\tau_{cor}}, \tag{9}$$

with the correlation time λ_{cor} is introduced we arrive at

$$\alpha \approx \frac{\lambda_{cor}^2}{H} \cdot \Omega, \tag{10}$$

a relation, which is often simplified to

$$\alpha \quad \approx \quad \Omega \cdot d, \tag{11}$$

i.e. $\lambda_{cor} \approx H \approx d$ is used, where d denotes the thickness of the convective layer.

Although relation (11) is in some sense conclusive, one has to hold in mind the oversimplification within the deduction. For example, (11) indicates by no means that $\alpha = 0$ if $u' = 0$. Especially, in galaxies there may be regions with reduced, or even vanishing turbulence; or, in contrast, there may be regions with exceptional strong turbulence.

For the models we consider in the following indeed an α-profile is used which is bound to the rotation, however, we will also take the freedom to consider, for example, an enhanced α in the central parts.

There is an additional point of interest, especially for considerations of galactic discs. We have till now not taken into account that the relation (5) takes no notice of the anisotropy of the turbulence in a rotating stratified object. Generally (5) has to be exchanged against the relation

$$\mathcal{E}_i \quad = \quad \alpha_{ij} \quad \overline{B}_j \tag{12}$$

with the tensor α_{ij}. Only in the case of isotropic homogeneous turbulence

$$\alpha_{ij} \quad = \quad \alpha\delta_{ij} \tag{13}$$

and so (5) is valid.

Even for galaxies the anisotropy may be important. (5) has to be exchanged by

$$\mathcal{E} \quad = \quad \alpha_{\perp}\mathbf{B}_{\perp} \quad + \quad \alpha_{\|}\mathbf{B}_{\|}, \tag{14}$$

where the symbols vertical $\perp$ and parallel $\|$ are related to the rotational axis of the galactic disc. A derivation on conditions modelling the turbulence in a galaxis by Rüdiger (1989) revealed significant differences, up to a factor 10, between $\alpha_{\perp}$ and $\alpha_{\|}$.

The strength of the induction action by the cyclonic turbulence will be described by the dimensionless parameter

$$C_\alpha = \mu\sigma\alpha_0 L. \tag{15}$$

It is clear that these two parameters (C_α, C_Ω) are not known with great accuracy. The value which is observationally best known is Ω_0, the characteristic scale of the differential rotation.

According to these uncertainties we will in the following not overinterpret the theoretical results. All reasonable estimates, however, show that in comparison with the results of the calculation of dynamo models the numerical values of (C_α, C_Ω) for galaxies lie beyond the border to self-excitation (Stix (1975), F. Krause (1987), Elstner et al. (1991)).

5 Dynamo Problem

5.1 The General Problem

The situation in rotating cosmical objects consisting of an electrically conducting fluid is described by the conservation of mass

$$\frac{\partial \varrho}{\partial t} + div \varrho \mathbf{u} = 0, \tag{16}$$

and momentum

$$\varrho(\frac{\partial \mathbf{u}}{\partial t} + (\mathbf{u} \cdot \nabla)\mathbf{u}) = -\nabla p + \varrho \mathbf{g} - 2\varrho \Omega \times \mathbf{u} + \mathbf{F} + \mathbf{j} \times \mathbf{B} + Div\ \tau, \tag{17}$$

and the induction equation

$$\frac{\partial \mathbf{B}}{\partial t} = \mathrm{curl}(\mathbf{u} \times \mathbf{B}) + \frac{1}{\mu\sigma}\Delta\mathbf{B}. \tag{18}$$

In addition we have an equation of state.

ϱ denotes the density, p the pressure, μ the permeability, $\mathbf{g}$ the gravity, Ω the angular velocity of the rotational motion, $\mathbf{u}$ the internal motion relative to the overall rotation. $\mathbf{F}$ denotes an assumed force which drives the internal motion, and τ the stress tensor.

Generally an equation describing the conservation of energy has to be added to (16), (17), (18), since in objects like the Sun or stars the internal motion, the convection, is driven by an energy transport process. However, in the context of galaxies, which we consider here, the turbulent motions of the interstellar medium are assumed to be powered by supernova explosions or stellar wind. To take this into account we have introduced the random forced $\mathbf{F}$ in equation (17).

An inspection of the system of partial differential equations (16), (17), (18) clearly shows that this system is highly nonlinear. The solutions we expect are of complex structure (since turbulent) and so they are beyond access even for the most powerful computers which are at present available.

Hence the solution of the dynamo problem is still in a state of development. Step by step the complexity has to be disentangled.

For the question of the origin of the magnetic field a natural simplification presents itself: Since the magnetic field $\mathbf{B}$ enters in the equation (17) for $\mathbf{u}$ via the nonlinear term $\mathbf{j} \times \mathbf{B}$, we can for $\mathbf{B} \approx 0$ neglect this term and consider simply the induction equation with $\mathbf{u}$ given. We can so decide on what conditions concerning the velocity field $\mathbf{u}$ a growth of a weak magnetic field will appear. We cannot decide questions concerning the magnitude and the structure of the finally formed field. In order to gain such information we have to analyze the complete system (16), (17), (18). Although a solution is not available, investigations during the recent years provided some insight which at least allow some conclusions (F. Krause and Meinel (1988), Brandenburg et al. (1989a), Rädler et al. (1990)).

5.2 The Kinematic Problem

In this case the velocity field is given and $\mathbf{B}$ has to be determined as a solution of the induction equation (16). In detail the problem may be posed as follows:

Given a region $\Re$ of an electrically conducting fluid with a prescribed velocity field $\mathbf{u} = \mathbf{u}(\mathbf{x})$. Let be the region $\Re$ embedded in the insulating space.

The magnetic field $\mathbf{B}$ fulfills inside of $\Re$ the equations

$$\frac{\partial \mathbf{B}}{\partial t} = \mathrm{curl}(\mathbf{u} \times \mathbf{B}) + \frac{1}{\mu\sigma}\Delta\mathbf{B} \tag{19}$$

$$div\mathbf{B} = 0, \tag{20}$$

outside of $\Re$

$$\mathrm{curl}\,\mathbf{B} = 0, \tag{21}$$

$$div\mathbf{B} = 0, \tag{22}$$

and the condition

$$\mathbf{B} \cong 0(\frac{1}{r^3}) \quad if \quad r \rightarrow \infty. \tag{23}$$

At the boundary of $\Re$ a smooth transition has to be guaranteed.

The assumption of an insulating space, say vacuum, outside of $\Re$ does not in any case correspond to the real situation. For example, a galaxy is embedded in a halo which also has non-zero conductivity. However, with the above assumption a problem is posed which is based on clear physical conditions.

We come here in touch with specific difficulties. As in the case of insulating outer space we have generally to take into account non-local boundary conditions, since the magnetic field is a far reaching quantity. Mathematical difficulties originate from these assumptions. Therefore, many authors use in investigations of the problems here considered local boundary conditions by simply assuming that the tangential component of the magnetic field vanishes. However, there is no physical justification, this is for mathematical simplicity only. A physical clearly posed problem with local boundary conditions is given by the assumption of ideal conductivity outside of $\Re$ (e.g. Meinel (1990)).

Let us now turn back to our problem. We assume a steady velocity field. Hence $\mathbf{B}$ can be represented by

$$\mathbf{B}(\mathbf{x}, t) = e^{\gamma t}\mathbf{B}(\mathbf{x}), \tag{24}$$

with the growth or decay rate γ. If the ansatz (24) is inserted into the system (19),...,(23) we arrive at an eigenvalue problem which generally has a denumerable set of solutions γ_n, $\mathbf{B}_n(\mathbf{x})$ with

$$Re\gamma_1 \geq Re\gamma_2 \geq ..., \quad Re\gamma_n \rightarrow -\infty \; if \; n \rightarrow \infty. \tag{25}$$

If

$$Re\gamma_1 \geq 0; \tag{26}$$

the system represents a dynamo, i.e. the induction action of the velocity field $\mathbf{u}$ prevents the magnetic field from decay due to dissipation by Joules heat.

We consider here dynamo excitation in galaxies, i.e. the region $\Re$ is a disc with finite radius R and thickness d. Especially, we assume with respect to all non-magnetic properties axisymmetry with respect to the axis of rotation and symmetry with respect to the central plane, the equatorial plane, of the disc. For the random velocity field these conditions have to be fulfilled in the average.

As a consequence, the eigensolutions $\mathbf{B}_n(\mathbf{x})$ reflect to some extend these symmetries. Generally a field $\mathbf{B}_n(\mathbf{x})$ is either antisymmetric (A) or symmetric (S) with respect to the central plane of the disc and depends on the azimuthal φ according to $e^{im\varphi}$, m = 0, 1, ... Correspondingly we denote the symmetry type of a magnetic field by Am or Sm (F. Krause (1971)).

Fields with m = 0 show axisymmetry with respect to the axis of rotation. So a dipole in the centre of the disc with its moment aligned to the rotational axis represents the type A0, even so an octupole. A quadrupole with analogue relations to the centre of the disc and the axis of rotation represents the type S0. They may appear combined with ring fields around the axis of rotation: two rings of opposite oriention for the A-type and one ring for the S-type. Then we have fields of axisymmetric spiral structure (ASS).

A dipole in the centre of the disc but with its moment in the central plane represents a field of type S1. If combined with an azimuthal field of the same symmetry we arrive at a field of bisymmetric spiral structure.

The fields $\mathbf{B}_n(\mathbf{x})$, which we obtaine by integrating the basic equation, adjust themselves into this scheme. Since they can be represented by a series of multipoles in the outer space, dipole or quadrupole will be the leading terms of the series.

From (25) we see that - even in the case of dynamo excitation - only a finite number of eigen fields $\mathbf{B}_n(\mathbf{x})$ will not decay with time. Generally, three fields of type A0, S0, S1 compete for taking the places $\mathbf{B}_1(\mathbf{x}), \mathbf{B}_2(\mathbf{x}), \mathbf{B}_3(\mathbf{x})$. In case all three have non-negative growth rates we cannot decide on the kinematic level which of them will dominate at the final state. This crucial point demands the solution of the nonlinear problem.

5.3 The Nonlinear Aspect

A solution of the dynamo problem by simply integrating the system (16), (17), (18) under physically resonable conditions is out of reach. In order to gain information concerning the behaviour of the solutions simple nonlinear models have been considered with the intention to find statements of some general character.

The consideration of the kinematic problem provides hints insofar as at the points where - in dependence on the induction action - the growth rates γ_n become positive, solutions $\mathbf{B}^{(n)}(\mathbf{x},t)$ bifurcate from the solutions $\mathbf{B} \equiv 0$ and form so the nonlinear extension of the eigenmodes. This nonlinear extension $\mathbf{B}^{(n)}(\mathbf{x},t)$ has the same spatial symmetry as the eigenmode $\mathbf{B}_n(\mathbf{x})$.

For sufficiently strong induction action a finite number N of solutions $\mathbf{B}^{(n)}(\mathbf{x},t)$, n = 1,..., N of the nonlinear problem exist. The point is now to find among the possible solutions $\mathbf{B}^{(n)}(\mathbf{x},t), n = 1,...,N$, that one which proves to be the attractor, i.e. the solution which corresponds to and describes the real physical situation.

Consideration of simple nonlinear dynamo models reveal (F. Krause and Meinel (1989)) that only the solution $\mathbf{B}^{(1)}(\mathbf{x},t)$ is stable, at least in the neighbourhood of the bifurcation point. This point separates the region where all magnetic fields decay from the region where even small seed fields will grow and form at the end the field $\mathbf{B}^{(1)}(\mathbf{x},t)$. All higher order fields $\mathbf{B}^{(n)}(\mathbf{x},t)$, n = 2,..., N, are first, i.e. in the neighbourhood of their bifurcation points, unstable.

For much stronger induction action $\mathbf{B}^{(1)}(\mathbf{x},t)$ may also lose its stability. The situation becomes complicated, for example, none of the solutions $\mathbf{B}^{(n)}(\mathbf{x},t)$, n = 1,..., N, may be stable (e.g. Brandenburg et al. (1990), Rädler et al. (1990)), no steady or regular behaviour will appear. General statements are not available.

The question raised at the end of the foregoing paragraph therefore finds an answer:

Only the marginal solution $\mathbf{B}^{(1)}(\mathbf{x},t)$ is stable and is thus the only one of physical interest.

The statement in this form has to be supplemented by some comments:
- The stability of the field $\mathbf{B}^{(1)}(\mathbf{x},t)$ is only guaranteed in a certain neighbourhood of the marginal bifurcation point,
- In the state of growth, i.e. when the magnetic field is small, a solution of the kinematic equation (19) can be represented by the series

$$\mathbf{B}(\mathbf{x},t) = \sum_{n=1}^{\infty} b_n \mathbf{B}_n(\mathbf{x}) \cdot e^{\gamma_n t}; \tag{27}$$

in this sense the higher order eigenmodes may play a certain role in case the field is still in the state of being developped with time.

Generally theoretical investigations are based on the view that even galaxies have had sufficient time to develop their magnetic fields into some final state. In this sense only such results, which are based on the marginal eigenmode, i.e. $\mathbf{B}^{(1)}(\mathbf{x},\tau)$, can be looked on as reliable.

Consequently, a theoretical analysis has to guarantee that indeed the marginal eigenmode has been found. Earlier investigations do not generally fulfill these requirements, especially in cases where not the full set of solutions has been considered (e.g. Fujimoto (1987)). Then the problem arises of "forgotten modes" (F. Krause (1989)); i.e. due to restrictive assumptions the marginal mode has not been found.

6 Disc Dynamo Models

6.1 General remarks

The mathematical treatment of discs is rather complicated. This is, for example, clearly to be seen at the investigations of Stix (1976) based on the considerations of oblate spheroids, where the convergence of the related series becomes increasingly bad with decreasing aspect ratio. In addition, the methods used do not allow an extension to the treatment of non-axisymmetric solutions. Most investigations were based therefore on certain approximations.

The simplest models treated so far consider an infinite slab where the differential rotation is given by a constant shear; a constant α but with opposite sign with respect to the central plane represents the action of the cyclonic turbulence (Parker (1971), Vainshtein and Ruzmaikin (1971),(1972), Moffatt (1978), Fujimoto (1987)). These models have been further developed by taking into account the large differences between the radial and the vertical scales which offer the possibility for an application of the WKB-method (Ruzmaikin et al. (1985), Baryshnikova and Shukurov (1987), Baryshnikova et al. (1987), Chiba and Tosa (1989), Tosa and Chiba (1990)). The models, although bound on certain assumptions of "locality", thus allow for a better modelling of parameters like differential rotation, α-effect, thickness of the layer etc. The same is true for models which attack the problem by expanding the relevant quantities in Bessel-Fourier-series (Stepinski and Levy (1990), Rädler and Wiedemann (1990)). Whether an infinite slab is actually a reliable approximation for a disc is a delicate question which deserves further attention (Soward (1978, 1991)).

Specific problems arise from the boundary conditions. In certain geometries, like the spheroidal one, the condition of vacuum outside can be fulfilled by expanding the solutions into series of the related harmonics. In this way also torus-dynamos have been considered by Deinzer et al. (1990) (see also Schmitt (1990)), which in some respect allow one to approximate the conditions in disc geometry. Difficulties arise in these cases with the consideration of non-axisymmetric solutions.

Models using the well-elaborated theory in spherical geometry do not meet these difficulties. The transition to discs is here attained by assuming flat dynamo active regions within the sphere (Brandenburg et al. (1990), Donner and Brandenburg (1991), Stepinski and Levy (1988)).

A step further is given by the galactic dynamo models without sharp boundaries (Elstner, Meinel and Rüdiger (1990)). These models evade the difficulties originating in the boundary conditions and allow the treatment of axisymmetric as well as non-axisymmetric solutions. The introduction of some artificial nonlinearity represents a trick, by which the numerical code automatically finds the marginal mode. Hence the danger of the "forgotten modes" will never appear.

Results of these investigations shall be presented here in more detail.

6.2 Galactic dynamo models without sharp boundaries

The main idea of these models is that the disc is embedded in a cylinder with much larger geometrical scales. The conductivity σ is non-zero in the whole cylinder, but the dynamo active processes, especially the α-effect, are non-zero in a flat disc situated in the central part of the cylinder.

A general analysis (Meinel (1989)) revealed that the boundary conditions which have to be posed on the cylinder are of minor influence. Meinel has shown that localized induction action can generate and maintain a localized field only, i.e. the field decays with increasing distance from the inducing region. This makes it possible to pose rather arbitrary conditions on the surface of the cylinder; in order to have a physically clear model infinite conductivity is assumed. Vacuum outside the disc is simply modelled by a drastically reduced conductivity.

For the induction processes is assumed a differential rotation according to

$$\Omega(r,z) = \left\{ \begin{array}{ll} \Omega_0 & \text{for} \quad r < r_\Omega \\ \Omega_0 \frac{r_\Omega}{r} & \text{for} \quad r > r_\Omega \end{array} \right\}; \tag{28}$$

and an α-effect given by

$$\alpha(r,z) = \alpha_0 f(z) \cdot g(r). \tag{29}$$

The angular velocity (28) does not depend on z. The r-dependence corresponds roughly to the rotation laws observed in galaxies: A rigid rotation inside the cylinder $r < r_\Omega$ and the so-called flat rotation curve for $r > r_\Omega$. In this way, apart from the choice of r_Ω, individual details observed at the different galaxies have been omitted.

With the α-effect the flatness of the object is described by the function $f(z)$, e.g. in models we present here a decrease according to $exp{-}z^2/z_\alpha^2$ is used. The same dependence but with a three times larger scale height is taken for the conductivity σ.

For the r-dependence of α according to (11) a proportionality with Ω is prescribed; however, (7) or (8) indicate that physical reasons exist which justify deviations. A great varity of models may so be calculated and their results compared with the observed magnetic fields in real objects (Elstner et al. (1991)).

The induction equation (19) has been integrated using a certain numerical code. By the formulation of the boundary conditions at the rotational axis and the central plane four different problems have been posed according to the four symmetry types S0, A0, S1, A1 of interest. The introduction of an artificial nonlinearity guarantees that at the temporal integration the marginal mode appears when the steady state is reached. The results of the four problems will then be compared and the marginal mode of the full problem determined.

6.3 Modelling of dynamo excited magnetic fields in discs

Above all it should be emphasized that nearly all calculations of physically resonable models show that the marginal mode is of type S0. Hence the theoretical prediction says: A disc, which with respect to the non-magnetic properties is axisymmetric, excites a magnetic field which is symmetric with regard to the central plane and axisymmetric with regard to the axis of rotation.

In this way dynamo theory can explain the fields as observed in M 31, IC 342 and others. These fields are of axisymmetric spiral structure (ASS) and can, depending on the strength of the differential rotation, show a rather strong toroidal component which provides the appearance of a ring field as in M 31 (fig. 1). The poloidal field in the halo is of quadrupolar character.

In fig. 11 the polarization map derived from a calculated magnetic field is represented. The dashes indicate the direction of the magnetic field component in the plane perpendicular to the line of sight, the lengths of the dashes are chosen proportional to the polarized intensity. The model rather well approximates the conditions observed in the Andromeda Nebulae.

In fig. 12 an A0-field is represented. This field is not marginal. There are not many differences to fields of type S0. A clearer decision between both symmetry types seems possible on the basis of rotation measure maps, which for A0-type fields can show four zeros (Elstner et al. (1991)). However, the widely accepted opinion that the magnetic fields observed in galaxies are symmetric with respect to the central plane is mainly based on the theoretical prediction that the marginal mode is preferentially excited.

In fig. 13 we present a polarization map of a model, where the α-effect in the centre is enhanced if compared with the profile of the angular velocity. We obtain an S0-field which is rather strong in the centre, in the direction of growing r a gap appears followed by the ring field. This model allows for an interpretation of the observed high non-thermal radiation from the central regions of galaxies as in M 31 (fig. 1).

If the α-effect is still more enhanced the models tend to excite an A0-type field as the marginal one. These fields are concentrated in the central

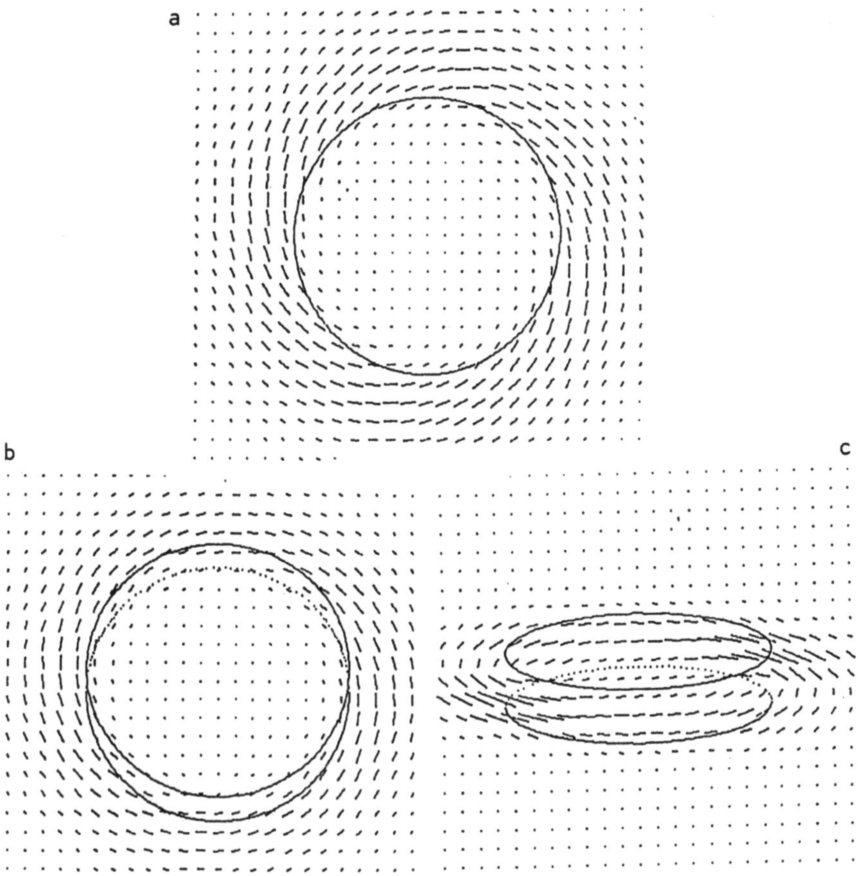

Fig. 11. Polarization map of an axisymmetric spiral field of type S0 (courtesy of D. Elstner, R. Meinel and R. Beck (1991)). a) face-on, b) inclination angle of 30°, corresponding to the galaxy IC 342; c) inclination angle of 74°, corresponding to the Andromeda nebulae. The solid line indicates the cylinder of the transition from rigid body rotation to flat rotation

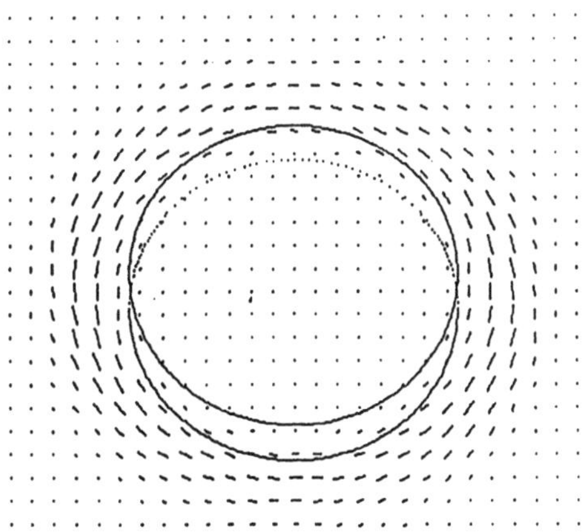

Fig. 12. Polarization map as fig. 11b, but of an A0-type magnetic field (courtesy of D. Elstner, R. Meinel and R. Beck (1991)). Note that there are not many differences between the figures

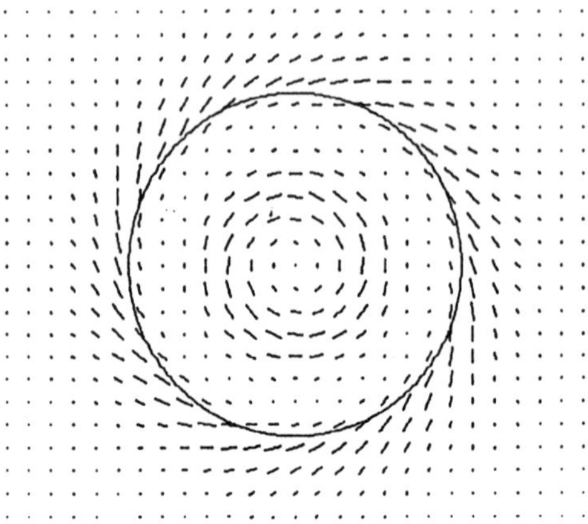

Fig. 13. A face-on polarization map of the S0-type field which is excited by a dynamo model with enhanced α-effect in the centre of the disc (courtesy of D. Elstner, R. Meinel and R. Beck (1991))

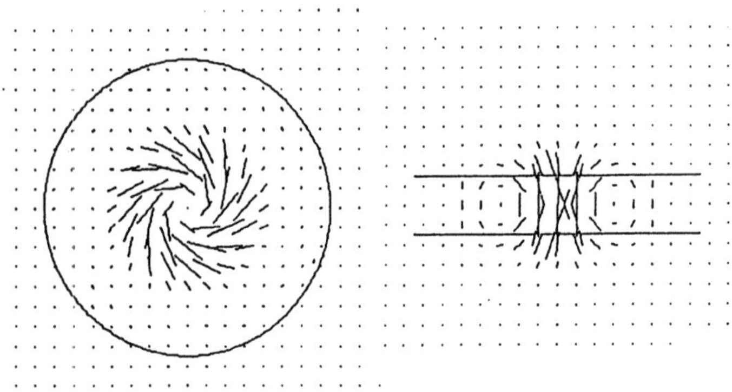

Fig. 14. Polarization maps of an A0-type field (courtesy of D. Elstner, R. Meinel and R. Beck (1991)), which is marginal due to an enhanced α-effect in the central region of the disc. The same model is represented in face-on and edge-on view. The edge-on view resembles that of our galactic centre

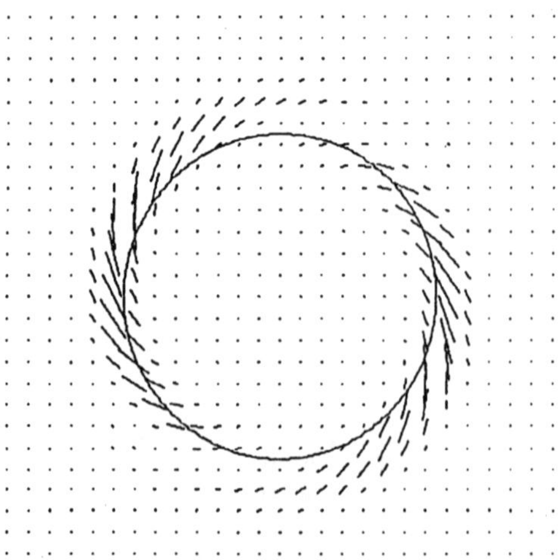

Fig. 15. Polarization map of a S1-type field (cortesy of D. Elstner, R. Meinel and R. Beck (1991)), i.e. a field of bisymmetric structure. This eigenmode is not marginal

regions. The dipole-type structure of the poloidal field apparently provides a better fit of the field observed in the centre of our galaxy (Reich et al. (1987)) than a field of type S0 (fig. 14).

Finally, we represent in fig. 15 a map of a calculated field of typ S1. This is also not marginal, hence dynamo theory cannot explain its existence. However, fig. 15 clearly shows that such bisymmetric structures also appear as eigenmodes of the dynamo equations.

7 Conclusions

This paper attempts to report on the exciting development which took place in the exploration of the magnetic fields in galaxies.

Although hardly any complete magnetic maps of galaxies exist, observations clearly reveal that in galaxies there exist magnetic fields which are extended over the whole object and possess its large scale spatial structure. In most cases this structure is an axisymmetric spiral one (ASS), a field of symmetry type S0, in accordance with the prediction of dynamo theory. This theoretical prediction rests on the view that only the marginally stable eigenmode as solution of the kinematic problem reflects the magnetic field that actually exists.

In no case do models which are axisymmetric with respect to their non-magnetic properties excite a field with bisymmetric structure (BSS), i.e. a field of S1-type, as the marginal mode. The observed BSS-field in M 81, for example, presents an enigma. Apart from the possibility of a primordial origin the theory presented here allows for two explanations:

- A non-axisymmetric influence (e.g. by a near companion as in the system M 81/M 82) makes the eigenvalues change their order in a way that the bisymmetric mode becomes marginally stable;
- A non-axisymmetric influence provides for a break of symmetry.

The first possibility was already anticipated by Krasheninnikova et al. (1990); however, it appears rather less probable according to numerical studies by Moss et al. (1990).

In the second case the excited field is not a genuine symmetry type, but a mixture of an axisymmetric and a non-axisymmetric part. The observational data of M 81 still allow a significant axisymmetric contribution.

Acknowledgement

We would like to thank D. Elstner and R. Meinel from the Astrophysikalischen Observatorium Potsdam and R. Beck and M. Krause from the Max-Planck-Institut für Radioastronomie Bonn for continued discussions about dynamos and magnetic fields in galaxies.

References

1. Alfven, H., Richtmayer, R.D. and Teller, E., 1949, Phys. Rev. **75**, 1169
2. Appenzeller, I., 1967, Publ. Astron. Soc. Pacific, **79**, 600
3. Baldwin, J.E., Pooley, G.G., 1973, Mon. Not. Roy. astron. Soc. **161**, 127
4. Baryshnikova, Y., Shukurov, A., 1987, Astron. Nachr. **308**, 89
5. Baryshnikova, Y., Ruzmaikin, A.A., Sokolov, D.D., Shukurov, A., 1987, Astron. Astrophys. **177**, 27
6. Beck, R., 1982, Astron. Astrophys. **106**, 121
7. Beck, R., Berkhuijsen, E.M., Wielebinski, R., 1978, Astron. Astrophys., **68**, L27
8. Beck, R., Berkhuijsen, E.M., Wielebinski, R., 1980, Nature, **283**, 272
9. Beck, R., Klein, U., Wielebinski, R., 1987, Astron. Astrophys. **186**, 95
10. Beck, R. and Gräve, R., 'Interstellar Magnetic Fields' 1987, Springer-Verlag
11. Beck, R., Kronberg, P.P., Wielebinski, R., 1990, "Galactic and Intergalactic Magnetic Fields", Kluwer Academic Publishers
12. Biermann, L., 1950, Z. Naturf., **5a**, 65
13. Biermann, L., 1952, Annalen der Physik **10**, 28
14. Bingham, R.G., McMullan, D., Pallister, W.S., White, C., Axon, D.J., Scarrott, S.M., 1976, Nature **256**, 463
15. Bracewell, R.N., Cooper, B.F.C., Cousins, T.E., 1962, Nature **195**, 1286
16. Brandenburg, A., Krause, F., Meinel, R., Moss, D., Tuominen, I., 1989, Astron. Astrophys. **213**, 411
17. Brandenburg, A., Krause, F., Tuominen, I., 1990, Geophys. Astrophys. Fluid Dyn. **50**, 85
18. Buczilowski, U.R., Beck, R. 1991, Astron.Astrophys. **241**, 47
19. Chiba, M., Tosa, M., 1989, Mon. Not. Roy. astr. Soc. **238**, 621
20. Cooper, B.F.C., Price, R.M., 1962, Nature **195**, 1084
21. Crutcher, R.M., Kazes, I., 1983, Astron. Asttrophys. **125**, L23
22. Davis, L., Greenstein, J.L., 1951, Astrophys. J. **114**, 206
23. Deinzer, W., Grosser, H., Schmitt, D., 1990, in: Galactic and Intergalactic Magnetic Fields (R. Beck, P.P. Kronberg, R. Wielebinski eds.) IAU. Printed in the Netherlands, 95
24. Donner, K.J., Brandenburg, A., 1990, Astron. Astrophys. **240**, 289
25. Elstner, D., Meinel, R., Rüdiger, G., 1990, Geophys. Astrophys. Fluid Dynamics **50**, 85
26. Elstner, D., Meinel, R., Beck. R., 1991, Astron. Astrophys. (in press)
27. Elvius, A., 1978, Astrophys. Space Sci. **55**, 49
28. Fermi, E., 1949, Phys. Rev., **75**, 1169
29. Fujimoto, M., 1987, in: Interstellar Magnetic Fields, (eds. R. Beck and R. Gräve), Springer-Verlag, Berlin, Heidelberg, 23
30. Gardner, F.F., Whiteoak, J.B., 1963, Nature **197**, 1162
31. Gioia, M., Gregorini, L., Klein, U., 1982, Astron. Astrophys. **116**, 164
32. Gräve, Beck, R., 1988, Astron. Astrophys. **192**, 66
33. Hiltner, W.A., 1958, Astrophys. J., **128**, 9
34. Horrelou, C., Beck, R., Klein, U., 1990, in: Galactic and Intergalactic Magnetic Fields (R. Beck, P.P. Kronberg, R. Wielebinski eds.), Kluwer Acad. Publ., Dordrecht, 211
35. Hummel, E., Lesch, H., Wielebinski, R., Schlickeiser, R., 1988, Astron. Astrophys. **197**, 29
36. Hummel, E., Beck, R., Dahlem, M., 1991, Astron. Astrophys. (in press)
37. Kiepenheuer, K.O., 1950, Phys. Res. Letters **79**, 738
38. Krasheninnikova, Y., Ruzmaikin, A.A., Sokolov, D.D., Shukurov, A., 1990, Geophys. Astrophys. Fluid Dyn. **50**, 131
39. Krause, F., 1967, Habilitationsschrift, Universität Jena
40. Krause, F., 1971, Astron. Nachrichten **293**, 187
41. Krause, F. and Rädler, K.-H., 'Mean Field Magnetohydrodynamics and Dynamo Theory' 1980, Pergamon Press
42. Krause, F., 1987, in: "Interstellar Magnetic Fields" (eds. R. Beck and R. Gräve), Springer-Verlag Berlin Heidelberg, 8

43. Krause, F., Meinel, R., 1988, Geophys. Astrophys. Fluid Dyn., **43**, 95
44. Krause, F., 1990, Geophys. Astrophys. Fluid Dynamics **50**, 67
45. Krause, F., Meinel, R., Elstner, D., Rüdiger, G., 1990, in: Galactic and Intergalactic Magnetic Fields (R. Beck, P.P. Kronberg, R. Wielebinski eds.), IAU Printed in the Netherlands, 97
46. Krause, M., Hummel, E., Beck, R., 1989a, Astron. Astrophys., **217**, 4
47. Krause, M., Beck, R., Hummel, E., 1989b, Astron. Astrophys., **217**, 17
48. Mathewson, D.S., Ford, V.L., 1970a, Astron. J. **75**, 778
49. Mathewson, D.S., Ford, V.L., 1970b, Mem. Roy. astr. Soc. **74**, 139
50. Mathewson, D.S., van der Kruitt, P.C., Brouw W.N., 1972, Astron. Astrophys., **17**, 468
51. Mayer, C.H., McCullough, T.P., Sloanaker, R.M., 1957, Astrophys. J., **126**, 468
52. Mayer, C.H., McCullough, T.P., Sloanaker, R.M., 1962, Astrophys. J., **67**, 581
53. Meinel, R., 1989, Astron. Nachr. **310**, 1
54. Meinel, R., 1990, Geophys. Astrophys. Fluid. Dyn. **50**, 79
55. Meinel, R., Elstner, D., Rüdiger, G., Krause, F., 1990, in: Galactic and Intergalactic Magnetic Fields (R. Beck, P.P. Kronberg, R. Wielebinski eds.) IAU Printed in the Netherlands, 115
56. Meyer, W.F., 1920, Lick Obs. Bull., **10**, 68
57. Moffatt, A.T., 1973, in: "Galaxies and the Universe" (eds. Sandage et al.) **9**, Chapter 7, Chicago Uni. Press
58. Moffatt, K.H., 1978, Magnetic Field Generation in Electrically Conducting Fluids, Cambridge University Press, Cambridge
59. Moss, D., Brandenburg, A., Tuominen, I., 1991, Astron. Astrophys. (in press)
60. Muliarchik, T.M., 1957, Izwest. Astropys. Inst. Alma Ata **5**, No 7
61. Öhman, Y., 1942, Stockholm Obs. Bull., **10**, 68
62. Parker, E.N., 1971, Astrophys. J. **163**, 255
63. Pooley, G.G., 1969, Mon. Not. Roy. astron. Soc. **144**, 101
64. Rädler, K.-H., Wiedemann, E., Brandenburg, A., Meinel, R., Tuominen, I., 1990, Astron. Astrophys. **239**, 413
65. Rädler, K.-H., Wiedemann, E., 1990, in: Galactic and Intergalactic Magnetic Fields (R. Beck, P.P. Kronberg, R. Wielebinski esd.) IAU, Printed in the Netherlands, 107
66. Reich, W., Sofue, Y., Inoue, M., Seiradakis, J.H., 1987, in: Interstellar Magnetic Fields, (eds. R. Beck und R. Gräve), Springer-Verlag, Berlin, Heidelberg, 146
67. Ruzmaikin, A.A., Sokolov, D.D., Shukurov, A.M., 1985, Astron. Astrophys. **148**, 335
68. Ruzmaikin, A.A., Shukurov, A.M. and Sokoloff D.D., 'Magnetic Fields in Galaxies' 1988, Kluwer Academic Publishers
69. Rüdiger, G., 1990a, Geophys. Astrophys. Fluid Dynamics **50**, 53
70. Rüdiger, G., 1990b, in: Galactic and Intergalactic Magnetic Fields (R. Beck, P.P. Kronberg, R. Wielebinski eds.), IAU, Printed in the Netherlands, 113
71. Sawa, T. and Fujimoto, M., 1986, Publ. Astron. Soc. Japan **38**, 132
72. Scarrott, S.M., White, C., Pallister, W.S., Solinger, A.B., 1977, Nature **265**, 32
73. Scarrott, S.M., Ward-Thompson, D., Warren-Smith, R.F., 1987, Mon. Not. Roy. astron. Soc. **224**, 299.
74. Schlüter, A., Biermann, L., 1950, Z. Naturforschung **5a**, 237
75. Schmidt, Th., 1970, Astron. Astrophys. **6**, 294
76. Schmidt, Th., 1976, Astron. Astrophys. Suppl. **24**, 357
77. Schmitt, D., 1990, Rev. in Modern Astronomy **3**, 88
78. Simard-Normandin, M., Kronberg, P.P., Button, S., 1981, Astrophys. J. Suppl., **45**, 97
79. Stepinski, T.F., Levy, E.H., 1988, Astrophys. J. **331**, 416
80. Stepinski, T.F., Levy, E.H., 1990, Astrophys. J. **362**, 318
81. Stix, M., 1975, Astron. Astrophys., **42**, 85
82. Soward, A., 1978, Astron. Nachrichten **299**, 25
83. Soward, A., 1991, Geophys. Astrophys. Fluid Dynamics (in press)
84. Tosa, M., Chiba, M., 1990, Geophys. Astrophys. Fluid Dynamics **50**, 147
85. Vainshtein, S.I., Ruzmaikin, A.A., 1971, Astron. J. (USSR), **48**, 902, [Sov. Astron. **15**, 714]

86. Vainshtein, S.I., Ruzmaikin, A.A., 1972, Astron. J. (USSR), **49**, 449, [Sov. Astron. **16**, 365]
87. Vallee, J.P., 1988, Astron. J. **95**, 750
88. Verschur, G.L., 1968, Phys. Ref. Lett. **21**, 775
89. White, M.P., 1978, Astron. Nachrichten **299**, 209
90. Westerhout, G., Seeger, Ch.L., Brouw, W.N., Tinbergen, J., 1962, Bull. Astron. Inst. Neth., **16**, 187
91. Wielebinski, R., Shakeshaft, J.S., Pauliny-Toth, I.I.K., 1962, Observatory, **82**, 158
92. Wielebinski, R., Shakeshaft, J.S., 1962, Nature **195**, 982
93. Wielebinski, R., Shakeshaft, J.S., 1964, Mon. Not. Roy. astron. Soc. **128**, 19
94. Wielebinski, R., 1990, in: "The Interstellar Medium in Galaxies" (eds. H.A. Thronson Jr., J.M. Shull) 349, Kluwer

Index of Contributors